Foundation of Digital Electronics and Logic Design

Foundation of Digital Electronics and Logic Design

Subir Kumar Sarkar
Asish Kumar De
Souvik Sarkar

Published by

Pan Stanford Publishing Pte. Ltd.
Penthouse Level, Suntec Tower 3
8 Temasek Boulevard
Singapore 038988

Email: editorial@panstanford.com
Web: www.panstanford.com

British Library Cataloguing-in-Publication Data
A catalogue record for this book is available from the British Library.

Foundation of Digital Electronics and Logic Design

ISBN 978-981-4364-58-4 (Hardcover)
ISBN 978-981-4364-59-1 (eBook)

Printed in the USA

Contents

Preface

The present treatise is meant to be a textbook for undergraduate students in electrical engineering, electronics and communication engineering, computer science, and information technology. It is also expected to be useful to the students of other disciplines studying basic courses on electronics. It covers the basic principles of digital electronics and logic design. This textbook is an outgrowth of the lectures we have delivered to our students over several years and covers the syllabi of many universities across the globe. It will help in bridging the gap between digital electronics and logic design. Topics have been illustrated with the help of many diagrams to make the students grasp the subject in a better manner. The subject matter is dealt with a clear and concise way with many examples, and a large number of problems have been worked out to make acquaint students with the applications of the principles and formulae they have encountered in the text. If the contents of the present book prove useful to those for whom it is intended, we will deem our effort amply rewarded. We will thankfully receive constructive suggestions for the improvement of the book. We acknowledge gratefully the encouragement given by our colleagues and students, whom we have taught over the past several years, for their thought-provoking questions, which have really helped us to clear our thoughts and have enabled us to develop simple explanations to complex-looking theories.

We wish to thank the editorial and production team at Pan Stanford Publishing for the meticulous processing of the manuscript. Subir Kumar Sarkar expresses his particular appreciation to his research scholars Mr. Suman Basu, Mr. Amit Jain, Mr. Pranab Kishore Dutta,

Mr. Bijoy Kantha, Mr. Rashmi Ranjan Sahoo, Mr. Bibhas Manna, Mr. Subhashis Roy, Ms. Gargee Bhattacharyya, Mrs. Saheli Sarkhel Ganguly and Mrs. Jayashree Bag for their skillful service in preparing the manuscript.

Subir Kumar Sarkar
Asish Kumar De
Souvik Sarkar

Chapter 1

Combinational Circuits

1.1 Introduction

The word digital implies that, information in the system is represented by values that are limited in number. The values hence used to represent the information are called quantized or discrete values. These values are processed by circuit components that maintain the limited set of values. A digital system performs more reliably and efficiently if it handles only two values ("0" and "1") that is only two states and such a digital system will be called a two state system or a binary system.

Quest for scientific knowledge to enhance our understanding of the world has led to all technological innovations responsible for the progress of civilization. Since, the invention of digital electronics/ integrated circuits a few decades ago, manufacturing of electronic systems has taken rapid strides in improving *speed, size,* and *cost.* For today's digital integrated chips, switching time is on the order of submicrons, transistor count is in the order of millions, and cost leads to the very pervasive introduction of integrated circuit (IC) chips to many aspects of modern engineering and scientific

Foundation of Digital Electronics and Logic Design
Subir Kumar Sarkar, Asish Kumar De, and Souvik Sarkar

ISBN 978-981-4364-58-4 (Hardcover), 978-981-4364-59-1 (eBook)
www.panstanford.com

endeavors including computations, telecommunication, aeronautics, genetics engineering and manufacturing, and so on. It is clear that the IC chips will play the role of a key building block in the information society of the 21st Century. Needless to say, it is the very large scale integration (VLSI) technology that has made this evolution a reality. The digital electronics together with VLSI technology has brought the power of mainframe computer to the laptop. The number of other inventions has changed the way human being live, think, communicate, interact, recreate and develop, individually, and as societies, as the digital circuit together with VLSI technologists have. This indicates that we have realized the tremendous potential and importance of the digital electronics in lives. The widespread uses of digital electronics have motivated us to decide who will be entrusted with design, development, test, and so on, so that we can extract the best possibilities from the digital system.

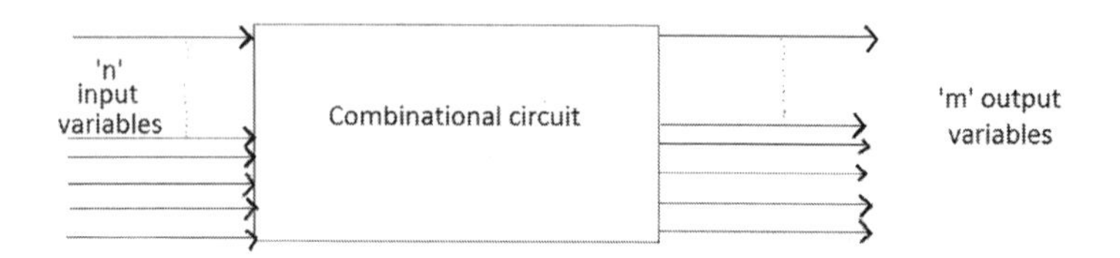

Figure 1.1 A digital circuit (combinational).

It is now appropriate time to give the ideas of analog and digital signals and systems: There are two broad categories of electronic circuits and systems—analog and digital circuit and systems. A continuous signal that can have any magnitude in a given range of time is called *analog signal* for example growth of a living creature such as a child and a plant. A system having analog circuit and/ or devices is called *analog electronics.* On the other hand, a signal having only two discrete values is called *digital signal.* The number of students in a class is an example digital signal. The branch of electronics that deals with the digital devices, circuits, and systems is called *digital electronics.* The digital circuits are mainly of two different types—combinational and sequential circuits. We shall describe the combinational digital circuit first and then the sequential circuits. A combinational circuit is a logic circuit in which the output of the circuit at any instant of time is depends totally on the inputs present at that time only. Figure 1.1 represents the block diagram of the combinational circuit: In general, a combinational circuit consists of "*n*" input variables, logic gates, and "*m*" output variables. The logic gates of the combinational circuits acknowledge signals from the inputs and generate the output signal. The combinational circuit in the process transforms binary information from the given input data to the required output data. For n input variables, the number of maximum input combination will be 2^n. It is important to remember that for each possible input combination, there is only one possible output combination. The detailed description of the circuit will be given in the due course. However, the design of combinational circuits starts from the verbal outline of the problem and ends in a set of Boolean function or in a logic circuit diagram.

1.2 Advantages of Digital System

The advantages of digital system are as follows:

1. Its basic components are highly reliable.
2. It is low cost.
3. It is small in size.
4. It is light weight.
5. It consumes less power.
6. It is faster in operation.
7. It is durable.
8. It requires number heating element, since it is based on field emission.
9. It is reasonably immune to noise.

1.3 Essential Characteristics of Digital Circuits

The advantages of a digital system are due to the advent of improved integrated circuit technology. Greater the probability of satisfying a digital circuit, greater will be its demand.

A digital circuit should satisfy certain essential properties, which are as follows:

1. The binary output signal of a digital circuit should be the prescribed function of the binary input or output (Fig. 1.2). This is called logical function.

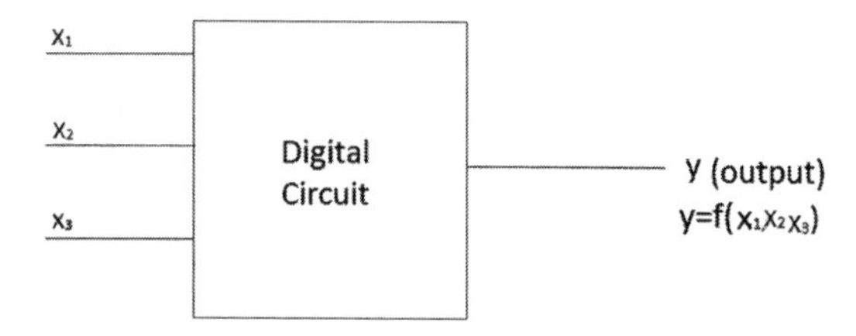

Figure 1.2 The block diagram of a digital system.

2. Both inputs and outputs of a digital circuit are discrete.
3. Quantization of amplitude within the normal operating range of the voltages is required. This requires nonlinearity in circuit operation. The specified range of voltage within the whole range will represent any one of the two states as shown in Fig. 1.3. The region of uncertainty between the two states should be as small as possible.
4.

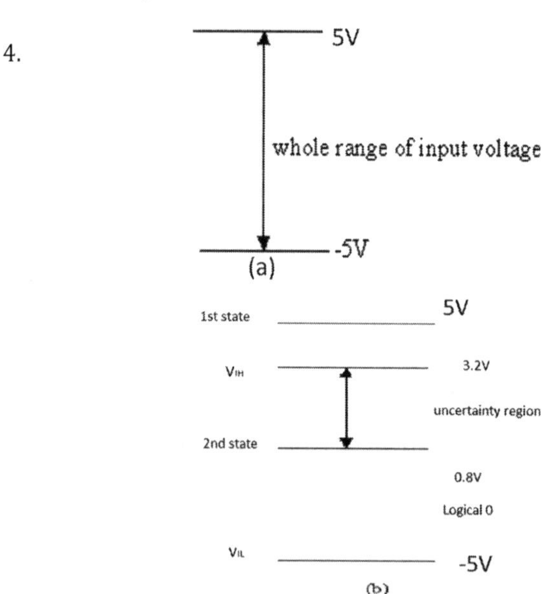

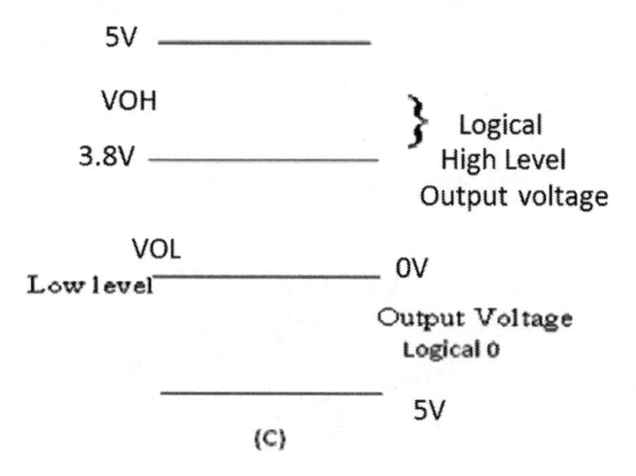

Figure 1.3 Schematic of logic levels.

5. The amplitude levels should be regenerated while passing through the digital circuit. This requirement dictates the presence of voltage transfer characteristics for the digital circuit as shown in Figs. 1.4 and 1.5.
6. Digital circuits are nonlinear circuits.

Output versus input transfer characteristics:

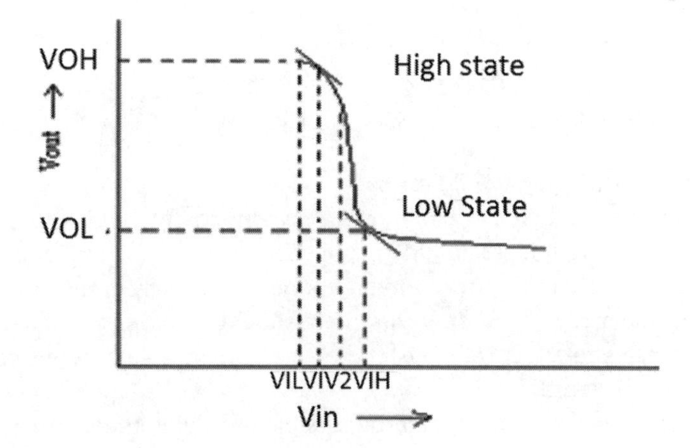

Figure 1.4 Input–output characteristics.

In between the logical states the magnitude of the low state voltage gain should be greater than unity.

$$\left|\frac{\Delta V_{\text{out}}}{\Delta V_{\text{in}}}\right| = 1 \tag{1.1}$$

$$\text{at } V_{\text{in}} = V_{\text{IH}}$$
$$\text{or } V_{\text{in}} = V_{\text{IL}}$$

V_{OH} = minimum value of high level output voltage
V_{OL} = maximum value of low level output voltage

$$\left|\frac{dV\int_{\text{out}}}{dV_{\text{in}}}\right| = 1 \tag{1.2}$$

$$\left|\frac{dV_{\text{out}}}{dV_{\text{in}}}\right| > 1$$

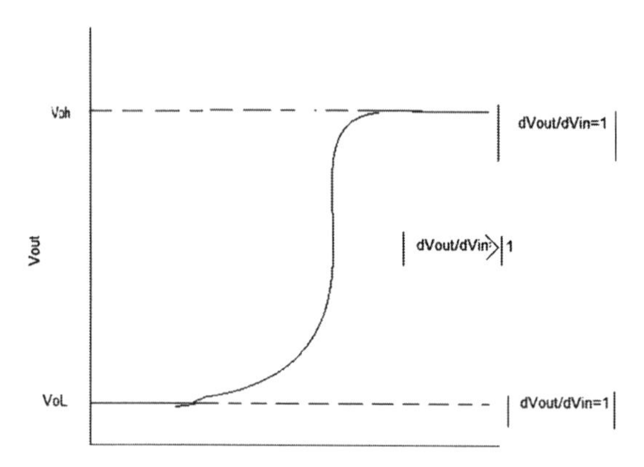

Figure 1.5 Schematic of input–output voltage variation.

Noninverting characteristics:

7. The useful digital circuit must have a directivity that is changes in the output level should not appear at any unchanging input of the same circuit (there should not be feedback from the output to the input). In other words, there should be implicit, unilateral cause-and-effect relationship between input(s) and output of a digital circuit.
8. Each digital circuit should be capable of driving the number of similar circuits. The number of similar circuits that a particular

digital circuit can drive without a false output is called the fan-out of the digital circuit. The number of independent inputs that a digital circuit can accept is called the fan-in of the digital circuit. Fan-in of this digital circuit is 3 and its fan-out is also 3 (Fig. 1.6).

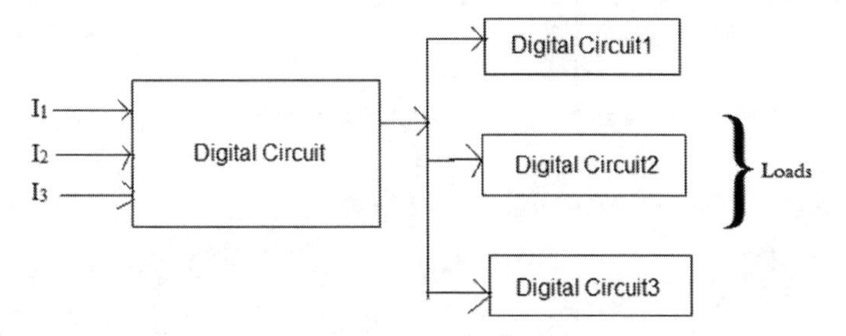

Figure 1.6 Digital circuits showing fan-in and fan-out.

1.4 Characteristic of an Ideal Digital Logic Element

An ideal digital circuit (Fig. 1.7) should have the following:

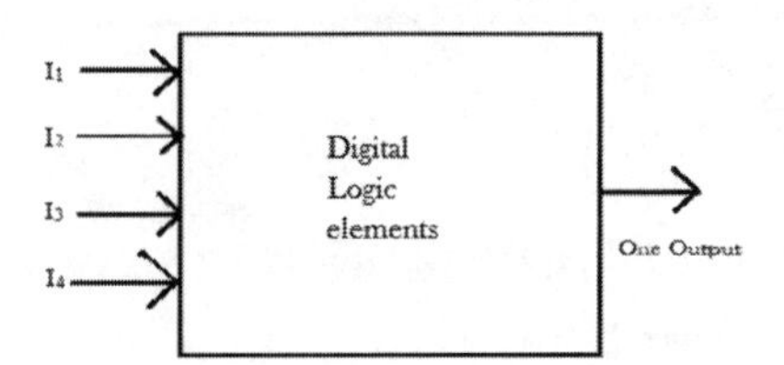

Figure 1.7 An ideal digital circuit.

1. *n* inputs.
2. one output.
3. power supply voltage input.
4. ground.
5. infinite input impedance.
6. zero output impedance (drive the current with maximum capacity).
7. negligible time delay (output appears almost instantaneously).

Figures 1.8 and 1.9 show the ideal and actual characteristics of a digital circuit, respectively.

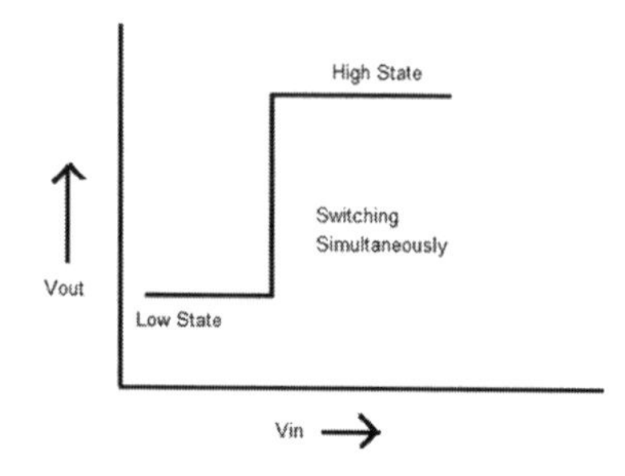

Figure 1.8 Ideal characteristic of digital system.

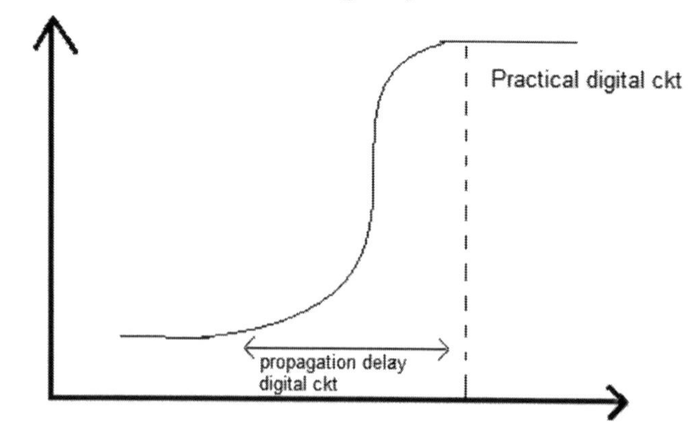

Figure 1.9 Practical digital circuits.

1.5 Definition of Truth Table and Various Logic Conventions

1.5.1 Logic Circuit

A logic circuit is an electronic circuit which operates on digital signals according to some logic function.

1.5.2 Logic Gate

A logic gate is a logic circuit whose output depends on the inputs in accordance with some logic rule.

1.5.3 Truth Table

One of the best statements in a tabular form that describes what a logic circuit can do or cannot do is called a truth table. The truth table is a formal specification sheet that describes the exact circuit behavior for all possible input combinations that is, it describes what its output will be for each combination of inputs. In a truth table there are 2^n rows and the number of columns will be one more than the number of input variables forming the logic function. The "n" is the number of input variables.

For example: If a logical function is $y = f(A, B, C)$ then $n = 3$ (A, B, and C).

Hence, the number of rows = 2^3 or 8 and number of columns is 3 + 1 or 4.

← Truth table for the AND gate

A	B	Y (output)
0	0	0
0	1	0
1	0	0
1	1	1

1.5.4 Logical Convention

There is specified voltage range corresponding to each 0 and 1. For a specific voltage values the positive and negative logic is demonstrated in Fig. 1.10.

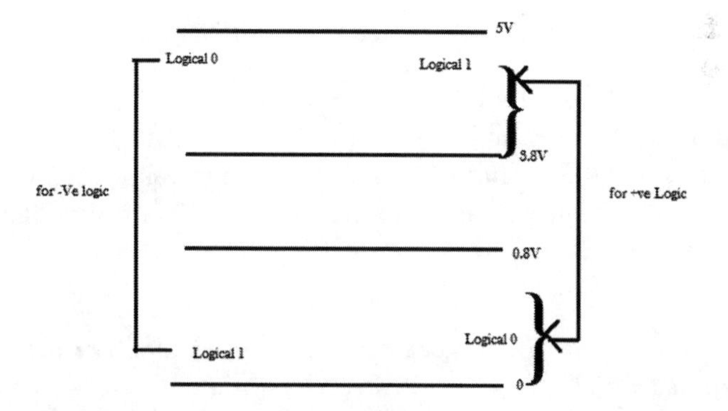

Figure 1.10 Illustration of the positive logic.

1.5.4.1 Positive logic

In a positive logic system, the more positive voltage value of the two is assigned the logical 1 and the other is assigned logical 0.

For example of the voltage values 0 V and 5 V. The 5 V is logical 1 and 0 V is logical 0. Similarly, for voltage ranges –1 V and –5 V, –1 V is logical 1 and –5 V is logical 0.

1.5.4.2 Negative logic

In a negative logical system, the more negative voltage value of the two available voltage values is assigned logical 1 and the other is assigned logical 0.

For example, if there are two voltages values, 0 V and 5 V, 0 V is assigned logical 1 and 5 V is assigned logical 0. Similarly for voltage values –1 V and –5 V, –1 is assigned logical 0, and –5 V is assigned logical 1.

In most cases, we will assume positive logic. The AND gate positive logic will not behave as AND gate in negative logic.

1.6 Number System

The data processed by the computer are numbers. A single number can be expressed in more than one way, using the same character–numerals and/or letters. The form a number will thus take depends on the choice of the number system.

Any nonnegative integer N may be written in the form:

$$N(b) = \{a_{n-1}b^{n-1} + a_{n-2}b^{n-2} + \cdots + a_1b^1 + a_0b^0\} \tag{1.3}$$

where "b" is the base (or radix) of the number system, "a" is the weight (or positional value), and n indicates how many digits (or positions) the number has. For a formal notation of the number N, the power of the base and the plus signs are deleted, leaving only the weights:

$$N(b) = a_{n-1}\, a_{n-2} \ldots a_1\, a_0 \tag{1.4}$$

with the base of the number system given as a subscript inside parenthesis.

1.6.1 Positional Number System

It is a system in which the value of each digit depends on its position in the number. The positional number systems are classified according to the base used. The number system most commonly used in everyday life is the decimal number system. Its base (radix) is 10. Computers use the binary number system for which the base is 2. The digits in the binary system are often called bits and the standard symbols used to represent the bits are 0 and 1.

Binary		Decimal
00	=	0
01	=	1
r	=	base of radix system

When $r = 10 \rightarrow$ we get decimal number which uses numbers 0, 1, 2, 3, 4, ..., 9.
When $r = 2$, we get binary number, that uses 0 and 1 that is $(r-1)$.
When $r = 8$, we get octal number that uses 0, 1, 2, ... $(r-1)$ that is 7.
When $r = 16$ we get hexadecimal that uses 0, 1, 2, ..., 9, A, B, C, D, E, F.
Any number in radix r or base r can be expressed as power of r.

Decimal number versus binary number:

$$(324)_{10} \quad = \quad 4 \times 10^0 + 2 \times 10^1 + 3 \times 10^2$$

$\downarrow$

Weight of 3 of $(324)_{10}$ is 100 "4" in the $(324)_{10}$ is the least significant bit or digit (LSB or LSD).

Left most digit or bit is most significant digit (MSD or MSB), such as 3 in $(324)_{10}$.

$$(1\quad 0\quad 1\quad 0\quad 1)_2 = (21)_{10} = 2^4 + 2^2 + 2^0 = 21$$

1.6.2 Generalized Approach of Number System

A number can be treated as a vector quantity as if we consider the sign. A number X in radix r is represented by vector form as follows:

$$(X)_r = X_{n-1}\, X_{n-2}\, X_{n-3} \ldots X_1 X_0 \qquad X_{-1} X_{-2} X_{-3} \ldots X_{-m}$$

$\uparrow$	$\uparrow$
Integer part of the number Including the sign bit	Fractional part of radix r

X_i is the i-th element of the vector representing a number.

The value of X_i is always in between $(0 \leq X_i \leq r-1)$ that is X_i must be an integer where r is called the base or the radix of the number system. X_{n-1} is called the sign bit of the number C. The radix point or decimal point elements or digits to the right hand side of the radix or the decimal point give the fractional part of the number and the digits to the left of the radix point represent the integer part of number. So the length of the number is $(n+m)$ digits. If $X_{n-1}=0$ then the number is positive.

However, if $X_{n-1}=(r-1)$ then the number is negative.

When $r = 10$, we get a decimal number system, which uses the digits 0–9.

When $r = 2$, we get the binary number system, which uses the digits, 0 and 1.

When $r = 3$, we get the trinary (ternary) number system, which uses the digits, 0, 1, and 2.

When $r = 4$, we get the quaternary number system, which uses the digits, 0, 1, 2, and 3.

With $r = 8$ we get octal number system, which uses the digits (0, 1, 2, 3, 4, 5, 6, and 7)

When $r = 16$, we get hexadecimal number system, which uses the digits (0, 1, 2, 3, 4, 5, 6, 7, 8, 9, A, B, C, D, E, and F).

Since 10 is not the single digit number, so it represents 10 (single digits). For negative number $X_{n-1}=1$ for the binary system.

$X_{n-1}=9$ for the decimal system
$X_{n-1}=7$ for the octal system
$X_{n-1}=\text{F}$ for the hexadecimal number system

Any number in a particular radix r can be represented in the following manner:

(a) sign and magnitude system
(b) the $(r-1)$s or digit complement system
(c) the r's complement system

Generally, the representations of the positive number in all the systems are identical.

For a positive number $X_{n-1}=0$ in radix r.

$X_{n-2}\,X_{n-3}\,\ldots\,X_{-1}\,X_{-2}\,\ldots\,X_{-m}\rightarrow$ represents the magnitude digits of the number in radix $r = |(X)_{\text{r}}|$

The decimal equivalent value of the magnitude of a number in radix r is given by:

$(X)_{\text{r}}|_{10}=X_{n-2}r^{n-2}+X_{n-3}r^{n-3}+\cdots+X_1r^1+X_0r^0+X_{-1}r^{-1}+X_{-2}r^{-2}+\cdots+X_{-m}r^{-m}$

The suffix of each number gives its weights (suffix r)

$$=\sum_{i=-m}^{n-2} X_i r^i$$

A binary number is represented as 01010.11

This gives the magnitude ($n = 5, m = 2$)
But decrease it by a factor of 2

$$X_3\, X_2\, X_1\, X_0 \cdot X_{-1}\, X_{-2}$$

$$0\ 1\ 0\ 1\ 0 \cdot 1\ 1$$

$\uparrow$

Sign bit

$$(X_r)_{10} = \sum_{I=-2}^{3} X_i r^i = X_{-2}2^{-2} + X_{-1}2^{-1} + X_0 2^0 + X_1 2^1 + X_2 2^2 + X_3 2^3$$

$$= \frac{1}{4} + \frac{1}{2} + 0 + 1 \times 2 + 0 \times 2^2 + 1 \times 2^3$$

$$= \frac{1}{4} + \frac{1}{2} + 2 + 8 = 10.75$$

So, if it is a positive number so the sign bit = 0.

1.6.3 Radix Conversion

Say, a number x in radix r has to be converted into a number of radix g.

$(X)_r = (I)_r(F)_r \Rightarrow$ The number x has got integer part and the fractional part of the number.

$(\bar{x})_g = (\bar{I})_g\, (\bar{F})_g$

To convert one number in radix r into a number the r number in radix g, we have to convert integer and fractional parts separately.

1.6.4 Integer Conversion

Let the integer part of the number be given by
$(X)_r = (I)_r\, (d_{n-2}d_{n-3} \ldots d_1 d_0)_r$ containing (n–1) digits.
The number in radix g that is the converted number.

$$(\bar{X})_g = (I)_g = (b_{n-2}\, b_{n-3} \ldots b_1 b_0) \tag{1.5}$$

Containing (n–1) digits.

For integer conversion, first find the decimal equivalent of the number in radix r (decimal equivalent of the number with radix r = decimal equivalent of the number with radix g).

Let $(I_r)_{10} = I$ = decimal equivalent value of the given number in radix r.

The decimal equivalent value of the converted number in radix g

$$(I)_g|_{10} = b_{u-2}g^{u-2} + b_{u-3}g^{u-3} + \cdots + b_1 g + b_0 \tag{1.6}$$

$$(I)_g|_{10} = b_{u-2}g^{u-2} + b_{u-3}g^{u-3} + \cdots + b_1 g + b_0 \tag{1.7}$$

$$= g\,(b_{u-2}g^{u-3} + b_{u-3}g^{u-4} + \cdots + b_1) + b_0 \tag{1.8}$$

From which $I = gQ_1 + b_0$ (1.9a)

where $Q_1 = (b_{u-2}g^{u-3} + b_{u-3}g^{u-3} + \cdots + b_2 g + b_1)$ (1.9b)

1.6.5 Procedure for Integer Conversion

For integer conversion we have to use repeated conversion.

1. First divide the I by g and after this get an integer quotient Q_1 and first remainder $r_1 = b_0$

$$g)\, I\, (Q_1$$
$$gQ_1$$
$$b_0$$

So, $I = gQ_1 + b$

where Q_1 is an integer quotient given by Eq. 3 and b_0 is the first remainder where $0 \le b_0 \le g{-}1$.

2. Again divide the quotient Q_1 by g to get to get the second integer quotient to get Q_2.

$$\text{So } Q_1 = gQ_2 + b_1$$

After this we get the second integer quotient Q_2 and second remainder $r_2 = b_1$.

And this is obtained by multiplying equation. (BLANK)

$$Q_1 = g\,(b_{u-2}g^{u-4} + b_{u-3}g^{u-5} + \cdots + b_2) + b_1 \tag{1.8}$$

$$Q_1 = gQ_2 + b_1 \tag{1.9}$$

$$\text{where} \quad Q_2 = (b_{u-2}g^{u-4} + b_{u-3}g^{u-5} + \cdots + b_2) \tag{1.10}$$

Repeat steps (1) and (2) until at say:

Step (1) you get an integer quotient $Q_i = 0$ and last remainder $r_i = b_{n-2}$ and thus, the remainder form in reverse order gives us the converted number in radix g.

Thus, integer number in radix (r)

$$= (I)_r = (b_{u-2} b_{u-3} b_{u-4} \ldots b_2 b_1 b_0)_g \tag{1.11}$$

The first remainder = b_0 = LSB bit of the converted number and last remainder.

b_{n-2} = MSB bit of the converted number.

$(I)_r = (5673)_{10}$ $r = 10$

Convert this into hexadecimal number of radix $g = 16$

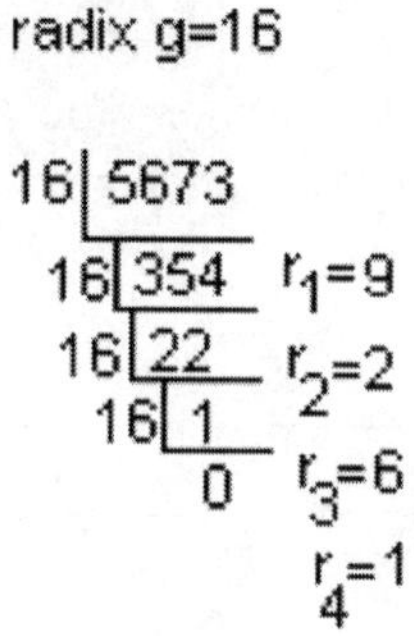

radix g=16

```
16| 5673
   16| 354   r1=9
    16| 22   r2=2
     16| 1
         0   r3=6
             r4=1
```

Here we stop as $Q_i = 0$

So, $(5673)_{10} = (1629)_{16}$

$$(1629)_{16} = 1 \times 16^3 + 6 \times 16^2 \times 2 \times 16^1 + 9 \times 16^0$$
$$= 4096 + 1536 + 32 + 9$$
$$= (5673)_{10}$$

$(578596)_{10} \rightarrow$ Convert it into binary

First convert the large decimal number into hexadecimal number, by repeated divisions by 16.

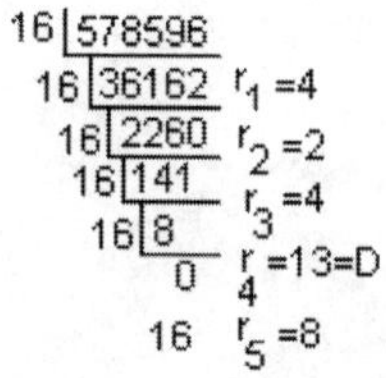

```
16|578596
  16|36162  r1 =4
   16|2260  r2 =2
    16|141  r3 =4
     16|8
        0   r4 =13=D
       16   r5 =8
```

$(578596)_{10} = (8D424)_{16}$

Represent hexadecimal digits by 4 bit binary

$(1000 \quad 1101 \quad 0101 \quad 0010 \quad 0100)_2$

If leading zeroes are there we can decimal to octal.

1.7 Logic Gates and Logic Circuits

Logic circuits make a series of decision to give or to obtain the logical answer to a problem for a given set of input conditions. It has to make some decisions.

To make such decisions the three basic logic gates are used are OR, AND, and NOT or inverter gate.

1.7.1 OR Gate

The output of OR = 1 when either input or all inputs are 1 and output = 0 when all inputs are 0 (Fig. 1.11). The truth table and definition is given in Fig. 1.11.

Table 1.1 OR gate

Definition	Truth table		
The output of the OR gate = 1 if any one or all inputs are 1 and output = 0 if the all inputs are zero	A(S_1)	B (S_2)	C
	0	0	0
	0	1	1
	1	0	1
	1	1	1

A and *B* are inputs and *y* is output.

Electrical Analogue:

When both the switches are open the current passes through the bulb and output is 1. If any one of the switch is open the current passes through the short circuit and output is 0.

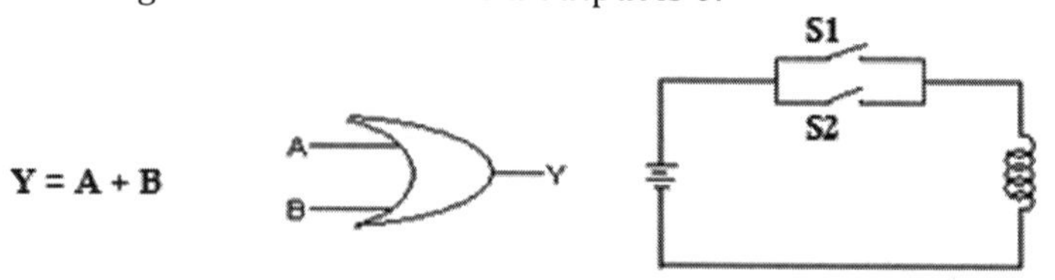

Figure 1.11 OR gate.

Number of input may be more than two but is limited by the fan-in capability (to be discussed later) of the OR gate.

1.7.2 AND Gate

The truth table is given in Table 1.2.

Table 1.2 AND gate

Definition	Truth table		
The output of AND gate = 1 when all inputs are 1 and output = 0 when any of the input = 0	A(S_1)	B (S_2)	C
	0	0	0
	0	1	0
	1	0	0
	1	1	1

Electrical Analogue:

When both the switches are closed the current passes through the bulb and output is 1. If any one of the switch is open the then the circuit becomes an open circuit and output is 0.

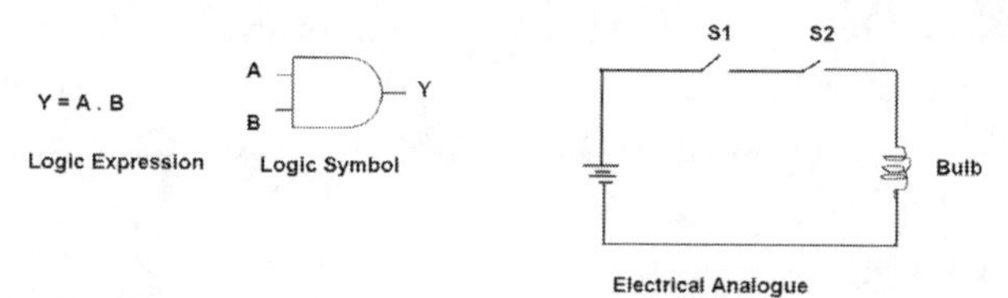

Figure 1.12 AND gate.

1.7.3 NOT Gate

This logic gate is having only one input and one output. Output of NOT gate is 1 when input is 0 and output is 0 when input is 1. *A* is input and *y* is output. The logical expression along with logical symbol, electrical analogue is given in Fig. 1.13. The truth table is given in Table 1.3. (Do the same for all basic gates)

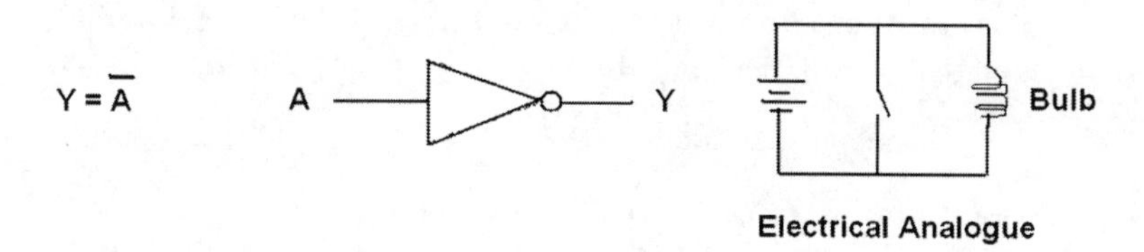

Figure 1.13 NOT gate.

Table 1.3 NOT gate

Definition	Truth table	
Output of NOT gate is 1 when input is 0 and output is 0 when input is 1	A	Y
	1	0
	0	1

The bar above the Boolean variable indicates invert of the logical variable.

Electrical Analogue:
When switch is open, the bulb glows that is input = 0, output = 1. The whole current will pass through the short circuited path and number current will pass through the bulb when switch is closed. Hence, output = 0.

1.7.4 NOR Gate

The output of NOR gate = 1 when all inputs are 0 and output = 0 when any of the input = 1. A and B are inputs and Y is the output in the present case. The logic symbol and the corresponding electrical analogue circuit is shown in Fig. 1.14. The corresponding truth table is given in Table 1.4.

Table 1.4 NOR gate

Definition	Truth table		
The output of NOR gate = 1 when all inputs are 0 and output = 0 when any of the input = 1.	$A(S_1)$	$B\ (S_2)$	C
	0	0	1
	0	1	0
	1	0	0
	1	1	0

Electrical Analogue:
When both the switches are open the current passes through the bulb and output is 1. If any one of the switch is closed the current will pass through the short circuited path and no current will pass through the bulb, hence output = 0.

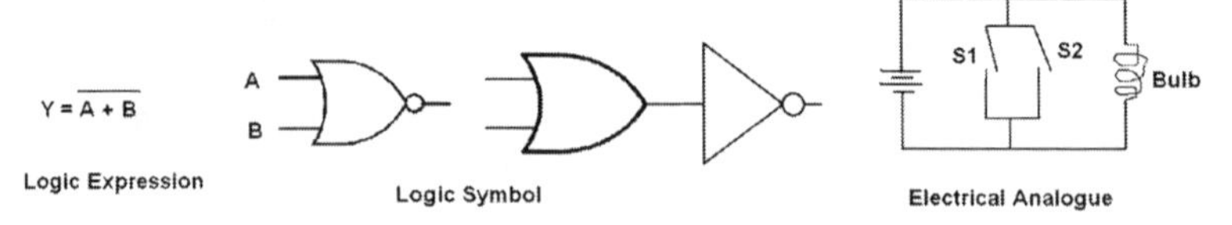

Figure 1.14 NOR gate.

Logical expression:

$$\therefore Y = \overline{A + B}$$

When both inputs are zero that is both switches are opened only then output = 1 will be obtained. Otherwise the bulb will not glow.

1.7.5 NAND Gate

The output of NAND gate = 0 when all inputs are 1 and output = 1 when any of the input = 1. A and B are inputs and Y is the output in the present case. The logic symbol and the corresponding electrical analogue circuit are shown in Fig. 1.15. The corresponding truth table is given in Table 1.5.

Table 1.5 NAND gate

Definition	Truth table		
	A(S_1)	B (S_2)	C
The output of NAND gate = 0 when all the inputs are 1 and output = 1 when any of the input = 1.	0	0	1
	0	1	1
	1	0	1
	1	1	0

Electrical Analogue:
When both the switches are closed the current will pass through the short circuited path and no current will pass through the bulb, hence output = 1. If any one of the switch is open the current will pass through the bulb and output = 0.

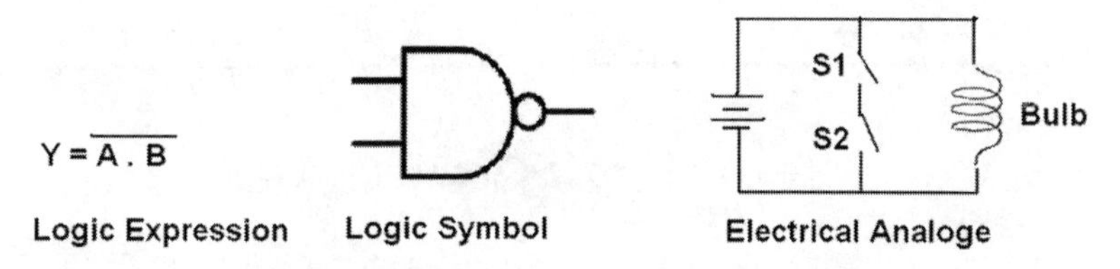

Figure 1.15 NAND gate.

NAND and NOR gates are called universal logic gates as we can realize any other gates by using these gates only.

The details of all logic gates is given in Table 1.6.

Table 1.6 Details of all the logic gates

Gate name	Definition	Truth table	Logical symbol	Logical expression	Electrical analogue
OR	The output of OR = 1 when either input or all inputs are 1 and output = 0 when all input are 0.	A B Y_{AND} 0 0 0 0 1 0 1 0 0 1 1 1	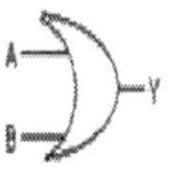 A, B, Y	$Y = A + B$ ""+"→ logical OR operator"	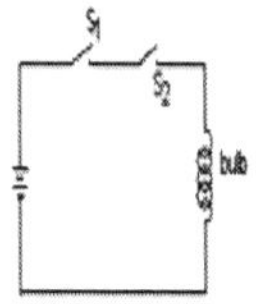 S_1, S_2, bulb
AND	The output of AND gate = 1 when all inputs are 1 and output = 0 when any of the input = 0	A B Y_{OR} 0 0 0 0 1 1 1 0 1 1 1 1	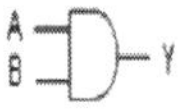 A, B, Y	$Y = A\,.\,B$ ""."→ logical AND operator"	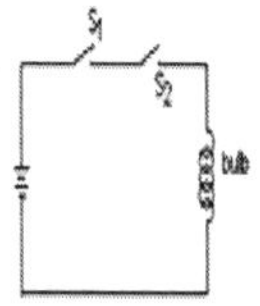 S_1, S_2, bulb
NOT	Output of NOT gate is 1 when input is 0 and output is 0 when input is 1	A Y_{NOT} 0 1 1 0	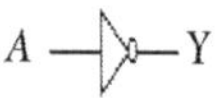 A, Y	$Y = \overline{A}$ "–"→ complement (the bar above Boolean variable) or A' → complement of A	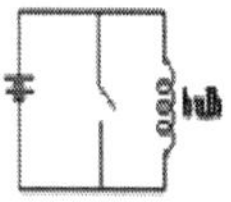 bulb
NAND	Output = 1 if at least one input = 0, Output = 0 if all inputs are 1	A B Y_{NAND} 0 0 0 0 1 1 1 0 1 1 1 0	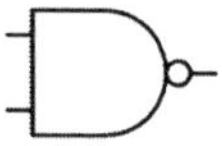	$Y = \overline{A.B}$	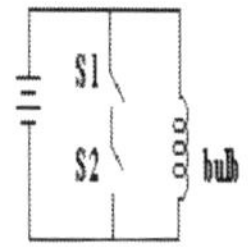 S1, S2, bulb
NOR	The output of NOR Gate = 0, if any one input is 1 and output = 1, if all inputs are zero.	A B Y_{NOR} 0 0 1 0 1 0 1 0 0 1 1 1	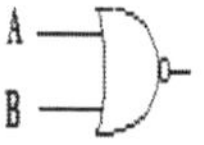 A, B	$Y = \overline{A + B}$	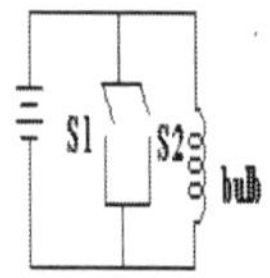 S1, S2, bulb

1.8 Logic Gates and Logic Circuits

1.8.1 X-OR Gate or Exclusive OR Gate

It has two or more inputs and only one output. All commercial XOR gates usually have two inputs.

The output of a XOR gate will be 1 if inputs are different, and the output will be 0 if all its inputs are 0 or all are 1. The truth table is given in Table 1.7.

Table 1.7 X-OR gate

Definition	Truth table		
	A(S_1)	B (S_2)	C
The output of a XOR gate will be 1 if inputs are different, and the output will be 0 if all its inputs are 0 or all are 1.	0	0	0
	0	1	1
	1	0	1
	1	1	0

We can connect a LED and if it glows then two inputs are same otherwise not.

Logical symbol of XOR gate:

Logical expression for XOR gate:

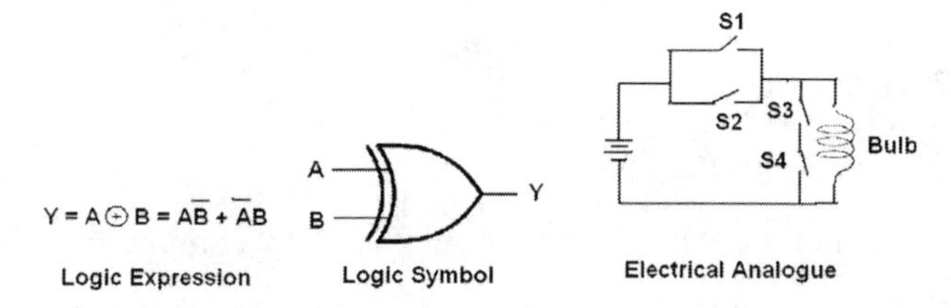

Figure 1.16 X-OR gate.

$$Y = A \oplus B = A\overline{B} + \overline{A}B$$

$\oplus \rightarrow$ XOR operator or it is also called Modulo 2 sum.

$$Y = A\overline{B} + \overline{A}B \quad (1.12)$$

$$Y = (A + B)\,(\overline{A} + \overline{B}) \quad (1.13)$$

$$Y = (A + B)\,\overline{AB} \quad (1.14)$$

$$Y = AB + \overline{A}\,\overline{B} \quad (1.15)$$

The four alternative expressions satisfy the logical XOR and the truth table is satisfied by all these, as shown in Table 1.8.

Table 1.8 XOR gate

A	B	$\overline{A}$	$\overline{B}$	$A\overline{B}$	$\overline{A}B$	Y
0	0	1	1	0	0	0
0	1	1	0	0	1	1
1	0	0	1	1	0	1
1	1	0	0	0	0	0

So expression (1) satisfies the truth table.

Electrical Analog of 2 input XOR gate:

Two switches are ganged at a time that is when one switch is operated other switch also operates.

S_1 and S_3 are ganged. That is S_1 and S_3 have to close simultaneously.

S_2 and S_4 are ganged. That is S_2 and S_4 have to close simultaneously.

When S_1, S_2, S_3, S_4 open then the bulb will not glow $\rightarrow$ 0, 0

$S_1\ S_3$	$S_2\ S_4$	Status of bulb
0	0	bulb do not glow = 0
0	1	bulb glows = 1
1	0	bulb glows = 1
1	1	bulb do not glow = 0

The circuit will be short circuited.

Application of XOR gate:

1. It is used in the arithmetic section of a digital computer.

Addition of two binary bits$\rightarrow$ sum bits carry bit will be formed the sum will follow the XOR gate.

2. Two input XOR gate can be used as a one bit inequality detector (refer Truth table).
3. It is used as a parity checker circuit by realizing multivariable XOR. (Ramp is tested in parity checker).

1.8.1.1 Parity

A given sequence of fixed bit pattern is said to have even parity, if the sequence contains even number of 1s or all 0s.

Say considering a 5 bits sequence

11111$\rightarrow$ odd parity
01101$\rightarrow$ odd parity
00000$\rightarrow$ even parity
01101$\rightarrow$ odd parity
01100$\rightarrow$ even parity

A given sequence of fixed bit pattern is said to have odd parity if the sequence contains odd number of 1s. How parity error can be detected? That is how parity checking is determined.

Transmitted side (TX) (4 bits sequence is used)	Extra Bit	Receiver side (RX)
1101	0	Odd Parity ←11010→ no error
1001	1	10010→ error in one bit
1101	0	11010→ no error
1001	1	10010→ error in one bit

At first we take up a 4 bits sequence. At the TX, we add one extra bit, so that overall sequence becomes either odd or even. Suppose 5 bits sequence of odd type is produced.

Receiving end knows in advance that what TX is sending is an odd pattern. The RX will test the parity. If it sees that it is still odd then number error otherwise error.

This type of error detection by parity checking is possible if there is not more than 1 bit error.

If the first bit in RX is 00010 still it is odd parity, so the error is number detected, whereas there are two errors

1. But, this error detection is important in the receiver. Since the probability of error of a digital system is very small as $P_e = 10^{-6}$ and hence error detection by parity checking is efficiency used or can be efficiency used for one bit error detection.

$P_e = \frac{1}{10^6} \rightarrow$ out of 10^6 bits, there is the probability, that only one bit is detected by this.

2. It can detect error, but it cannot correct error.

For n bit parity checker circuit we require n—1 number of 2 input XOR gates. For 5 bit four 2 input XOR gates are required (Fig. 1.17).

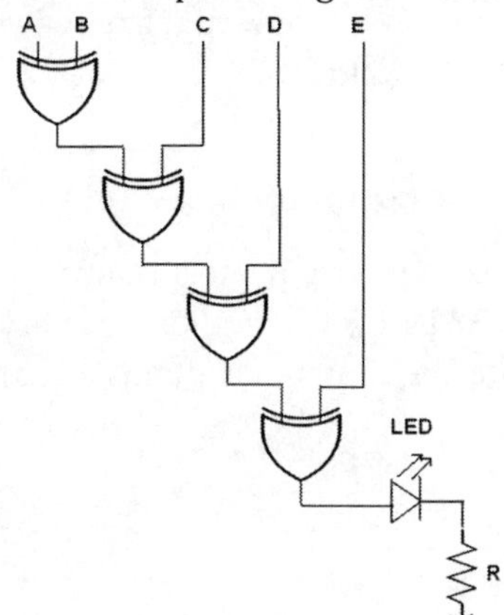

Figure 1.17 Five bit parity checker circuit.

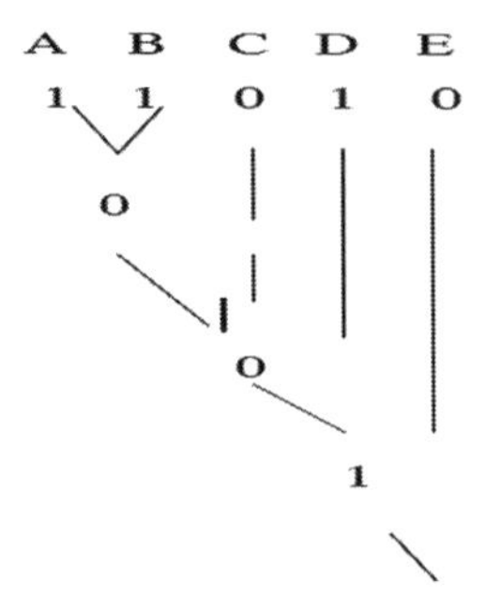

So, by using 4, 2 input XOR gate, we can realize 5 input XOR gate. This is called realization of multivariable XOR gate. An application of XOR gate is demonstrated in Fig. 1.18.

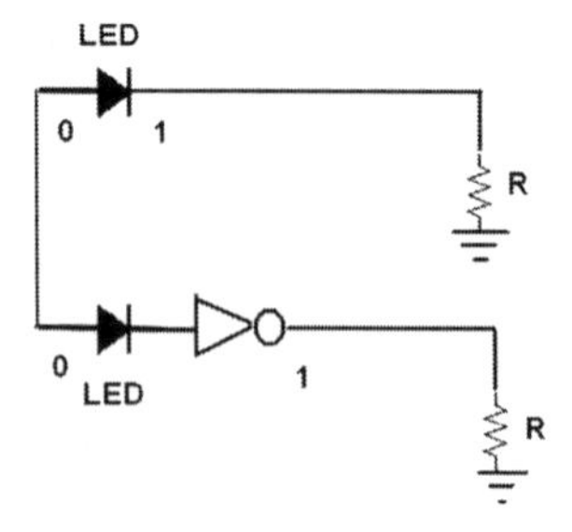

Figure 1.18 Application of XOR gate.

If light emitting diode (LED) is forward biased then it emits light. If LED glows then 5 bit input sequence is having odd parity and if LED does not glow, 5 bit input sequence have even parity.

If 1 = 5 V say then LED will glow. If we want LED to glow, input must have odd number of 1s. If the input has even number of 1s then the output circuit gets modified to

1.8.2 XNOR Gate or Equality Detector

It has got two or more inputs and one output.

The output of XNOR gate = 0, if inputs contains odd number of 1s. Output of XNOR gate = 1, if inputs contains all zeroes even number of 1s.

Logical symbol for XNOR or equality detector is shown in Fig. 1.19. The truth table of the XNOR gate is given in Table 1.9. The XOR gate followed by inverter gate is XNOR gate (Fig. 1.19).

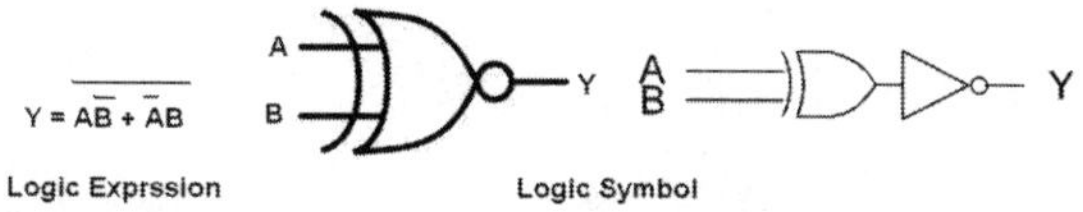

$Y = \overline{A\bar{B} + \bar{A}B}$

Logic Exprssion

A B Y A B Y

Logic Symbol

Figure 1.19 XNOR gate.

Table 1.9 XNOR gate

Definition	Truth table		
The output of XNOR gate = 0 if inputs contains odd number of 1's. Output of XNOR gate = 1 if inputs contains all zeroes even number of 1's.	*A*	*B*	*Y*
	0	0	1
	0	1	0
	1	0	0
	1	1	1

Logical expression for XNOR

$$Y = A \Theta B$$

Θ→XNOR operator or equality operator. It is same as $Y = A\Theta B$. The XNOR gate can be used as a one bit equality detector, one bit digital comparator.

$$\begin{aligned} Y_{\text{XOR}} &= A\bar{B} + A\bar{B} \\ &= (A+B)\,(\bar{A}+\bar{B}) \\ &= (A+B)\,\bar{A}\bar{B} \\ &= \overline{(A+B)\,\bar{A}\bar{B}} \end{aligned}$$

$$\begin{aligned} Y_{\text{XNOR}} &= \overline{A\bar{\bar{B}} + \bar{\bar{A}}B} \\ &= \overline{(A+B)\,(\bar{A}+\bar{\bar{A}})} \\ &= \overline{(A+B)\,\bar{\bar{A}}\bar{\bar{B}}} \\ &= \overline{\overline{(AB+\bar{A}\bar{B})}} = AB + \bar{A}\bar{B} \end{aligned}$$

1.9 Enable Inputs

In Fig. 1.20, if $S = 0$, this will perform its expected function, but $S = 1$ will inhibit the operation.

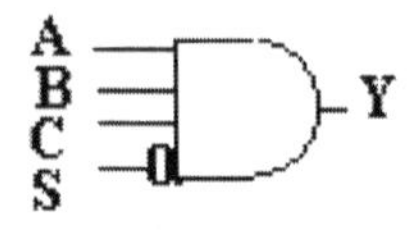

Figure 1.20 Logical symbol of enable input operation.

$$Y = A.B.C.\overline{S}$$

S input is called enable or strobe input when $S = 0$.

S input is called inhibit input when $S = 1$.

When $S = 0$, the output Y depends upon the value of A, B, and C and then perform its intended function. When $S = 1$, the Y is zero always being independent of A, B, C, and under this condition $S = 1$, the gate is called anticoincident gate and the gate performs inhibit operation. So, it is disabling the operation.

1.10 OR Gate Using Diode Resistor Logic

Two input OR gate using diode resistor logic (in positive logic system) along with a logical symbol of OR gate is shown in Fig. 1.21.

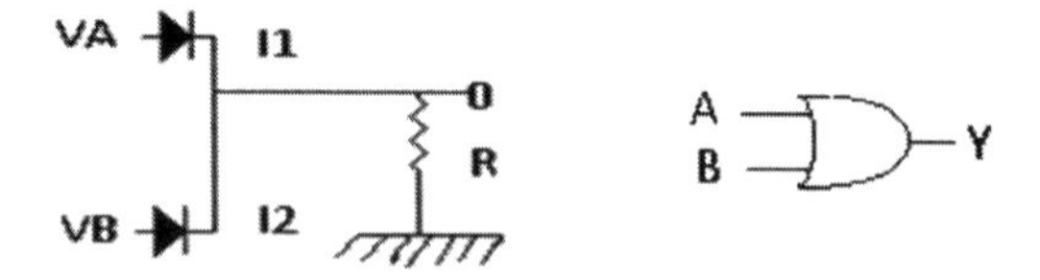

Figure 1.21 Two input OR gate using diode resistor logic.

We assume, let the two diodes be ideal one. $R_f = 0$ = forward resistance. V_A and V_B are positive logic inputs. The assigned voltage values to V_A and V_B are given in Table 1.10.

Table 1.10 Assigned voltage values for the diode circuit of two input OR gate

V_A	V_B	V_0
0	0	0
0	5 V	5 V
5 V		5 V
5 V	5 V	5 V

as both diodes are reverse biased as $I_1 = 0, I_2 = 0$

$I = I_1 + I_2 = 0$

$V_0 = IR = 0 \times R = 0$

In case, D_1 is reverse biased and D_2 is forward biased and $I_2 = I$ (by applying KCL)

$$I_2 = \frac{V_A - V_0}{R_f} = \frac{V_0}{R} = I$$

$$V_0\left(\frac{1}{R} + \frac{1}{R_f}\right) = \frac{V_A}{R_f}$$

$$V_0 = \left(\frac{RV_A}{R+R_f}\right) = \frac{5\text{ V}(R)}{(R+0)} = 5\text{V}$$

as ideally $R_f = 0$

when $V_A = 5$ V and $V_B = 5$ V both the diodes are forward biased.

So, $I_1 + I_2 = I_0$

$$\frac{V_A - V_0}{R_f} + \frac{V_B - V_0}{R_f} = \frac{V_0}{R}$$

As the voltage drop across the diode is taken also (ideally)

So
$$V_0\left\{\frac{1}{R} + \frac{2}{R_f}\right\} = \frac{V_A - V_B}{R_f}$$

$$V_0 = \frac{(V_A + V_B)R}{2R + R_f} = \frac{(5+5)R}{2R}$$

Then this circuit satisfies the 2 input OR gate

V_A	V_B	V_C
0	0	0
0	1	1
1	0	1
1	1	1

In the negative logic 0 V = 1, 5 V = 0 as 0 V is more negative than 5 V.

As $R_f = 0$ for ideal diode $V_0 = 5$ V

If 0 V = 0

5 V = 1

1.11 Realization of an Inverter Using Transistor

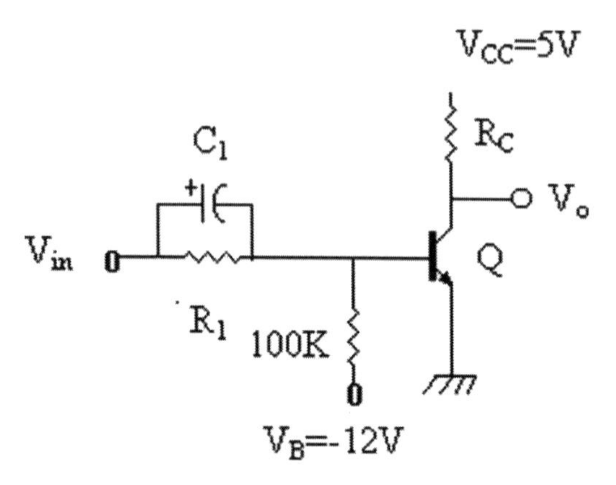

Figure 1.22 Inverter circuit.

An inverter circuit implemented using transistor is shown in Fig. 1.22. $V_B = -12$ V is applied at the base of Q via a 100 K resistance so that the transistor will operate either in the cut off region or in the saturation region. The Q is not to operate in active region, as the collector and emitter current may vary; also the output voltage is not fixed. When Q is on, the output voltage is $V_0 = V_{\text{CE Sat}} \approx 0.2\ V =$ Logical 0.

When Q is at cut off V_0 = 5 V. The C_1 is used to improve the transient response of the transistor (The base charge is removed by the capacitor).

V_{in} Q V_0

0 off 5 V = 1→Base is at lesser potential compared to emitter.

5 V on 0.2 V = 0 → transistor is in saturation.

V_{in}	V_o	
0	1	5 V = sate '1'
1	0	0 V = sate '0'

1.12 Boolean Algebra and Its Postulates

This was developed by George Boole and constitutes mathematical basic and finds its importance after the advent of digital computer.

Boolean algebra is an algebra of logic forms the mathematical basis on which the logic design is based. It is used for the description, synthesis, and analysis of the binary logical function. It is based upon the concept that any logical statement can be designated true or false. The logical statement cannot be assigned any value other than 1 and 0.

$$Y = A + \overline{B}C$$
$$Y = f(A,B,C)$$

Whatever the binary values of *A, B, C, Y* can assume only two values 0 and 1 for any set of combinations. In Boolean algebra it is based upon several postulates for explaining the various steps.

Definition: It is an algebra comprising a set B (B; +, 1,–, 0, 1) that contained at least two elements 0 and 1 together with the operation (described by "+" symbol), complement operation (described by "—" symbol), such that if *x, y, z* are three elements within {B} $x + y$ that is, logical OR of *x* and *y*. The logical AND of *x* and *y* is *x.y* complement of x also be in {B}. Together with the three operations the logical OR operation denoted by "+", logical AND operation is denoted by (.), logical complement denoted by | or such that if *x, y,* and *z* are three elements within B then $x + y$ that is logical OR of *x* and *y*, the *x. y* that is logical AND of *x* and *y* and $\overline{x}$ or $x^{/}$ that is logical complement of *x* are also in B. (Set B).

Axioms or postulates of Boolean algebra:

Idempotent property	$x + x = x$ (Logical OR operator)
Commutative property	$x + y = y + x$
Associative property	$x + (y + z) = (x + y) + z$
Distributive property	$x.(y + z) = x.y + x.z$
Absorptive property	$x + y.x = x\,[x.(1 + y) = x]$
Complement	$\overline{x}$

If *x* be any element in B that is $x \in B$ then there must exist an element $x^{/}$ or $\overline{x}$, which is called the complement of *x* where $x^{/}$ or $\overline{x}$ also $\in B$ then $x + \overline{x} = 1$, $x.\overline{x} = 0$

(G) The zero or null element and one universal element. There exists a unique element 0 (called 0 element) where $0 \varepsilon B$. For each an every element x where $\alpha \varepsilon B$. Then $0 + x = x + 0 = 0$ (OR)

$$0.x = x.0 = 0 \text{ (AND)}.$$

There exists a unique element 1 (called or universal element) where 1εB, for each an every element xεB we must have $x + 1 = 1 + x = 1$

$$x.1 = 1.x = x$$

$$x + y \cdot z = (x + y)(x + z)$$

R.H.S.

$$\begin{aligned}(x+y)(x+z) &= x.x + y.x + x.z + y.z \\ &= x + x.z + x.y + y.z \\ &= x(1 + z + y) + y.z \\ &= x.1 + y.z \\ &= x + y.z \\ &= L.H.S. \\ &= \text{(Distributive property)}\end{aligned}$$

The simplification and minimization are different. In some case, they may be same.

1.13 Demorgan's Theorem in Dual Form

Dual form = one form is the dual of the other.

1.13.1 Sum into Product

Statement: This states that the complement of the logical sum of several Boolean variables is equal to the logical product of the complement of the individual variable for two inputs Demorgan's theorem is expressed as

1. $$\overline{A + B + C + \cdots} = \overline{A} \cdot \overline{B}\overline{C} \ldots \qquad (1.16)$$

1.13.2 Product into Sum

Statement: This state that logical complement of the product of several variables is equal to the logical sum of the complement of the individual variable.

2. $$\overline{A.B.C \ldots} = \overline{A} + \overline{B} + \overline{C} + \cdots \qquad (1.17)$$

For two inputs Demorgan's theorem is expressed as

$$\overline{A+B} = \overline{A}.\overline{B} \tag{1.18}$$

$$\overline{A.B} = \overline{A}+\overline{B} \tag{1.19}$$

The realization corresponding to POS and SOP form of the Demorgan's theorem is given in Table 1.11. Pictorial form of Demorgan's theorem is shown in Fig. 1.23.

Table 1.11 Realization of the truth table of POS and SOP form

A	B	$\overline{A+B}$	$\overline{A}$	$\overline{B}$	$\overline{A}\cdot\overline{B}$	$\overline{A\cdot B}$	$\overline{A}+\overline{B}$
0	0	1	1	1	1	1	1
0	1	0	1	0	0	1	1
1	0	0	0	1	0	1	1
1	1	0	0	0	0	0	0

So, $\overline{A+B} = \overline{A}.\overline{B}$

And $\overline{A.B} = \overline{A}+\overline{B}$ (proved)

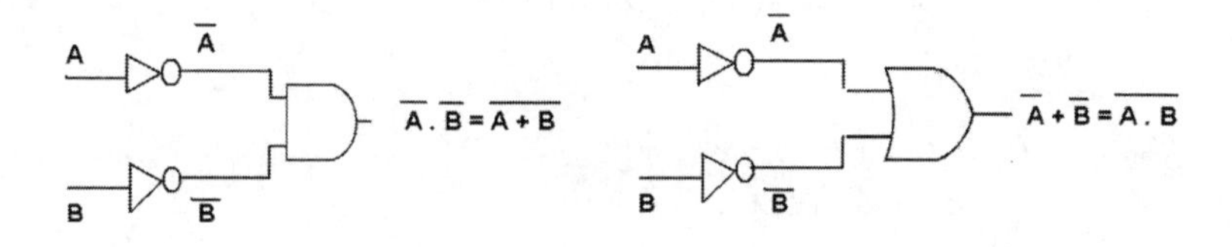

Figure 1.23 Pictorial form of Demorgan's theorem.

1.14 Simplification of Boolean Expressions by Using Boolean Algebra

$$L = AB\cdot C + AB\overline{C} + \overline{A}\cdot B\cdot C + \overline{A}\cdot B\cdot\overline{C}$$
$$L = AB(C+\overline{C}) + \overline{A}B\,(C+\overline{C}) \text{ as } C+\overline{C} = 1$$
$$L = (A\cdot B)\,(1) + \overline{A}\cdot B\,(1)$$
$$L = AB + \overline{A}\cdot B$$
$$L = B\,(A+\overline{A}) = B(1) = B, \text{ as } A+\overline{A} = 1.$$

The actual output is independent of C and A. To realize this simplification terms of the AND, NOR gate.

So, actually it requires 3 input four AND gates, four NOT gate, and one 4 input OR gate as shown in Fig. 1.24.

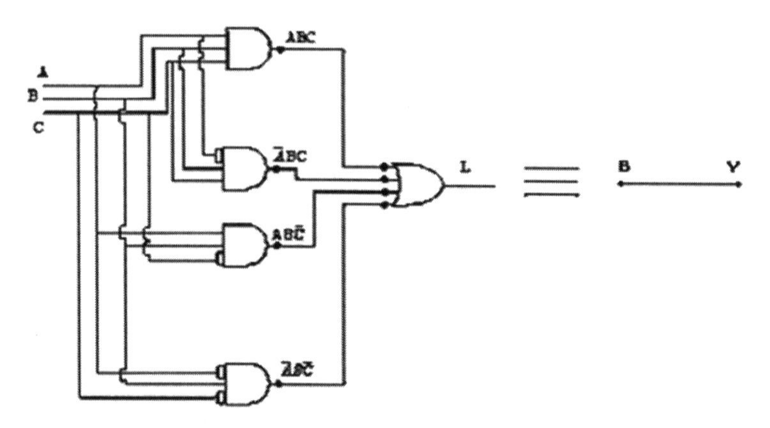

Figure 1.24 Simplifications of Boolean algebra.

By simplifying using Boolean algebra, we may not get the minimized expression (expression which can be realized by using minimum number of gates or expressions).

$$L = A.B + A.\bar{B}.C + \bar{A}.B.\bar{C}$$
$$L = B(A + \bar{A}\bar{C}) + A\bar{B}.C$$
$$L = B(A + \bar{A})(A + \bar{C}) + A\bar{B}C \text{ (by distributive property)}$$
$$L = B(A + \bar{C}) + A\bar{B}.C$$
$$L = B.A + B.\bar{C} + A\bar{B}.C$$

Simplified to some extent as 2nd term has two terms.

$$= (AB + A\bar{B}C) + B\bar{C}$$
$$= A(B + \bar{B}C) + B\bar{C}$$
$$= A(B + \bar{B}) + (B + C) + B\bar{C} \quad \text{(by distributive property)}$$
$$= A(B + C) + B\bar{C}$$
$$= AB + AC + B\bar{C}$$

1.15 Logical Expression in SSOP and SPOS Form (Min and Max Term Form)

Logical expression can be expressed in the following two forms:

(i) standard sum of products (SSOP)
(ii) standard product of sum (SPOS)

Let a logical expression which is a function of three variables may be expressed as:

$$f(A,B,C) = A + B\overline{C} \tag{1.20}$$

A given logical expression is said to be in standard sum of product form.

If each term contains the entire input variables in product form (either normal form or complemented form or combination of them) and the resulting expression of logical function is the sum of such product term.

If

$$f(A,B,C) = A.B.\overline{C} + \overline{A}.BC + \overline{A}B\overline{C} + ABC \tag{1.21}$$

where each term contains the product of all the three variables and resulting expression is sum of all terms. Then only it is SSOP. But (1) is not SSOP. So, we have to convert (1.20) into (1.21).

The converting a given logical expression in to SSOP form:

$$f(A,B,C) = A + B\overline{C} = A.1.1 + 1.B.\overline{C}$$

$$f(A,B,C) = A(B + \overline{B})(C + \overline{C}) + (A + \overline{A})(B.\overline{C})$$

$$f(A,B,C) = A(B.C) + A\overline{B}C + AB\overline{C} + A\overline{B}\overline{C} + \overline{A}.B.\overline{C}$$

(Using distributive property of B and A)
If a particular term is more than once, omit one term.

$$f(A,B,C) = A.B.C + \overline{A}.B.\overline{C} + AB\overline{C} + A\overline{B}C + A\overline{B}\overline{C} + AB\overline{C}$$

$$f(A,B,C) = ABC + AB\overline{C} + A\overline{B}C + A\overline{B}\overline{C} + \overline{A}B\overline{C}$$

Eq. (A) is the SSOP form for the given logical function.

$$f(A, B, C) = U + V + W + X + Y$$

where

$$U = ABC$$

$$V = AB\overline{C}$$

$$W = A\overline{B}C, X = A\overline{B}\,\overline{C}, Y = \overline{A}B\overline{C}$$

$f(A, B, C) = 1$, if $U = 1$, or $V = 1$, or $W = 1$,
or $X = 1$,or $Y = 1$, irrespective of other values.
If, $U = V = W = X = Y = 1$ then also $f(A, B, C) = 1$
If $U = 1$ then $A = B = C = 1$.

$$
\begin{aligned}
f(A, B, C) &= 1, \text{ if } V = 1, A = B = 1, C = 0 \\
,, &= 1, \text{ if } W = 1, A = C = 1, B = 0 \\
,, &= 1, \text{ if } X = 1, A = 1, B = C = 0 \\
,, &= 1, A = C = 0, B = 1
\end{aligned}
$$

Represent uncomplemented variable by 1 and complemented variable by 0.

$$
\begin{aligned}
f(A, B, C) &= 1.1.1 + 1.1.0 + 1.0.1 + 1.0.0 + 0.1.0 \\
&= 7 + 6 + 5 + 4 + 2
\end{aligned}
$$

Generally written as $f(A, B, C) = \sum m\,(2, 4, 5, 6, 7)$ (in ascending order)

This representation in Eqn. (B) is called the min term representation for the given logical function.

From Eq. (B) the truth table can be directly found out, once min term form for a given logical function is obtained.

Min term representation of a given logical function is shown in Table 1.12.

Table 1.12 Min term representation of a given logical function

Decimal	Inputs			$f(A, B, C)$
	A	B	C	
0	0	0	0	1
1	0	0	1	1
2	0	1	0	1
3	0	1	1	0
4	1	0	0	1
5	1	0	1	0
6	1	1	0	1
7	1	1	1	1

So, $f(A, B, C) = \sum m\,(0, 1, 2, 4, 5, 7)$ = min term form

Standard product to sum form (SPOS)

$$f(A,B,C) = A + B\overline{C} \tag{1.22}$$

same expression given logical function is said to have standard product of sum, if each term contains the sum of all the input variables and the resultant logical expression is the product for such term.

If $$f(A,B,C) = \left(\overline{A} + B + C\right)(A + \overline{B} + C)(A + B + C) \tag{1.23}$$

It is a SPOS

Now to convert (20) to (21) SPOS form.

$$f(A,B,C) = A + \overline{BC} = (A + B)\ (A + \overline{C}) = (A + B + 0)\ (A + \overline{C} + 0)$$
$$= (A + B + C\overline{C})(A + \overline{C} + B\overline{B})$$
$$= (A + B + C)(A + B + \overline{C})(A + \overline{C} + B)(A + \overline{C} + \overline{B})$$
$$= (A + B + C)(A + B + \overline{C})(A + \overline{C} + \overline{B})...$$

Eq. (C) is the standard product of sum form.
Say from Eq. (C) $f(A,B,C) = uvw$
where $u = A + B + C, v = A + B + \overline{C}, w = A + \overline{B} + \overline{C}$
This can be represented by a AND gate and $f(A,B,C) = 0$ if $u = 0$ $(A = B = C = 0)$
So, we substitute 0 for uncomplemented variables and vice versa

$$,, = 0 \text{ if } v = 0\ A = B\ C = 1$$
$$,, = 0 \text{ if } w = 0\ A = 0\ B = C = 1$$

$$f(A, B, C) = (0 + 0 + 0)\ (0 + 0 + 1)\ (0 + 1 + 1)$$
$$= \Pi M\ (0, 1, 3) \qquad (1.24)$$

The values within the brackets of equation D will have output = 0 otherwise = 1 and similarly from truth table max term form can be found. The Max term representation of a given logical function is demonstrated in Table 1.13.

Table 1.13 Max term representation of a given logical function

Decimal	*A*	*B*	*C*	*f* (*A*, *B*, *C*)
0	0	0	0	0
1	0	0	1	0
2	0	1	0	1
3	0	1	1	0
4	1	0	0	1
5	1	0	1	1
6	1	1	0	1
7	1	1	1	1

$f(A,B,C) = AB + B\overline{C}$. In order to convert into max term form, first convert to SPOS form

$$= AB(C + \overline{C}) + (A + \overline{A})B\overline{C} = ABC + AB\overline{C} + AB\overline{C} + \overline{A}B\overline{C} = ABC + AB\overline{C} + \overline{A}B\overline{C}$$

Because, it is easier to convert to the min term form and then convert to max term form. So, whichever is suitable we convert it into that first min term or max term.

$$=111 + 110 + 010 = \sum m\ (2, 6, 7)$$
$$f(A,B,C) = \prod M\ (0, 1, 3, 4, 5)$$

as those terms absent in min term are present in the max term form (This is an intelligent alternative way of getting max term form).

$$f(A, B, C) = \prod M\ (1, 3, 5) = (001)(011)(101)$$
$$(A + B + \overline{C})\ (A + \overline{B} + \overline{C})\ (\overline{A} + \overline{B} + \overline{C})$$
$$\text{So } f(A, B, C) = \sum m\ (0, 2, 4, 6, 7)$$

If min term form is given:

$$f(A, B, C) = \sum m\ (0, 1, 6)$$
$$= 000 + 001 + 110$$
$$= \overline{A}\overline{B}\overline{C} + \overline{A}\overline{B}C + AB\overline{C}$$

as zero replaced by complemented variable and 1 by the un-complemented variable.

Now let us explain a little bit about min and max terms.

1.15.1 Min Terms

If a logical function of multiple variables is expressed in the form of sum of product terms and each product term contains all the variables either in the normal form or complemented form or combination of both the forms then each term is called the min term and the function under such condition is said to be standard sum of product form (SSOP).

Example: Let $y = f(A, B, C)$ be a logical function of three variables (A, B, and C).

Suppose $y = BC + AB\overline{C} + A\overline{B}C + \overline{A}BC + A\overline{B}\overline{C} + \overline{A}\overline{B}C + \overline{A}B\overline{C} + \overline{A}\overline{B}\overline{C}$

Then y is in SSOP form and each term of it is called min term.

1.15.2 Max Term

If a logical function of multiple variables is expressed in the form of product of sum terms and each product term contains all the variables either in the normal form or complemented form or combination of both the forms then each term is called the max term and the function under such condition is said to be standard product of sum form (SPOS).

Example:

$$y = (A + B + \overline{C})\,(A + B + \overline{C})\,(A + \overline{B} + C)\,(\overline{A} + B + C)$$
$$(A + \overline{B} + \overline{C})\,(A + \overline{B} + \overline{C})\,(\overline{A} + B + C)\,(\overline{A} + \overline{B} + \overline{C})$$

Each term is called max term and the logical function is then said to be standard product of sum form.

The main difference between min and max term is that the normal logical variable is taken as logical one in case of min term $(A = 1, = 0)$ but it is taken as logical zero in the case of max term $(A = 0, = 1)$.

Universal gates: The two gates namely NAND and NOR are called universal gates as we can realize all the other gates from those two gates separately.

So, as any logic gate can be realized by using either NAND or NOR gates only and hence NAND and NOR are called universal gates.

1.16 NAND as a Universal Gate

Any gate that can realize all logical functions is said to be universal gate.

1.16.1 NOT Gate from NAND

Logical expression

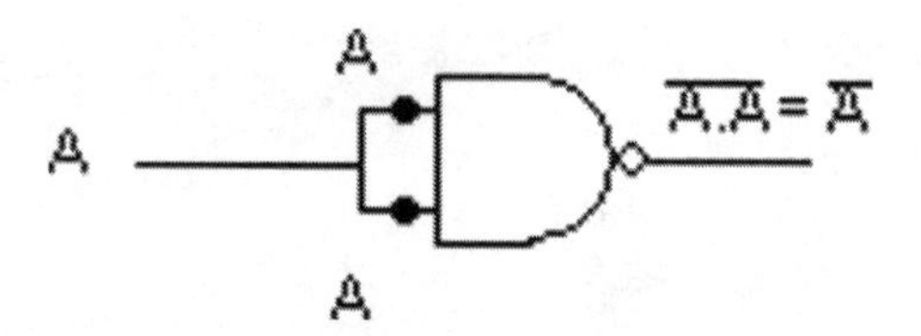

Figure 1.25 NOT gate from NAND gate.

The logic circuit of a NOT gate from a NAND gate is shown in Fig. 1.25.

So, a NAND gate acts as a NOT gate with 1 input and 1 output. Higher input (4 input) NAND can also be used to realize NOT gate as shown in Fig. 1.26.

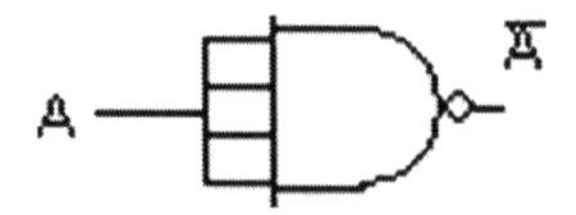

Figure 1.26 NOT gate from four input NAND gate.

1.16.2 OR Gate from NAND

Logical expression from OR

$$Y = A + B$$

Making complement of a complement

$$Y = A + B = \overline{\overline{A + B}}$$

from Demorgan's theorem.

$$Y = A + B = A.B$$

The corresponding logic circuit is given in Fig. 1.27.

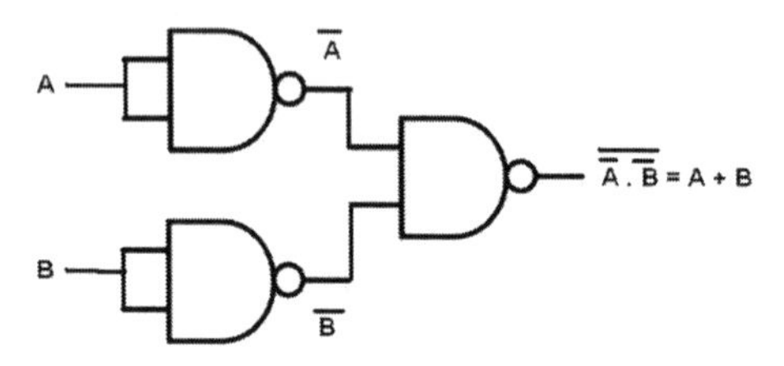

Figure 1.27 OR gate from NAND gate.

1.16.3 AND Gate from NAND Gate

Logical expression for AND

$$Y = A.B$$
$$Y = \overline{A.B}$$

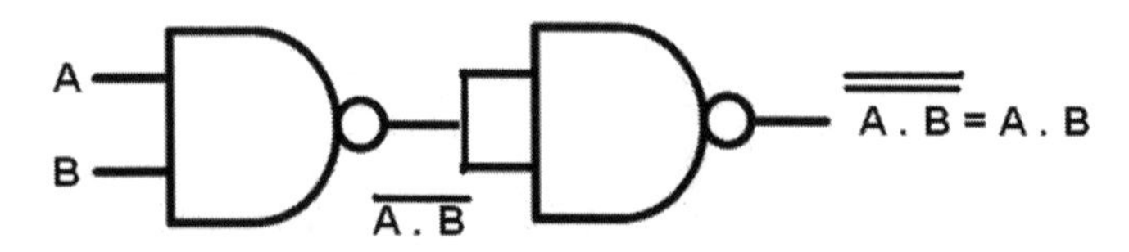

Figure 1.28 AND gate from NAND gate.

The logic circuit is given in Fig. 1.28.

1.16.4 NOR Gate from NAND Gate

Logical expression

$$Y = \overline{A + B}$$

$$Y_{\text{NOR}} = \overline{\overline{A}\,\overline{B}}$$

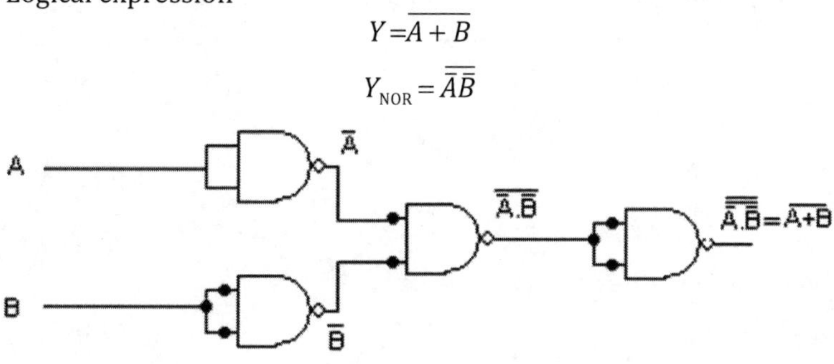

Figure 1.29 NOR gate from NAND gate.

The logic circuit is depicted in Fig. 1.29. So, we require 4 NAND gates.

1.16.5 XOR Gate from NAND Gate

Logical expression

$$Y = A.\overline{B} + \overline{A}\,B$$

Taking complement on both sides

$$\overline{Y} = \overline{A.\overline{B} + \overline{A}\ B}$$

Now Demorganizing the right hand side,

$$\overline{Y} = \overline{A.\overline{B}} + \overline{\overline{A}\ B}$$

Again taking complement on both sides

$$\overline{Y} = \overline{\overline{A.B} \cdot \overline{AB}} = \overline{C.D}$$

$$\text{Let } \overline{A.\overline{B}} = C$$

$$\text{Let } \overline{A.\overline{B}} = D$$

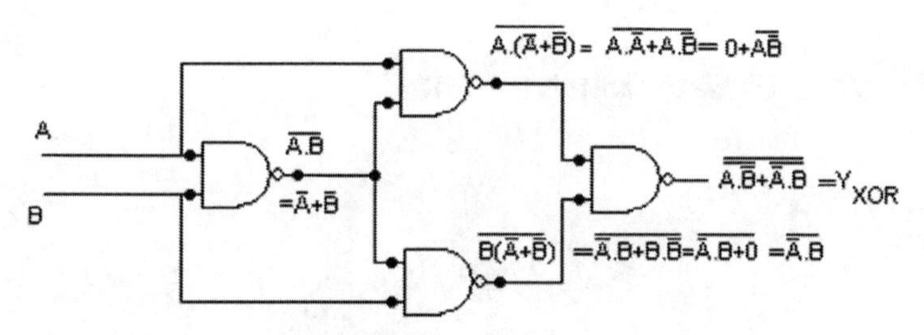

Figure 1.30 XOR gate from NAND gate.

The logic circuit is depicted in Fig. 1.30.

1.16.6 XNOR Gate from NAND Gate

Logical expression for XNOR or equality detector:

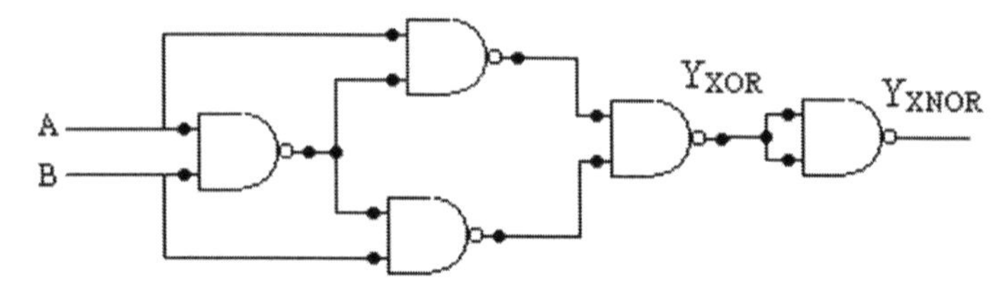

Figure 1.31 XNOR gate from NAND gate.

$$Y = \overline{A\overline{B} + \overline{A}B} = \overline{Y_{XOR}}$$

The output of XOR gate to another NOT gate.

So, we require 5 two input NAND gate to realize XNOR gate as shown in Fig. 1.31.

1.17 NOR as a Universal Gate

1.17.1 NOT Gate from NOR Gate

$$Y = \overline{A}$$

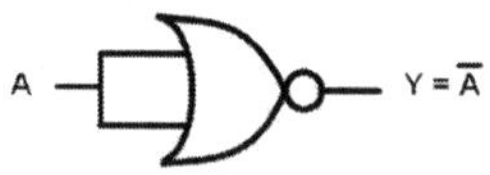

A — Y = $\overline{A}$

Figure 1.32 NOT gate from NOR gate.

The logic circuit is depicted in Fig. 1.32.

$$Y = \overline{A + A} \text{ (as } A+A = A\text{)}$$

1.17.2 AND Gate from NOR Gate

$$Y = A.B$$

$$Y = \overline{\overline{AB}}$$

$$Y = \overline{\overline{A} + \overline{B}}$$

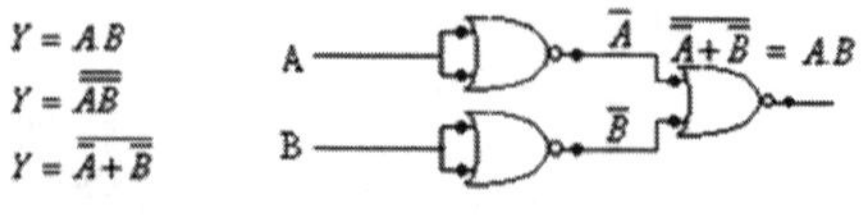

$Y = AB$

$Y = \overline{\overline{AB}}$

$Y = \overline{\overline{A} + \overline{B}}$

A, B, $\overline{A}$, $\overline{B}$, $\overline{\overline{A}+\overline{B}} = AB$

Figure 1.33 AND gate from NOR gate.

The logic circuit is shown in Fig. 1.33.

1.17.3 OR Gate from NOR Gate

$$Y = A + B$$

$$Y = \overline{\overline{A + B}}$$

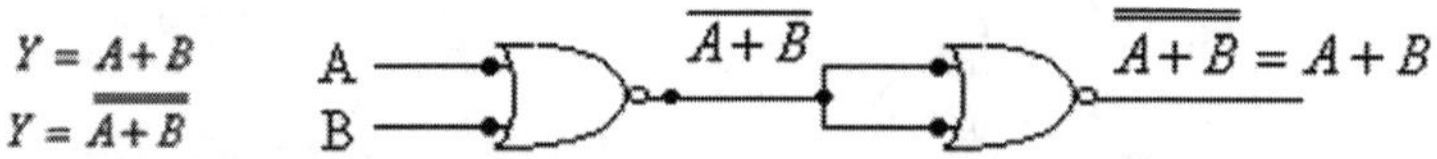

$Y = A+B$ A B $\overline{A+B}$ $\overline{\overline{A+B}} = A+B$

$Y = \overline{\overline{A+B}}$

Figure 1.34 OR gate from NOR gate.

The logic circuit is shown in Fig. 1.34.

1.17.4 NAND Gate from NOR Gate

$$Y = \overline{A.B} = \overline{A} + \overline{B}$$

$$Y = \overline{A.B} = \overline{\overline{\overline{A} + \overline{B}}}$$

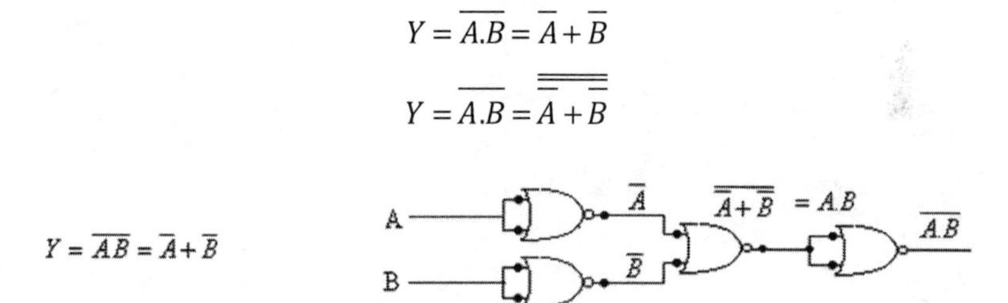

Figure 1.35 NAND gate from NOR gate.

The logic circuit is shown in Fig. 1.35.

1.17.5 XOR Gate from NOR Gate

Assume in the form of product of sum form POS.

$$Y = (A + B)(\overline{A} + \overline{B}) = \overline{A}B + A\overline{B}$$

$$\overline{Y} = \overline{(A + B)(\overline{A} + \overline{B})} = \overline{(A + B)} + \overline{(\overline{A} + \overline{B})}$$

$$Y = \overline{\overline{Y}} = \overline{\overline{A + B} + \overline{\overline{A} + \overline{B}}} = \overline{C + D},\ \ C = \overline{A + B}, D = \overline{\overline{A} + \overline{B}}$$

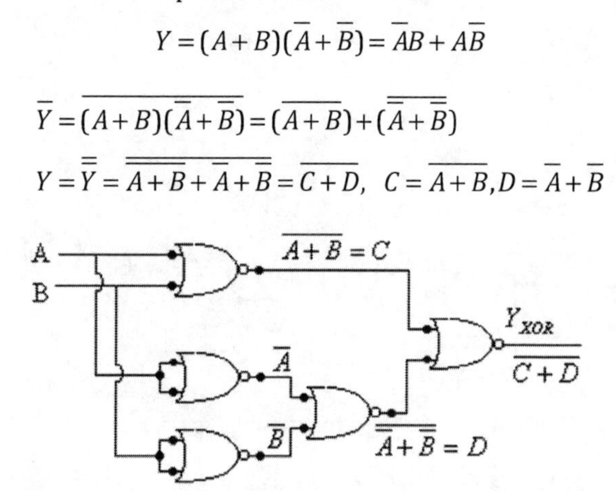

Figure 1.36 XOR gate from NOR gate.

So 5 two input NOR gate is required to realize 2-input XOR gate as shown in Fig. 1.36 or it can be realized in this way (Fig. 1.37).

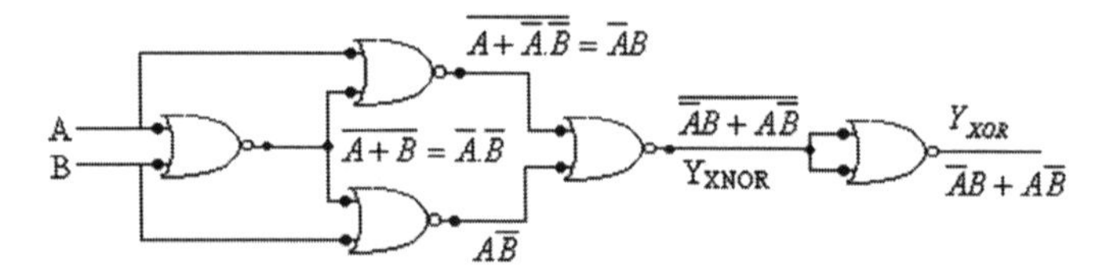

Figure 1.37 XOR gate from NOR gate in other way.

The realization of XOR gate is more economical with NAND gate. The realization of XNOR gate is more economical with NOR gate. (as 4 NOR gates are only required)

1.18 AND-OR Logic

For example-1

$$Y = AB + B\overline{C}$$

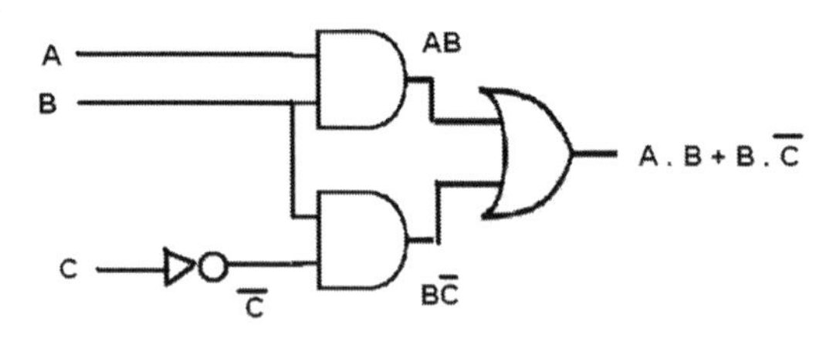

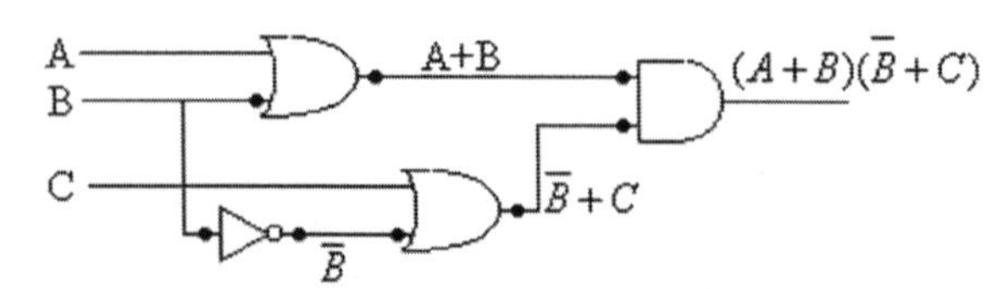

Figure 1.39 Realization of a logic function using OR and AND logic.

But none (1) and (2) are not SSOP or SPOS. (Standard sum of product SSOP or Standard product of sum SPOS)

If we want to realize SSOP or SPOS then we have:

1. by using either NAND or NOR (universal gate)
2. by NAND gate and by NOR gate (given a choice)

Rule: Sum of product form should be realized by using NAND gates only whereas the product of sum form can be better realized by NOR gates only.

$$Y = AB + B\overline{C}$$

$$Y = \overline{\overline{AB + B\overline{C}}}$$

$$Y = \overline{\overline{AB}.\ \overline{B\overline{C}}}$$

Realization by NAND gates is shown in Fig. 1.40.

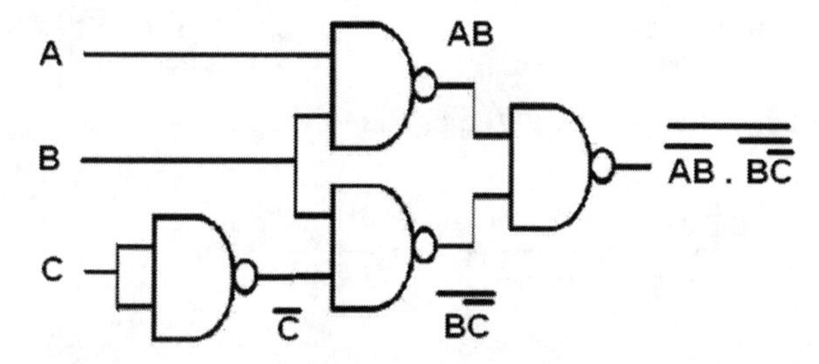

Figure 1.40 Realization of a logic function in SOP form using NAND gate.

Realization by NOR gates is shown in Fig. 1.41.

$$Y = (A+B)(\overline{B}+C)$$

$$\overline{Y} = \overline{(A+B)(\overline{B}+C)}$$

$$\overline{Y} = \overline{(A+B)} + \overline{(\overline{B}+C)}$$

$$\overline{\overline{Y}} = Y = \overline{\overline{A\_B} + \overline{\overline{B}+C}}$$

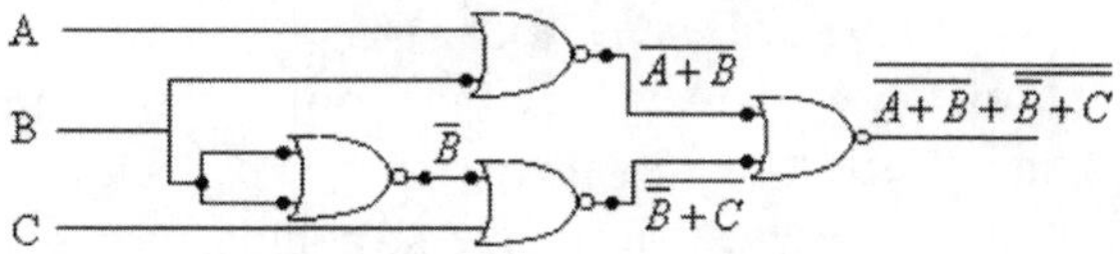

Figure 1.41 Realization of a logic function in SOP form using NOR gate.

1.19 Gray Code and Binary to Gray Code Conversion

Now, we will discuss about K map method of simplification of the logical expressions. But K map utilizes a special coding scheme which requires some elaboration of gray code as K map uses it.

1.19.1 Gray Code

It is a reflected nonweighted binary code.

1.19.2 Binary Code

Binary number 1101 $= 1 \times 2^3 + 1 \times 2^2 + 0 \times 2^1 + 1 \times 2^0 = 8 + 4 + 0 + 1 = 13$

So, binary code is a weighted code but gray code is nonweighted.

Table 1.14 Gray code representation of decimal numbers

Decimal code	Binary code	Gray code
0	000	000
1	001	001
2	010	011
3	011	010
4	100	110
5	101	111
6	110	101
7	111	100

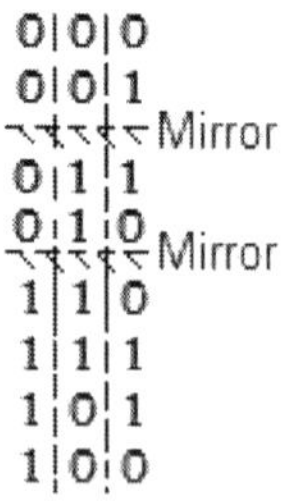

0|0|0
0|0|1
Mirror
0|1|1
0|1|0
Mirror
1|1|0
1|1|1
1|0|1
1|0|0

That is why gray code is called reflected code.

In the gray code any two successive numbers differ by one bit only. This is called unit distance property of a gray code that is only one bit change between successive number 5–6 whereas, more than one bit (5 = 101, 6 = 110) have changed.

This unit distance property of the gray code is utilized in the Karnaugh Map for minimizing a given logical function. Representation of the decimal numbers by binary code and grey code is given in Table 1.14.

1.19.3 Binary to Gray Code Conversion

1. 111 (binary of decimal 7) ... 100 (gray code of decimal 7) Most significant bits will remain same for gray code.
2. Perform exclusive-OR between the successive bits starting MSD. (MSDs next bit both are "1" in binary have XOR is 0. Similarly, the 2nd and 3rd bits).

Decimal	Binary	Gray Code
57	111001	100101

1.20 Karnaugh Map

It is a graphical method of minimizing a given logical function or given result table. The K map is a diagram in which each term of a logical function or each row of a truth table occupies a specific area in the diagram. The importance of the K map lies in the fact that the manner in which the areas are chosen so as to minimize a given logical function by usual inspection. For n number of input variables (for a logical function) the K map will contain 2^n number of cells or squares cells or squares are arranged in a matrix of rows and columns. The rows and columns are numbered by using gray code.

1.20.1 Limitation

It can minimize a logical function if $n \leq 6$.

For a K map the number of input variables, $n \leq 6$, for $n \geq 6$, the other method of minimization are used. In K map, a given area or cell or squares can have either 1s or 0s but not both.

Table 1.15 Truth table of the given logic function

Decimal	A	B	Y
0	0	0	1
1	0	1	0
2	1	0	1
3	1	1	0

$$f(A, B) = \Sigma m\ (0, 2).$$

To simplify the truth table shown in Table 1.15 for $n = 2$, number of cells or squares $= 2^2 = 4$.

We use

2 rows or 2 columns

or

4 columns and 1 row

1. 2 rows or 2 columns

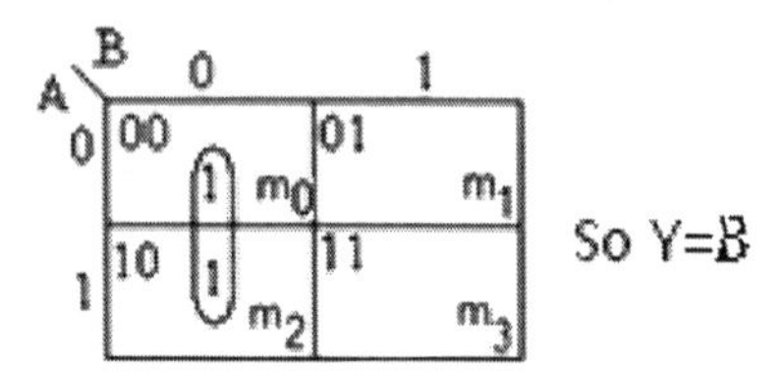

2. 4 columns and 1 row

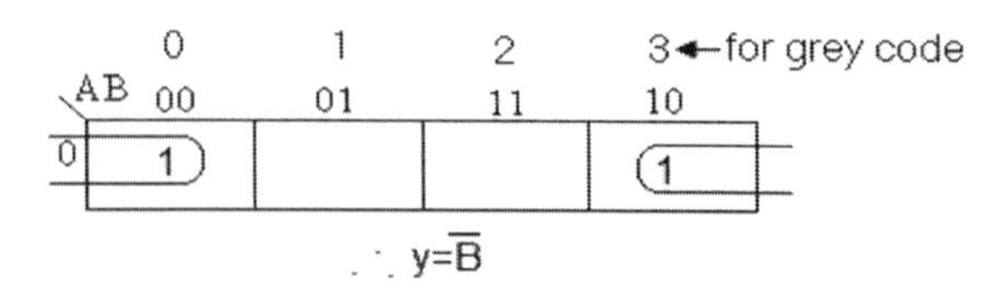

The gray code is used for column number as unit (blank) property is utilized here.

Any two cells in K map will be called an adjacent cell.

Two or more cells are said to be adjacent cells if 1, 2, 3 variables can be eliminated between them or in other words, two cells will be called adjacent cells if the two cells differ only in one bit value.

That is 1 and 1 in graph can be continued.

$$Y = m_0 + m_2 = \overline{A}\overline{B} + A\overline{B} = 00 + 10 \leftarrow \text{in binary}$$
$$= \overline{A}\overline{B} + A\overline{B} = \overline{B}(A + \overline{A}) = \overline{B}(\because A + \overline{A} = 1)$$

$\therefore$ $Y = \overline{B}$ is independent of A. This is what the truth table is showing after the minimization.

So realizing it—

B—▷o—Y   B—[NAND]—$Y=\overline{B}$ B—[NOR]—Y

If the truth table is;

A	B	Y
0	0	1
0	1	0
1	0	0
1	1	1

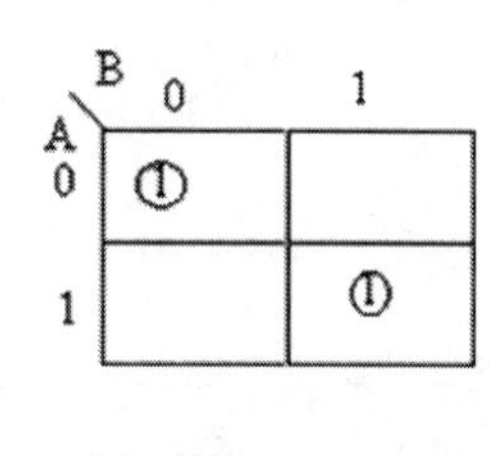

We cannot combine these two cells. So, they are not adjacent cells as the two cells differ by more than 1 bit. So, we encircle them individually.

Let us consider, Y as in the following:

$$Y = 00 + 11$$
$$Y = \overline{A}\overline{B} + AB$$
$$Y = \overline{A \oplus B}$$

(When we plot 1 in the K map the resulting expression will be in the form of SOP)

So, Y is the XNOR logic. So, for all expressed in min term form.

Now let us, try with max terms. For plotting in $f(A, B) = \prod \text{M}\ (1, 3)$ K map.

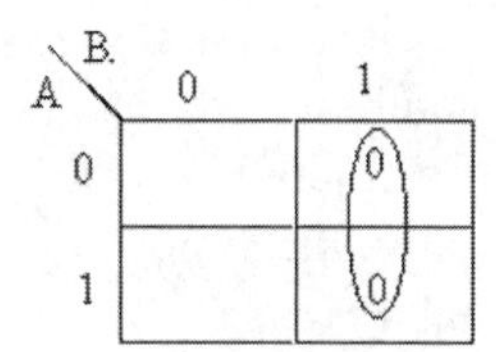

Whenever, it is in the form of max term, plot 0 in K map.

So, we can combine the two cells.

$$\begin{aligned} f(A, B) &= (\text{M}_1)\ (\text{M}_3) \\ &= (0\ 1)\ (1\ 1) \rightarrow \text{in binary} \\ &= \overline{\text{B}} \end{aligned} \qquad (1.25)$$

Examples:

$f(A, B, C) = \sum m\ (0, 1, 3, 5, 7)$ minimize this by K map.

As number of variables is three, eight cells are required = 2^3 conversion is four columns and two rows.

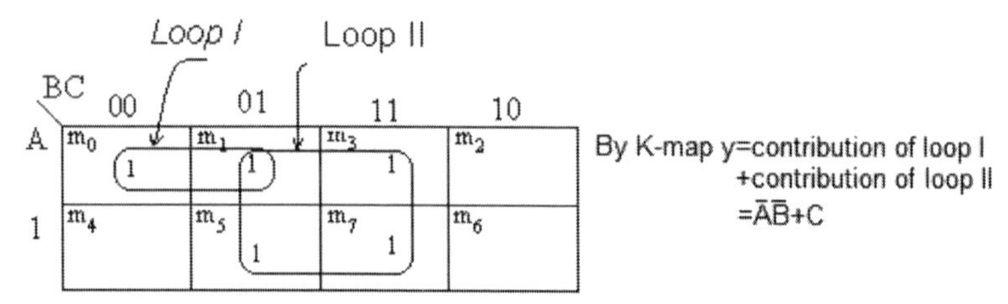

Column variables identified by BC always m_1 m_2 m_3 all written with respect to binary numbers. So, minimization proceeds as: (Analytical ways)

$$f(A,B,C) = (m_0 + m_1) + \{(m_1 + m_3) + (m_5 + m_7)\}$$
$$= (000 + 001) + \{(001 + 011) + (101 + 111)\}$$
$$= (\overline{A}\overline{B}\overline{C} + \overline{A}\overline{B}C) + \{(\overline{A}\overline{B}C + \overline{A}BC) + (A\overline{B}C + ABC)\}$$
$$= AB(C + C) + \{AC(B + B) + AC(B + B)\}$$
$$= \overline{A}\overline{B} + \{\overline{A}C + AC\}$$
$$= \overline{A}\overline{B} + C\{A + \overline{A}\}$$
$$= \overline{A}\overline{B} + C$$

So, two adjacent cells when combined eliminate one variable that is in result (1.25) where (A, B) is eliminated to $\overline{B}$ only. Four adjacent cells when combined two variables are eliminated (2nd loop II). In general, we can say if we combine 2^n adjacent cells, n variables are eliminated.

Three cells cannot combine, combination of cells always in multiples of 2^n (2, 4, 8, 16, ...). But six cells, 12 cells (multiple of 2) cannot be combined for simplification.

Three variables Karnaugh Map:

Let us take an example of three variables K map.

$y = f(A, B, C) = \sum m\ (1, 3, 5, 7)$→As it has three variables hence there are $2^3 = 8$ cells which can be arranged in two rows and four columns.

A \ BC	00	01	11	10
0	m_0	m_1 1	m_3 1	m_2
1	m_4	m_5 1	m_7 1	m_6

The combining 2^n adjacent cells/squares will eliminate n number of variables, here $n = 2$ as $2^2 = 4$ cells are adjacent.

$y = f(A, B, C) = C$ by visual inspection variables A and B are eliminated.

$$y = f(A, B, C) = \sum m(0, 1, 2, 3)$$

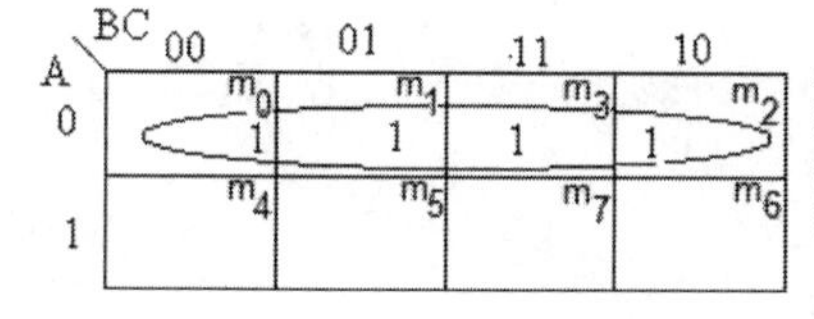

As no. of adjacent cells are four 2^2 Hence n=2, so two variables are eliminated. From the encircled loop y=$\bar{A}$ (B and C are eliminated)

$y = f(A,B,C) = A$. As both B and C can be eliminated from the columns

Example: $y = f(A, \bar{B}, C) = \sum m\ (0, 2)$ as the two cells becomes adjacent cells by folding it.

A \ BC	00	01	11	10
0	1			1
1				

$$y = f(A,B,C) = \overline{A}\overline{C}$$

Example: $y = f(A, B, C) = \sum m\ (0, 6)$

A \ BC	00	01	11	10
0	1			
1				1

None of the cells will combine with each other and both the cells are called Essential Prime Implicit. Final minimized expression will be:

$$y = f(A,B,C) = (000 + 110) = \overline{A}\overline{B}\overline{C} + AB\overline{C}$$

From the few examples illustrated above we can predict the rules for taking combination of adjacent cells.

Rules for taking combination of adjacent cells:

To select the combination, the rules are as follows: (In K map)

1. In each combination of cells or squares containing 1s or zeroes, there should be at least one cell, which is not involved in any other combination.
2. All the cells or squares containing 1s or 0s should be combined in such a way so that all the cells or squares can be combined by using minimum possible number of combination (that is we have to accommodate more number of 1s or 0s in each combination as possible so that the number of such combinations should be as small as possible.

For a four variable K map:

$$y = f(A, B, C, D) = \sum m\ (1, 5, 6, 7, 11, 12, 13, 15)$$

It has $2^4 = 16$ cells→ with 4 rows and 4 columns as shown in Fig. 1.42.

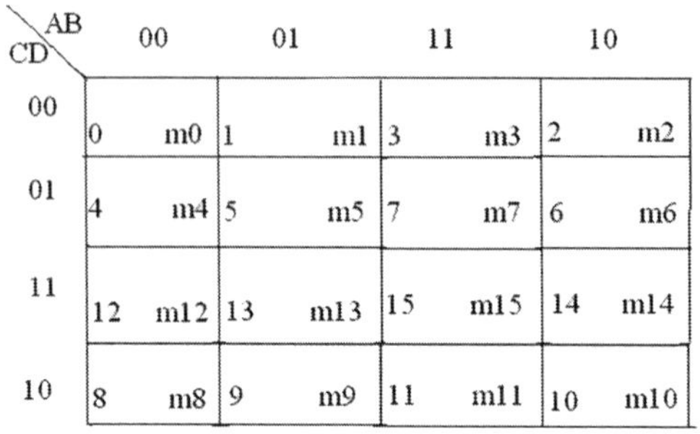

CD \ AB	00	01	11	10
00	0 m0	1 m1	3 m3	2 m2
01	4 m4	5 m5	7 m7	6 m6
11	12 m12	13 m13	15 m15	14 m14
10	8 m8	9 m9	11 m11	10 m10

Figure 1.42 K Map 1 (A four variable K Map).

A four variable boolean function is given as $y = f\ (A, B, C, D) = \sum m$ (1,5,6,7,11,12,13,15). The corresponding K Map is shown in Fig. 1.43.

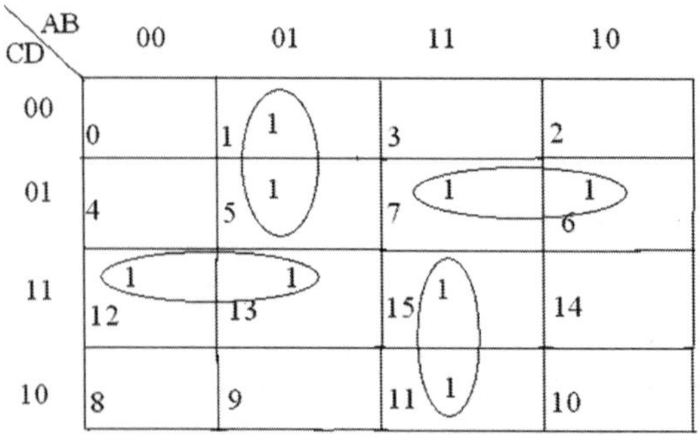

CD \ AB	00	01	11	10
00	0	1 1	3	2
01	4	5 1	7 1	6 1
11	12 1	13 1	15 1	14
10	8	9	11 1	10

Figure 1.43 K Map 2.

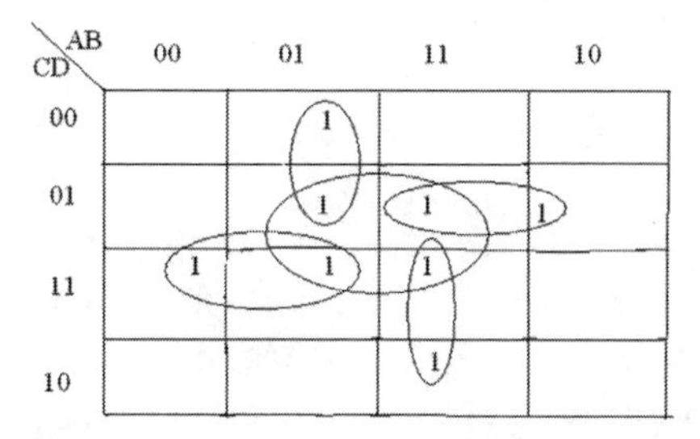

Figure 1.44 K Map 3.

Another four-variable function is given by

$$f(A,B,C,D) = \overline{A}\overline{C}D + \overline{A}BC + AB\overline{C} + ACD$$

The corresponding K Map is given in Fig. 1.45.

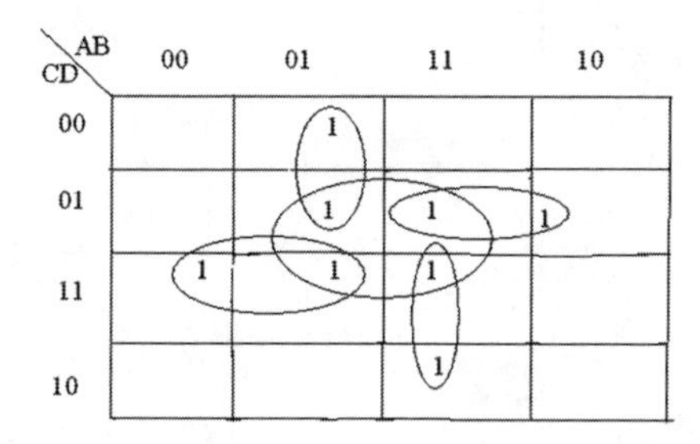

Figure 1.45 K Map 4.

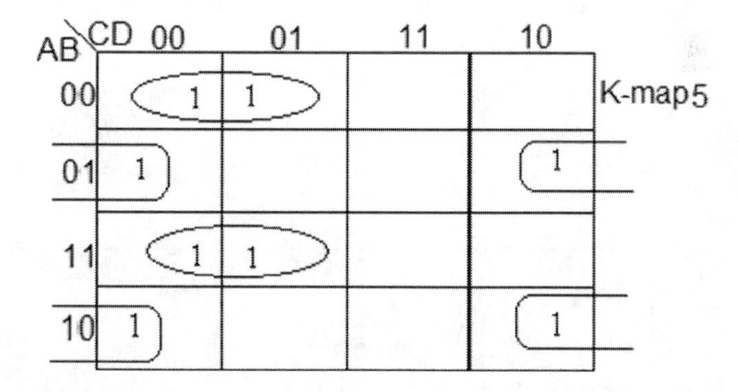

Figure 1.46 K Map 5.

The logical expression for this K Map is given below.

$$y = f(A,B,C,D) = 000- + 01-0 + 110- + 10-0$$
$$= \overline{A}\overline{B}\overline{C} + \overline{A}B\overline{D} + AB\overline{C} + A\overline{B}\overline{D}$$

$$y = f(A,B,C,D) = \sum m(2,3,8,10,12)$$
$$y = f(A,B,C,D) = 1-00+001-+10-0$$
$$= A\overline{C}\,\overline{D} + \overline{A}\,\overline{B}C + A\overline{B}\,\overline{D} \quad (1.26)$$

So, this is final minimum expression. But minimized expression may be more than one. We take that minimum expression and can be realized for minimum number of gates. The K Map for this expression is shown in Fig. 1.46.

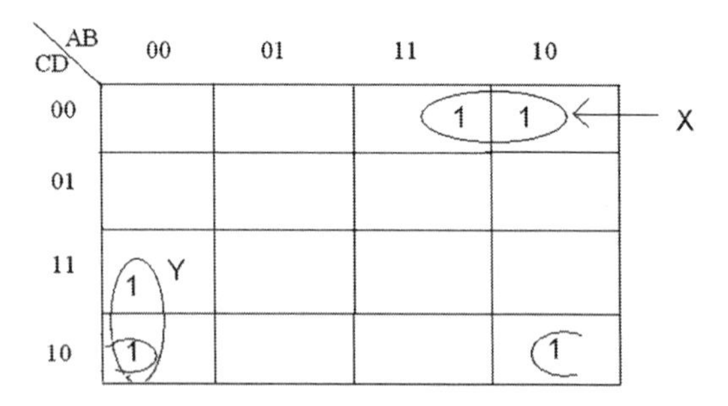

Figure 1.47 K Map 6.

Another solution to the K Map shown in Fig. 1.47 is also possible and it has been shown in Fig. 1.48.

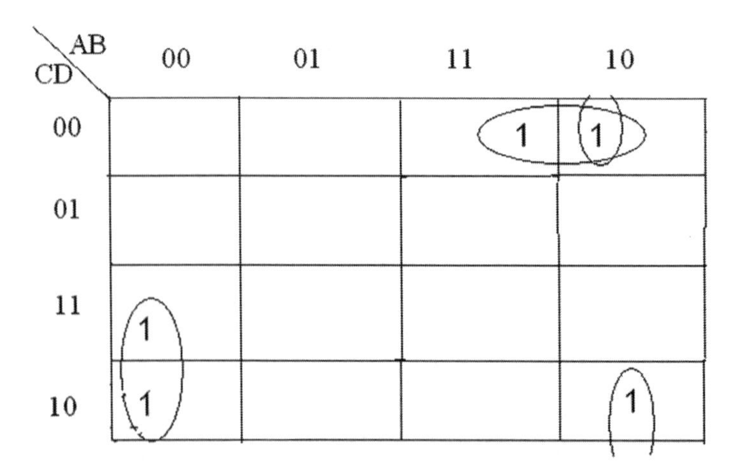

Figure 1.48 K Map 7.

The expression for this K Map is given as

$$y = f(A,B,C,D) = \Sigma m(2,3,8,10,12)$$

$$y = f(A,B,C,D) = 001- + 1-00 + -010$$

$$= \bar{A}\bar{C}D + \bar{A}\bar{B}C + \bar{B}C\bar{D} \qquad (1.27)$$

The last term in Eqs. (1.26) and (1.27) are same, so any one of them can be considered as the final minimized expression as the other 1st two terms are same.

1.20.2 Plotting Zeros (Max Term Representation)

(Instead of 1) So max term expression will come as shown in Fig. 1.49.

AB \ CD	00	01	11	10
00	0 (0)	1	3	2 (0)
01	4	5	7	6
11	12	13	15	14
10	8 (0)	9	11	10 (0)

From K.map.8
y=f(A,B,C,D)=ΠM(0,2,8,10)
All the four zeroes can be combined by folding twice.
y=f(A,B,C,D)=(-0-0)=(B+D)

Figure 1.49 K Map 8.

K Map is given in Fig. 1.50.

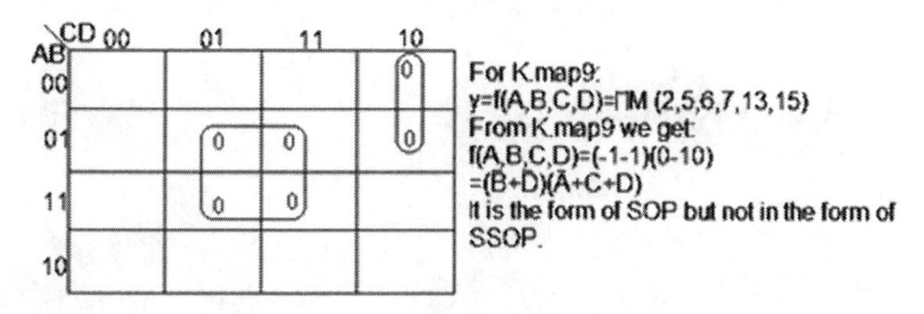

Figure 1.50 K Map 9.

1.20.3 Five Variable K Map

There will be 32 squares (2^5). It is plotted by two adjacent K-maps. (K-map11 & K-map12) as demonstrated in Fig. 1.51.

Example: $f(A, B, C, D, E) = \sum m\ (0, 1, 2, 3, 16, 17, 18, 19)$

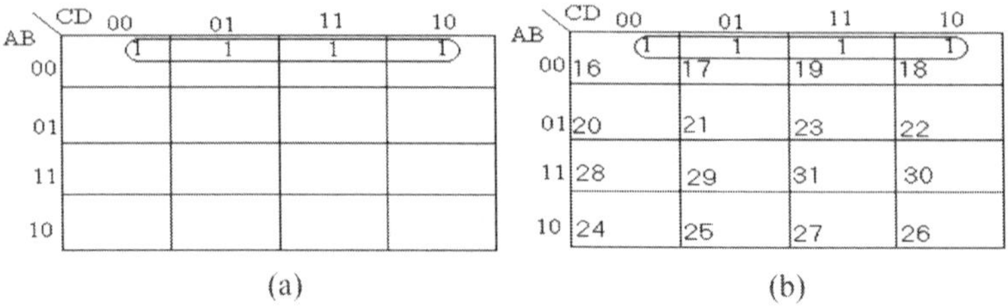

AB \ CD	00	01	11	10
00	1	1	1	1
01				
11				
10				

(a)

AB \ CD	00	01	11	10
00	16 1	17 1	19 1	18 1
01	20	21	23	22
11	28	29	31	30
10	24	25	27	26

(b)

Figure 1.51 (a) K Map 10 and (b) K Map 11.

From Fig. 1.51(a), minimized expression is

$$\begin{matrix} & ABCDE \\ y_1 = & (000--) = \overline{A}\,\overline{B}\,\overline{C} \ (\because A = 0) \end{matrix}$$

From Fig. 1.51(b), minimized expression is

$$\begin{matrix} & ABCDE \\ y_2 = & (100--) = A\overline{B}\,\overline{C} \ (\because A = 1) \end{matrix}$$

So, final minimized expression is

$$y = y_1 + y_2 = \overline{A}\,\overline{B}\,\overline{C} + A\overline{B}\,\overline{C} = \left(\overline{A} + A\right)\overline{B}\,\overline{C}$$

$$\therefore y = \overline{B}\,\overline{C}$$

That is three variables are eliminated.

1.20.4 Six Variable K Map

Let us take an example

$y = f(A, B, C, D, E, F) = \sum m\ (0, 1, 2, 3, 32, 33, 34, 35, 48, 49, 50, 51)$

For six variables K map we should have $2^6 = 64$ cells or squares. This is arranged by using four variable map as shown in Fig. 1.52.

From Fig. 1.52(a), minimized expression.

$$\begin{matrix} & ABCDEF \\ y_1 = & (0000--) = \overline{A}\,\overline{B}\,\overline{C}\,\overline{D} \ (\because A = 0, B = 0) \end{matrix}$$

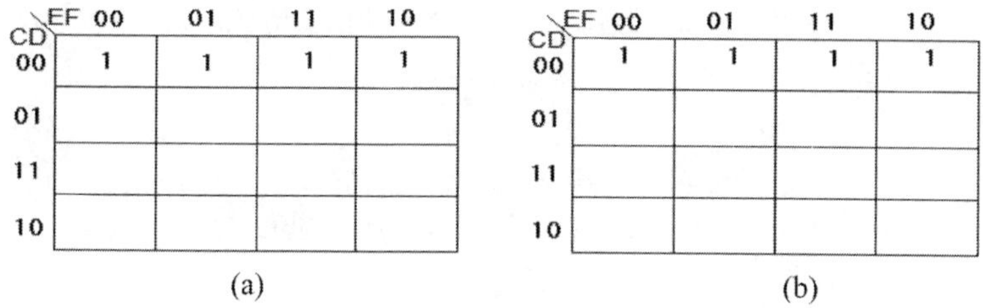

CD \ EF	00	01	11	10
00	1	1	1	1
01				
11				
10				

(a)

CD \ EF	00	01	11	10
00	1	1	1	1
01				
11				
10				

(b)

Figure 1.52 (a) K Map 12 and (b) K Map 13.

From Fig. 1.52(b), minimized expression:

$$\begin{array}{c} ABCDEF \\ y_2 = (0100--) = \overline{A}B\overline{C}\overline{D} (\because A = 0, B = 1) \end{array}$$

From Fig. 1.52a, minimized expression:

$$\begin{array}{c} ABCDEF \\ y_3 = (1000--) = A\overline{B}\overline{C}\overline{D} (\because A = 1, B = 0) \end{array}$$

From Fig. 1.52b, minimized expression:

$$\begin{array}{c} ABCDE \\ y_4 = (1100--) = AB\overline{C}\overline{D} (\because A = 1, B = 1) \end{array}$$

Final minimized expression:

$$y = y_1 + y_2 + y_3 + y_4$$

$$y = f(A, B, C, D, E, F) = \overline{A}\overline{B}\overline{C}\overline{D} + \overline{A}B\overline{C}\overline{D} + A\overline{B}\overline{C}\overline{D} + AB\overline{C}\overline{D}$$

$$= \overline{A}\overline{C}\overline{D}(\overline{B} + B) + A\overline{C}\overline{D}(\overline{B} + B) = \overline{A}\overline{C}\overline{D} + A\overline{C}\overline{D} = (\overline{A} + A)\overline{C}\overline{D} = \overline{C}\overline{D}$$

So, four variables are eliminated from $2^4 = 16$ variables combination (the first row of all squares). This is the maximum number of input variables for which K map is used. For minimizing the logical function by K map we should keep in mind that K map can be applied for minimization of logical function if logical function is given either in min term form or max term form or if the truth table for a logical function is given or if logical expression is given (but the number of input variables is less than or equal to 6).

Simplify or minimize by the K map of the following logical function.

$$y = f(A,B,C,D) = \bar{A}\bar{B}\bar{C} + \bar{A}\bar{B}D + CD$$

Convert the logic function to SSOP or SPOS and then express it in terms of max or max term function and then apply the K map.

$$= \bar{A}\bar{B}\bar{C}(D+\bar{D}) + \bar{A}\bar{B}(C+\bar{C})D + (A+\bar{A})(B+\bar{B})CD$$

$$= \bar{A}\bar{B}\bar{C}D + \bar{A}\bar{B}\bar{C}\bar{D} + \bar{A}\bar{B}CD + \bar{A}\bar{B}\bar{C}D + ABCD + \bar{A}BCD + A\bar{B}CD + \bar{A}\bar{B}CD$$

$\rightarrow$ SSOP form

$= 0001, 0000, 0011, 0001, 1111, 1011, 0111, 0011$

$f(A, B, C) = \sum m\ (0, 1, 3, 7, 11, 15)$

= min term form

$= \prod M\ (2, 4, 5, 6, 8, 9, 10, 12, 13, 14)$

= max term form

Now draw the K map plot

1. for min term form shown in Fig. 1.53 and
2. for max term form shown in Fig. 1.55.

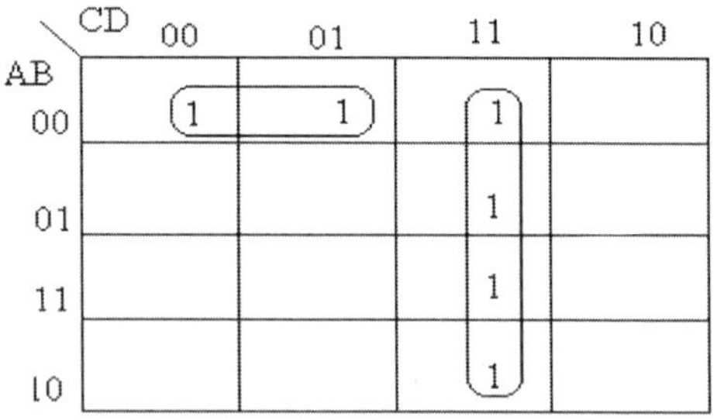

AB \ CD	00	01	11	10
00	1	1	1	
01			1	
11			1	
10			1	

Figure 1.53 K Map 14.

$$f(A,B,C,D) = (000-) + (--11) = \bar{A}\bar{B}\bar{C} + CD \quad (\text{SOP form})$$

So, final minimized expression in SOP form (for realization by NAND gate, we always express the minimized expression in SOP form). Taking complement on both sides:

$$\bar{f}(A,B,C,D) = \overline{CD + \bar{A}\bar{B}\bar{C}} = \overline{CD}.\overline{\bar{A}\bar{B}\bar{C}}$$

Again complementing both sides:

$$y = f\left(A,B,C,D\right) = f(A,B,C,D) = \overline{\overline{\overline{CD}.\overline{\overline{A}}.\overline{\overline{B}}.\overline{\overline{C}}}}$$

Realization by NAND gate is shown in Fig. 1.54.

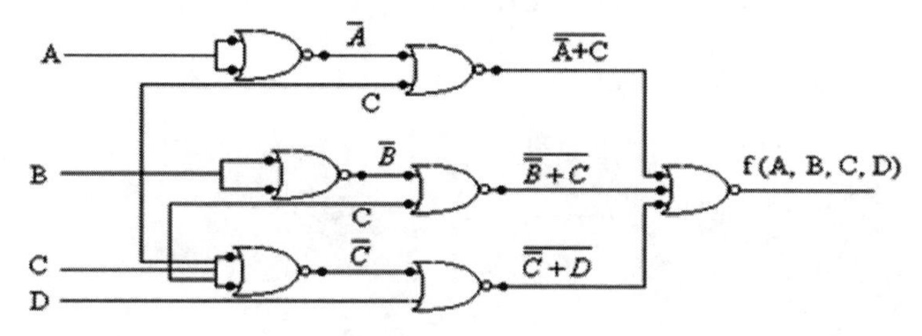

Figure 1.54 Realization of a logic function using NAND gate.

For realization of the function we require five two input NAND gates and one three input NAND gate. Some NAND gate in the form of chips is given in Table 1.16.

Table 1.16 Specification of ICs

7410	7400	7402	7432	7408	7486	7404	7420	7430
Triple three input NAND gate	Quad two input NAND gate	Quad two input NOR gate	Quad two input OR gate	Quad two input AND gate	Quad two input XOR gate	Hex inverter gate	Double four input NAND gate	Single twelve input NAND gates

Now, plotting with the max term in the form of POS and can be realized by NOR gates.

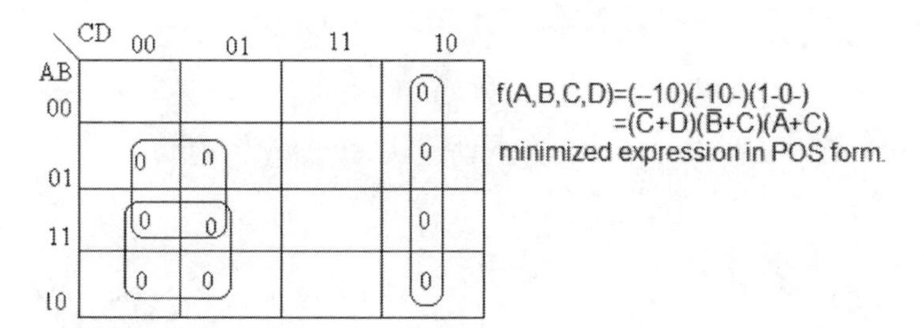

Figure 1.55 K Map 15.

Realizing it by NOR gate

$$\overline{f}(A,B,C,D)=\overline{(\overline{C}+D)(\overline{B}+C)(\overline{A}+C)}=\overline{\overline{C}+D}+\overline{\overline{B}+C}+\overline{\overline{A}+C}$$

$$y=\overline{\overline{f}}(A,B,C,D)=\overline{\overline{\overline{C}+D}+\overline{\overline{B}+C}+\overline{\overline{A}+C}}==\overline{u+v+w}$$

$$where,\, u=\overline{\overline{C}+D}, v=\overline{\overline{B}+C}, w=\overline{\overline{A}+C}$$

Realization by NOR gates is shown in Fig. 1.56.

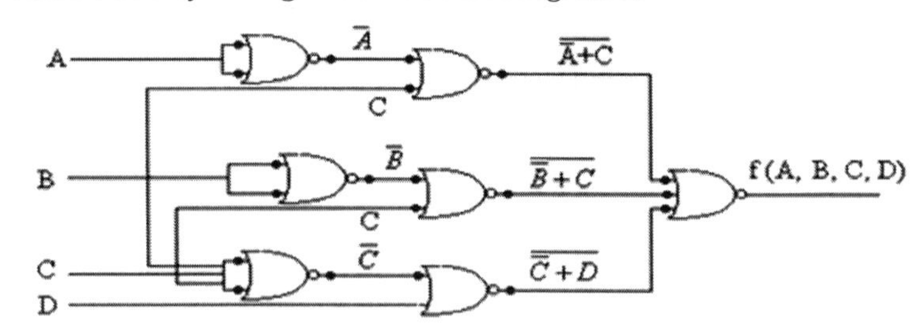

Figure 1.56 Realization of a logic function using NOR gate.

We do not have three input NOR gate in lab, we have two inputs.

$$Y=\overline{A+B+C}=3 \text{ input NOR gate expression.}$$

How we realize it by using two input NOR is shown in Fig. 1.57.

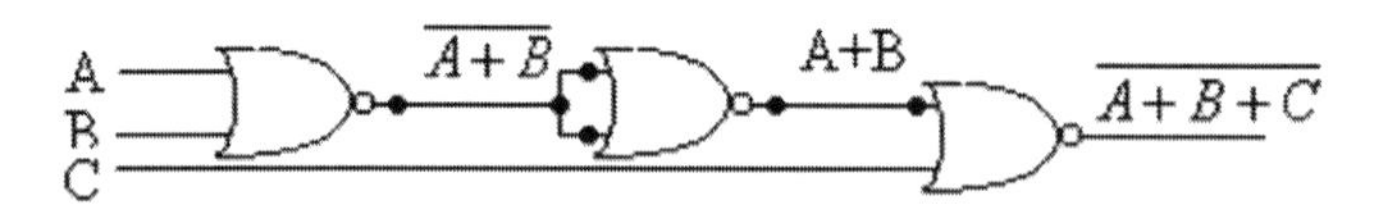

Figure 1.57 Realization of a 3 input NOR gate using 2 input NOR gates.

Why is a logic function called min term function or max term form?

$$f(A,B,C)=AB\overline{C}+\overline{A}BC$$

This is SSOP and it is min term form.

$\overline{A}\overline{B}C=$ consider the single min term of the given logic function.

Plotting this single min term in K map

A \ BC	00	01	11	10
0				
1				1

For each logic minterm, the K-Map will be filled up by minimum no. of 1's. So it is called minterm, we have only one 1 in the K-Map.

Example:

if $f\ (A,B,C) = (A + B + \bar{C})\ (A + B + C) = (000)(100) = \Pi\text{M}\ (0,4)$
$Q.f\,(A, B, C) = (A + B + C)\ (A + B + C)$

$A + B + C$ is zero when all three variables are zero. The expression is in the form of SPOS.

(0,4)

A \ BC	00	01	11	10
0	0	1	1	1
1	0	1	1	1

So, for each max term, there are maximum number of 1s, so max term form. So, maximum number of 1s occupies the cells of a K map in max term form.

Minimize the following logical function by using three variables K map for a four variable function.

$$f\,(A, B, C, D) = \sum m\ (0, 2, 3, 5, 6, 7, 9, 11, 13)$$

So, we have to plot two 3 variable K maps. First search for any cell in Fig. 1.58(a) and Fig. 1.58(b) which have got the same position with respect to BCD and which do not have any other combination. *X* and *Y* are combined between the two K maps, BCD position.

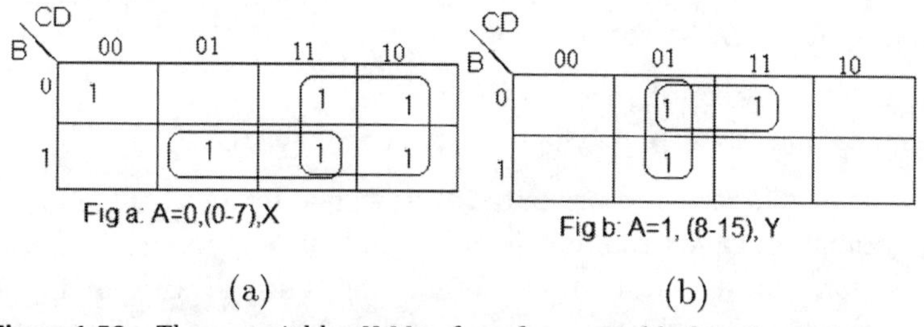

Figure 1.58 Three-variables K-Map for a four-variable function.

So, we have four combinations and we have four terms.

$$f\,(A, B, C, D) = (0\text{–}1\text{–}) + (00\text{–}0) + (10\text{–}1) + (\text{–}101)$$
$$= (\bar{A}\,C) + (\bar{A}\,\bar{B}\,\bar{D}) + (\bar{A}\,BD) + (BCD) \qquad (1.28)$$

We can verify by using four variables K map.

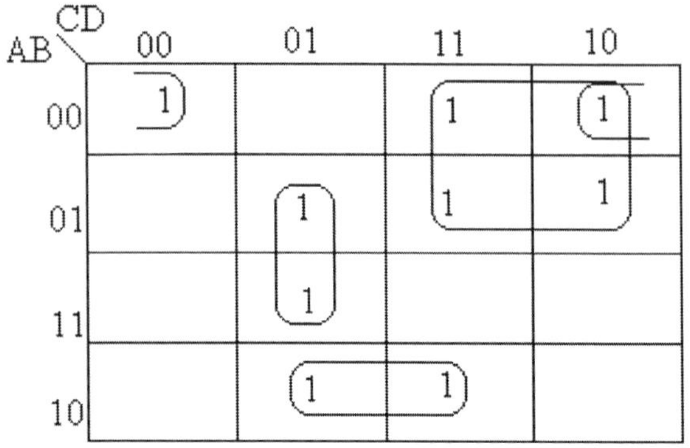

AB \ CD	00	01	11	10
00	1		1	1
01		1	1	1
11		1		
10		1	1	

$$f(A,B,C,D) = (0-1-)+(00-0)+(10-1)+(-101)$$
$$= \left(\bar{A}C\right)+\left(\bar{A}\bar{B}\bar{D}\right)+\left(A\bar{B}D\right)+\left(B\bar{C}D\right) \quad (1.29)$$

$f(A, B, C, D) = (0–1–) + (00–0) + (–101) + (10–1)$

which is same as Eq. (1.28).

1.21 Completely and Incompletely Specified Logic Functions

Logical functions are of two types:

1. completely specified logical function
2. incompletely specified logical function

A logical function whose output is specified for all possible input combination called completely specified logical function. A logical function whose output may not be specified for certain input combinations/conditions or for which a certain input combinations may never occur is called incomplete specified logical function. A table showing examples of completely and incompletely specified logic functions is shown in Fig. 1.59.

Inputs			Output
A	B	C	F
0	0	0	0
0	0	1	1
0	1	0	0
0	1	1	1
1	0	0	0
1	0	1	0
1	1	0	X
1	1	1	X

Figure 1.59 Completely and incompletely specified logic functions. For the last two conditions the output is not fixed.

As generally, an unknown quantity is represented by X, which is called the do not care condition, because the output may be 0 or 1 but it is not specified. This is an example of Eq. (2).

1.22 Minimization of Incompletely Specified Logic Functions

Another thing is that, certain input combinations are not fixed for which same multiple output logical function is one which may have more than one output for each possible input combination for example of multiple output logical function.

The 9's complement circuit (in binary) is a multiple output incompletely specified logical function (Fig. 1.60).

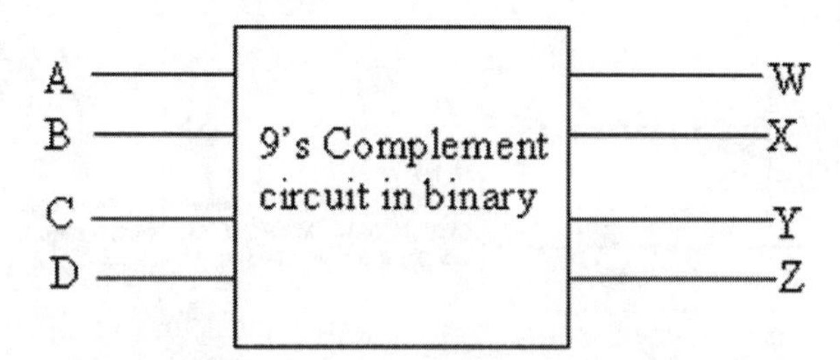

Figure 1.60 Block diagram of an incompletely specified logic function.

The 9's complement circuit should be such so that bit binary output + 4 bit binary input = $(9)_{10} = 1001$

Truth table for 9's complement circuit is given in Table 1.17.

Table 1.17 Truth table for 9's complement circuit

Decimal	A	B	C	D	W	X	Y	Z	
0	0	0	0	0	1	0	0	1	9
1	0	0	0	1	1	0	0	0	8
2	0	0	1	0	0	1	1	1	7
3	0	0	1	1	0	1	1	0	6
4	0	1	0	0	0	1	0	1	5
5	0	1	0	1	0	1	0	0	4
6	0	1	1	0	0	0	1	1	3
7	0	1	1	1	0	0	1	0	2
8	1	0	0	0	0	0	0	1	1
9	1	0	0	1	0	0	0	0	0
10	1	0	1	0	X	X	X	X	
11	1	0	1	1	X	X	X	X	
12	1	1	0	0	X	X	X	X	
13	1	1	0	1	X	X	X	X	
14	1	1	1	0	X	X	X	X	
15	1	1	1	1	X	X	X	X	

These are don't care conditions. Do not care conditions for any input greater than 9, output cannot be obtained.

$$w=\sum m(0,1)+\sum d(10,11,12,13,14,15)=f(A,B,C,D)$$

$w = 1$ when input combination is binary equivalent of decimal 0 or decimal 1 and w = do not care when input combination is binary equivalent of decimal 10, 11, 12, 13, 14, and 15.

For a four bit binary 9's complementary circuit input > 9 is not possible or output for input > 9 is do not care. So, we put the do not care conditions for all *WXYZ* in the O/P.

$$X=\sum m\ (2, 3, 4, 5)+\sum d\ (10, 11, 12, 13, 14, 15)=f\,(A, B, C, D)$$
$$Y=\sum m\ (2, 3, 6, 7)+\sum d\ (10, 11, 12, 13, 14, 15)=f\,(A, B, C, D)$$
$$Z=\sum m\ (0, 2, 4, 6, 8)+\sum d\ (10, 11, 12, 13, 14, 15)$$

For getting the minimized expression for *W, X, Y, Z* we have to use four variable Karnaugh Map demonstrated in Figs. 1.61–1.65.

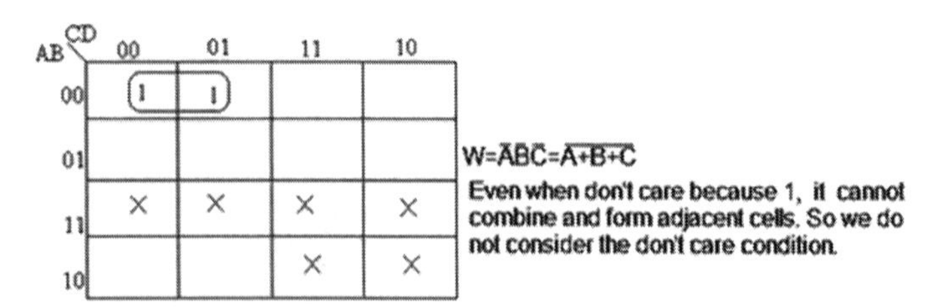

Figure 1.61 K Map for W.

K-map for X without including the don't care conditions.
$X=(001\text{-})+(010\text{-})=\bar{A}\bar{B}C+\bar{A}B\bar{C}$
But expression for X can be further simplified if we take don't care conditions into our condition.

Figure 1.62 K Map for X without incorporating do not care conditions.

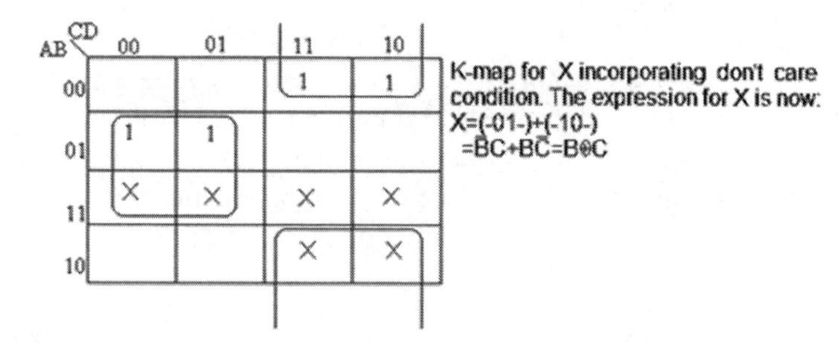

Figure 1.63 K-Map for X incorporating do not care conditions.

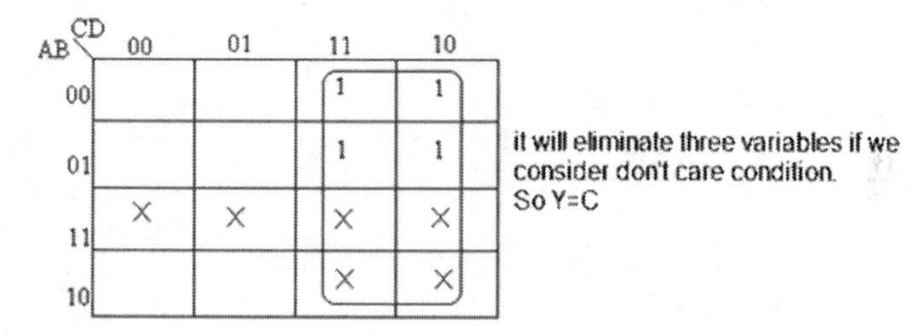

Figure 1.64 K Map for Y.

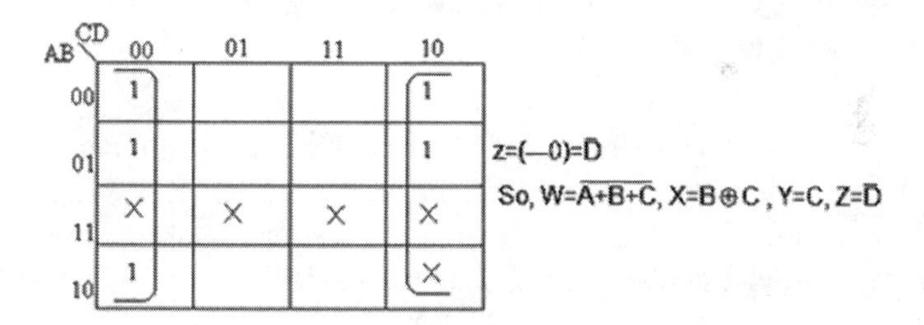

Figure 1.65 K Map for Z.

Realizing this by logic gates as shown in Fig. 1.66.

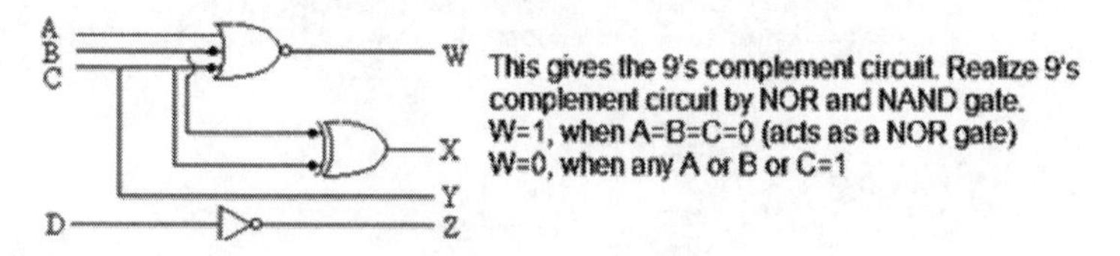

Figure 1.66 9's Complement circuit of the logic function.

And X is a XOR gate as it satisfies XOR truth table and Y is same as C and Z is complement of D (verified from Truth table).

This is an example of incompletely specified function. So, minimize such expression by K map, we have to include the Do not care condition, in which case the minimized expression is reduced to a much simpler form. At the same time it is also an example of multiple output logical function as more than one output occurs.

1.23 K Map Consideration

1. Include all the combination of the cells or square as a term in the minimized expression.
2. If a cell or square cannot be combined with any other cell then it is called the essential prime implicants. This should also be included in the final minimized expression.
3. If there is more than one minimized expression, take one whose cost of realization is minimum as the final minimized expression.

Limitations:

(a) It is a graphical method.

(b) It can be used when input variable is limited upto six only.

1.24 Digital Arithmetic Half Adder/ Half Subtractor

In digital system, addition and subtraction are the fundamental arithmetic operations as multiplication can be done by repeated addition and division can be done by repeated subtraction. As multiplication and division can be implemented by:

$$4 \times 3 = 12 \Rightarrow 4 + 4 + 4$$

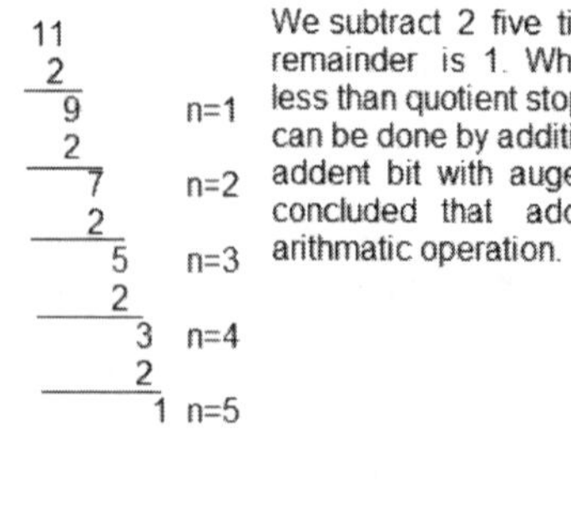

We subtract 2 five times so quotient is 5 and remainder is 1. When remainder becomes less than quotient stop at that point. Subtraction can be done by addition of two's complement of addent bit with augend bit. Hence, it can be concluded that addition is the fundamental arithmatic operation.

Simplest adder is called the HALF adder, it is a circuit which adds 2 binary bit at a time (Fig. 1.67). Half adder circuit contains 2 input bits. The Augend bit and Addend bit two outputs sum (S) and carry (C).

The truth table of a half adder is shown in Table 1.18.

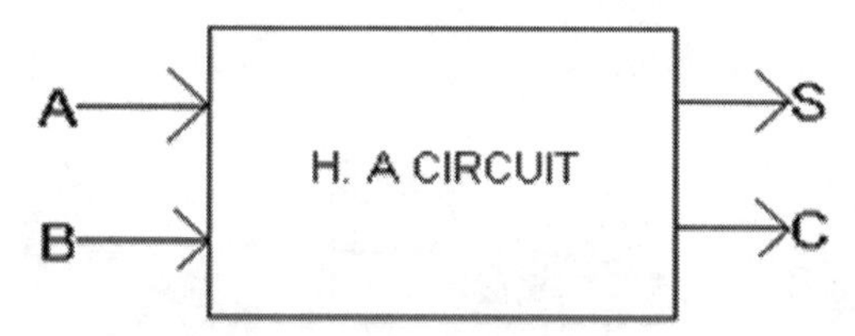

Figure 1.67 Block diagram of a half adder circuit.

So, it is a multiple output function.

Addition of two binary bits are defined as:

Augend bit A→0	0	1	1	Why (1+1) gives S=0 and C=1?
Addend bit B→0	1	0	1	As (1+1)=2 and 2 in binary is
S=0	S=1	S=1	S=0	represented by 10 hence (1+1)
C=0	C=0	C=0	C=1	gives S=0 and C=1.

Table 1.18 Truth table for half adder

Decimal	A	B	S	C
0	0	0	0	0
1	0	1	1	0
2	1	0	1	0
3	1	1	0	1

satisfies XOR logic and C satisfies AND logic. So XOR gate is applied in the Arithmetic section of Digital Electronics. (When both inputs are same logic output is 0 otherwise 1) Gate is applied in the Arithmetic section of Digital Electronics. From the truth table of half adder it is seen that

$$S = \sum m\ (1, 2)$$
$$C = \sum m\ (3)$$

K Map for carry is shown in Fig. 1.68 and the logic circuit of a half adder is depicted in Fig. 1.69.

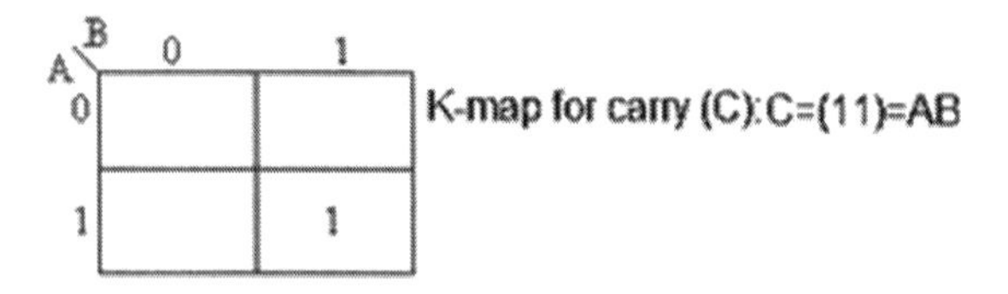

Figure 1.68 K Map for carry.

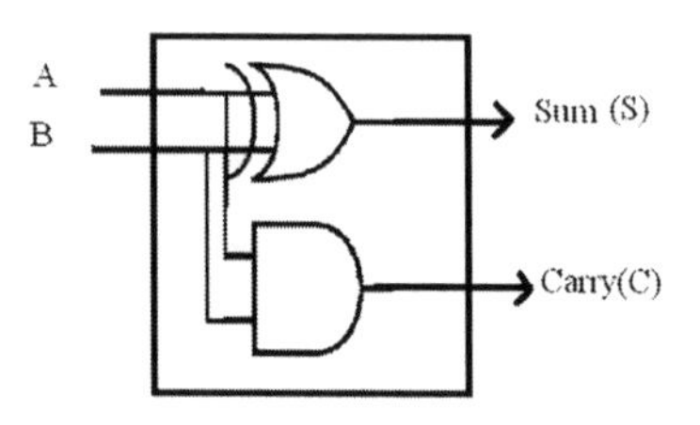

Figure 1.69 Logic circuit of a half adder.

The logic circuit of a half adder using NAND gates is shown in Fig. 1.70.

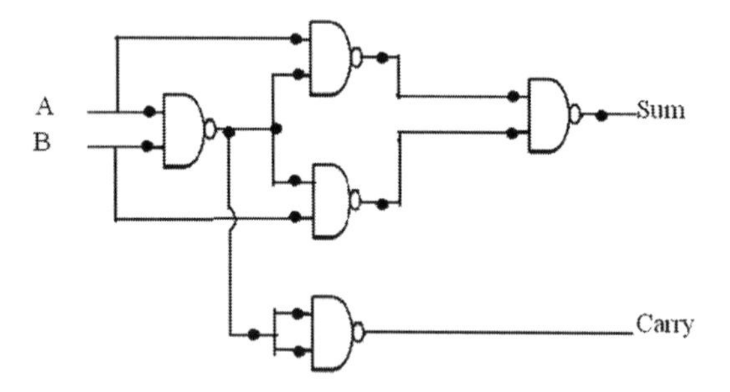

Figure 1.70 Half adder with NAND gates.

One and a half 7400 ICs are required, as 7400 is a quad 2 input NAND gate and here we require 6 gates.

1.24.1 Subtraction of Two Variables or Half Subtractor

It is a circuit which subtracts one binary bit called subtratend bit from the other bit called minuend. Half subtractor has got two output—one is called the difference (D), other is called the borrow (B).

Truth table data for half subtractor is given in Table 1.19

Table 1.19 Truth table of half subtractor

Decimal	*A*	*B*	*D*	β
0	0	0	0	0
1	0	1	1	1
2	1	0	1	0
3	1	1	0	0

D follows XOR logic $D = A \oplus B$ and β follows AND operation together with NOT operation.

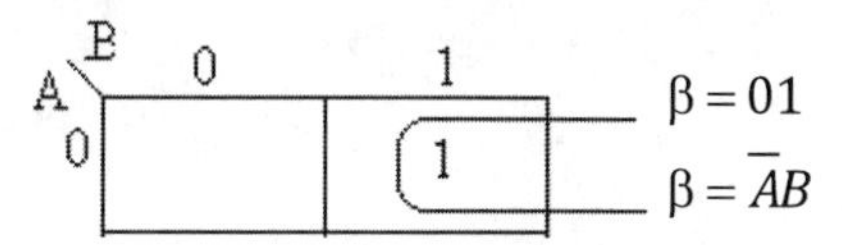

Realizing by the NAND gate (Fig. 1.71).

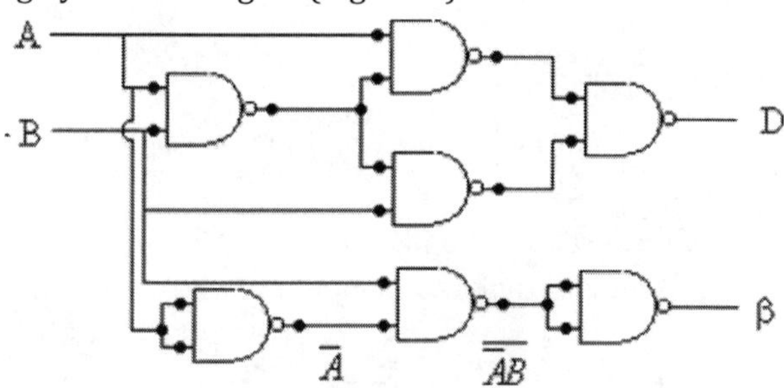

Figure 1.71 Half subtractor with NAND gates.

So, the half subtractor has a more complicated circuit as it requires one more NAND gate. We can use HS and HA by using 1 control.

HALF adder/half subtractor realization by using a control input *M* is shown in Fig. 1.72.

For $M = 0$ we get a half adder circuit

For $M = 1$ we get a half subtractor circuit

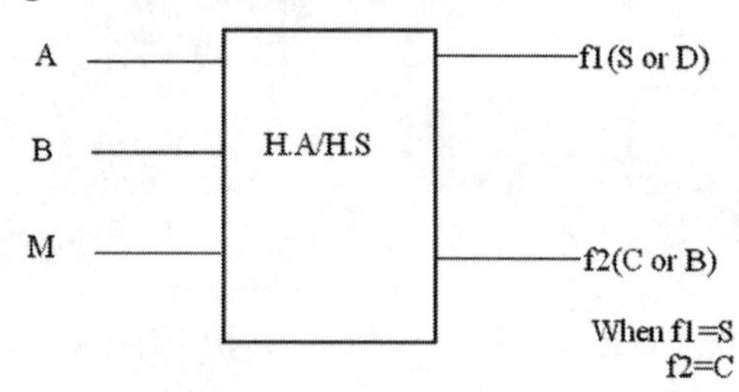

Figure 1.72 Block diagram of a half adder/half subtractor with a control input *M*.

Truth table data for H.A./H.S. is given in Table 1.20

Table 1.20 Truth table for this can be written as follows

Decimal	M	A	B	f_1	f_2
0	0	0	0	0	0
1	0	0	1	1	0
2	0	1	0	1	0
3	0	1	1	0	1
4	1	0	0	0	0
5	1	0	1	1	1
6	1	1	0	1	0
7	1	1	1	0	0

$$f_1\,(M, A, B) = \Sigma m\,(1, 2, 5, 6)$$

$$f_2\,(M, A, B) = \Sigma m\,(3, 5)$$

f_1 satisfies the XOR logic between A and B. Output is zero when even number of 0 or 1. So it is independent of M.

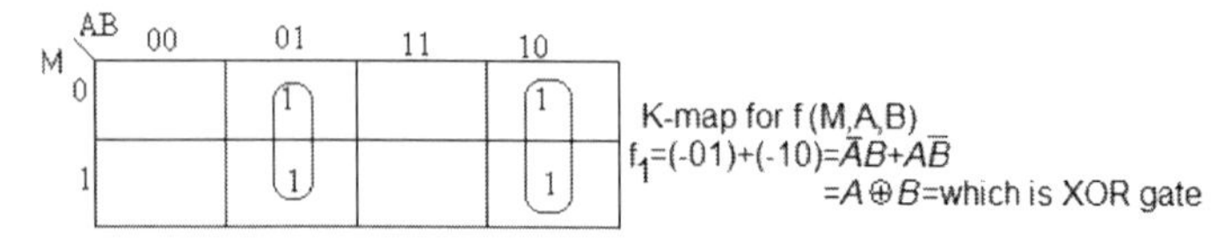

Since, S and D both satisfy the XOR logic, so f_1 whether it is S or D, will satisfy the XOR logic and independent of M.

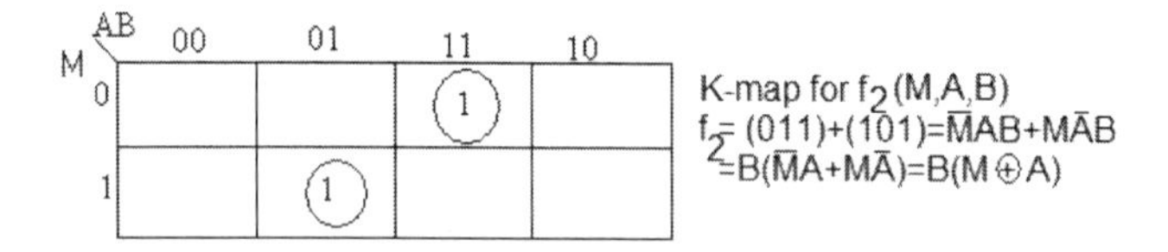

Realizing f_1 and f_2 with a control input M is demonstrated in Fig. 1.73.

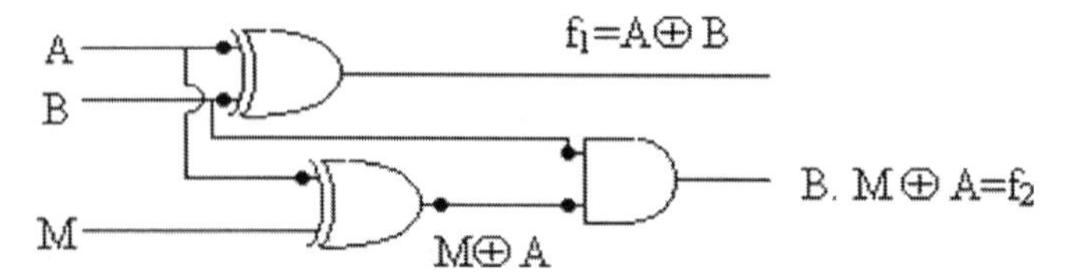

Figure 1.73 Logic circuit realization of the functions $f_1 = (M, A, B)$ and $f_2 = (M, A, B)$.

When switch M is grounded (Logic = 0) f_1 and f_2 will satisfy half adder output.

If M is connected to +5 V (Logic = 1) f_1 and f_2 will satisfy half subtractor output.

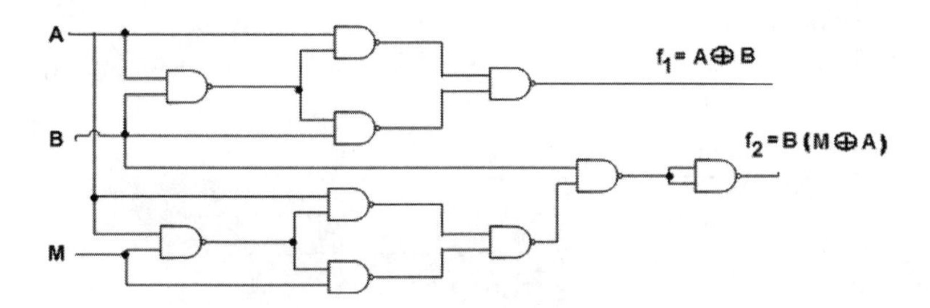

Figure 1.74 Logic circuit realization of the functions $f_1 = (M, A, B)$ and $f_2 = (M, A, B)$ using NAND gate.

Ten NAND gates are required for this (Fig. 1.74).

For multiple bit addition and subtraction HA and HS are not suitable.

That is

0 1 1

0 1 1

0 (Since there is a carry, 3 bit addition have to be done which is not possible with half adder.

1.25 Full Adder and Full Subtractor

It adds three binary bits at a time. The two binary bits are called augend bit and addend bit and the other bit is called the carry bit from the previous stage (Fig. 1.75).
here A = augend bit and B = addend bit

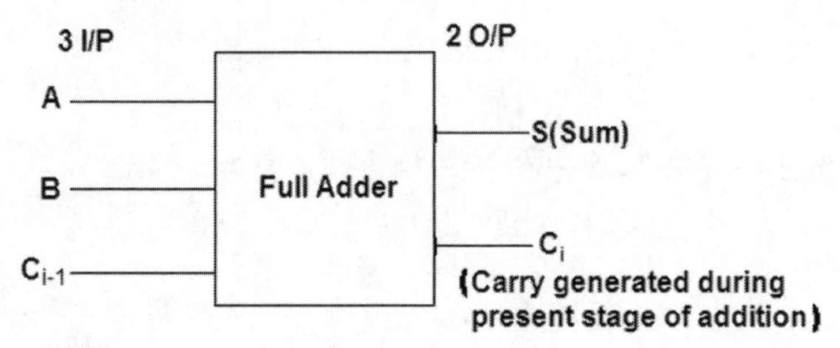

Figure 1.75 Block diagram of a full adder.

C_{i-1} = the carry bit generated from the previous stage that is just previous to present stage.

So full adder circuit has multiple output and this is a multiple output function.

The corresponding function of the full adder can be verified using the data given in Table 1.21

Table 1.21 Truth table of a full adder

	Three inputs			Two outputs	
Decimal	A	B	C_{i-1}	S	C_i
0	0	0	0	0	0
1	0	0	1	1	0
2	0	1	0	1	0
3	0	1	1	0	1
4	1	0	0	1	0
5	1	0	1	0	1
6	1	1	0	0	1
7	1	1	1	1	1

First add B with C_{i-1} and then add it with A to get the final sum and carry indicated by the logical expressions S and C, respectively.

$$S = \sum m\ (1, 2, 4, 7)$$
$$C = \sum m\ (3, 5, 6, 7)$$

The sum satisfies XOR logic (odd number of 1s as I/P) $S = 1$ and even number of 1s $S = 0$.

$$S = \mathrm{A} \oplus \mathrm{B} \oplus \mathrm{C}_{i-1.}$$

Now verify the result with the K map.

A \ BC_{i-1}	00	01	11	10
0		(1)		(1)
1	(1)		(1)	

But none of the cells can be included in group and hence,

$$\begin{aligned} \mathrm{S} &= 001 + 010 + 100 + 111 \\ &= \bar{A}\bar{B}C_{i-1} + \bar{A}B\bar{C}_{i-1} + A\bar{B}\bar{C}_{i-1} + ABC_{i-1} \end{aligned} \quad (1.30)$$

If we expand Eq. (1), we get Eq. (2).

Expanding Eq. (1), we have:

$$S = A \oplus B \oplus C_{i-1}$$

Let $\mathrm{u} = \mathrm{A} \oplus \mathrm{B}$

$$S = u \oplus C_{i-1}$$
$$= \bar{u}C_{i-1} + \overline{uC}_{i-1} \quad (1.31)$$

and

$$u = (A\overline{B} + \overline{A}B)$$
$$\bar{u} = \overline{A\overline{B} + \overline{A}B} = \overline{A\overline{B}}.\overline{\overline{A}.B}$$
$$\left(\overline{A}+B\right)\left(A+\overline{B}\right) \quad (1.32)$$

By Demorgan's theorem we get

$$= A\overline{A} + AB + \overline{A}\overline{B} + B\overline{B}$$
$$0 + \overline{A}\overline{B} + AB + 0$$
$$\text{So, } \bar{u} = \overline{A}\overline{B} + AB \quad (1.33)$$

Using Eqs. (3), (4) in (5)

$$S = (\overline{A}\overline{B} + AB)C_{i-1} + (A\overline{B} + \overline{A}B)\ \overline{C_{i-1}}$$
$$= \overline{A}\overline{B}C_{i-1} + ABC_{i-1} + A\overline{BC}_{i-1} + \overline{A}B\overline{C_{i-1}}$$

A \ BC	00	01	11	10
0			1	
1		1	1	1

$$C_i = (AC_{i-1} + AB + BC_{i-1})$$
$$C_i = (-11) + (1-1) + (11-)$$
$$= BC_{i-1} + AC + AB \quad (1.34)$$

$C_i \rightarrow$ can be realized by NAND gates. Full adder can also be realized by NAND gates (Fig. 1.76).

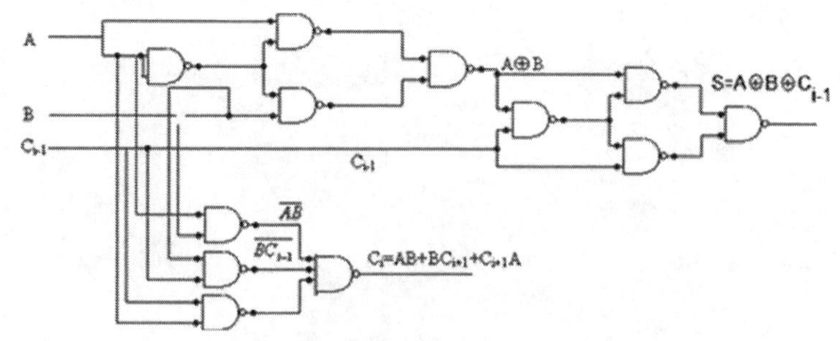

Figure 1.76 Logic circuit of a full adder with NAND gates.

$$\overline{C_i} = \overline{AB + BC_{i-1} + AC}$$
$$= \overline{AB}.\overline{BC_{i-1}}.\overline{AC}$$
$$C_i = \overline{\overline{C_i}} = \overline{\overline{AB}.\overline{BC_{i-1}}.\overline{AC}}$$

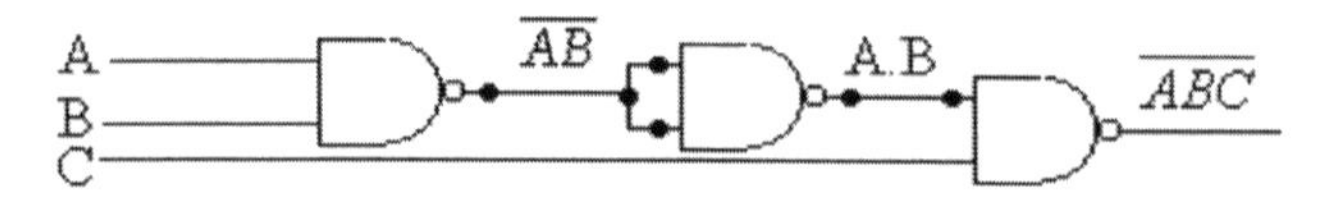

Full adder circuit can be realized by three half adder circuits (Fig. 1.77).

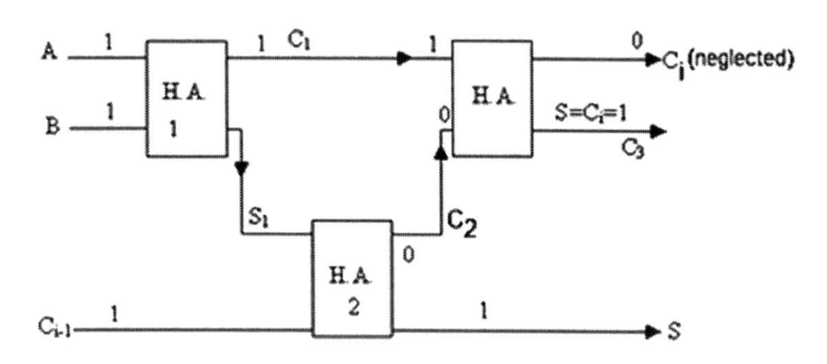

Figure 1.77 Realization of a full adder using three half adders.

Each half adder requires six gates. For this half adder circuit 18 gates are required which is not the minimized case as number = 14.

Full subtractor is a circuit which uses three binary bits at a time for subtraction (Fig. 1.78). A full subtractor circuit will contain three inputs of which one is minuend bit (A), the other is Subtrahend bit (B) and the remaining input is called the borrow input from previous stage. It has got two outputs as difference D and final borrow output β_i.

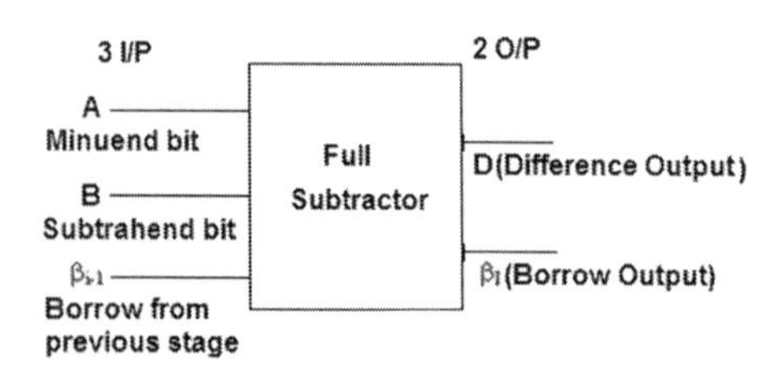

Figure 1.78 Block diagram of a full subtractor.

It is also an example of multiple output logical function.

The corresponding function of the full subtractor can be verified using the data given in Table 1.22

Table 1.22 Truth table of a full subtractor

A	B	β_{i-1}	D	β_i
0	0	0	0	0
0	0	1	1	1
0	1	0	1	1
0	1	1	0	1
1	0	0	1	0
1	0	1	0	0
1	1	0	0	0
1	1	1	1	1

$D = \Sigma m\ (1, 2, 4, 7) \rightarrow$ same as that for sum.

$\beta_i = \Sigma m\ (1, 2, 3, 7)$

D satisfies XOR logic so $D = A \oplus B \oplus \beta_{i-1}$.

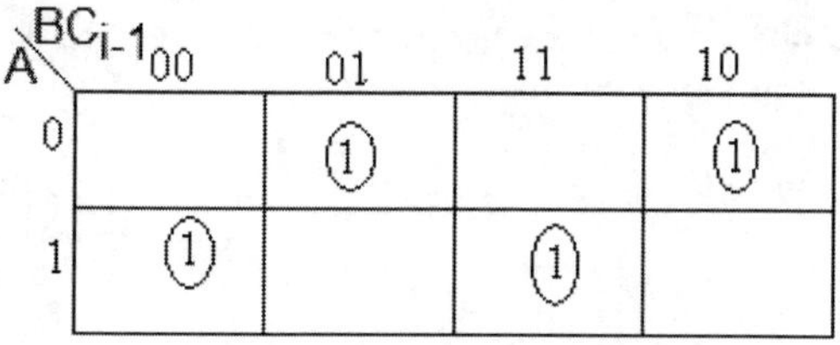

A \ BC$_{i-1}$	00	01	11	10
0		(1)		(1)
1	(1)		(1)	

$$D = 001 + 010 + 100 + 111$$

$$= \overline{A}\,\overline{B}\beta_{i-1} + \overline{A}B\overline{\beta_{i-1}} + A\overline{B}\overline{\beta_{i-1}} + AB\beta_{i-1}$$

$$= A \oplus B \oplus \beta_{i-1}$$

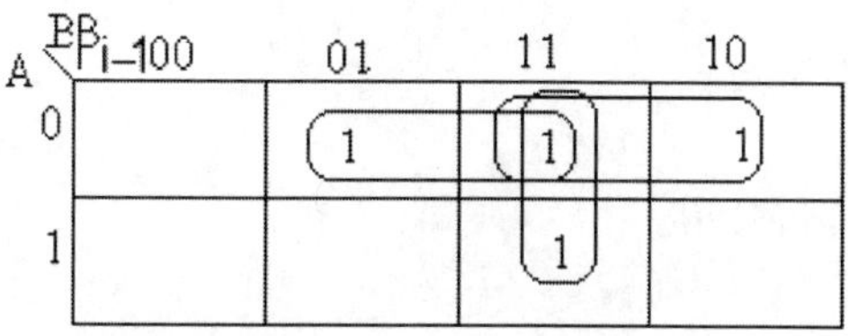

A \ Bβ_{i-1}	00	01	11	10
0		1	1	1
1			1	

$$\beta = (0-1) + (01-) + (-11)$$

$$= \overline{A}\beta_{i-1} + \overline{A}B + B\beta_{i-1}$$

$$\overline{\beta_i} = \overline{\overline{\overline{A}}\beta_{i-1} + AB + B\beta_{i-1}}$$

$$= \overline{\overline{\overline{A}}\beta_{i-1}}.\overline{AB}.\overline{B\beta_{i-1}}$$

$$\beta_i = \overline{\overline{\beta_i}} = \overline{\overline{\overline{\overline{A}}\beta_{i-1}}.\overline{AB}.\overline{B\beta_{i-1}}}$$

Realize it by NAND gates (full subtractor circuit) using 2-input NAND gates only (Fig. 1.79).

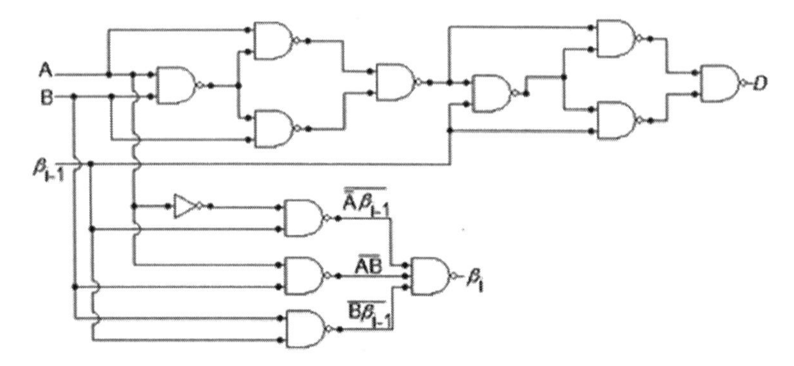

Figure 1.79 Realization of a full subtractor using NAND gate.

Full subtractor circuit can be realized by using three half subtractor circuit as shown in Fig. 1.80.

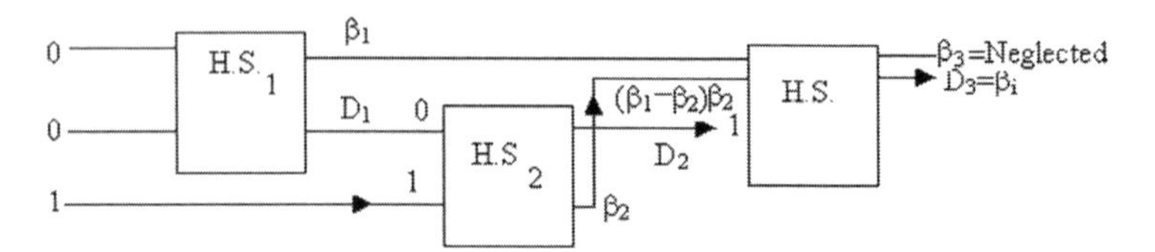

Figure 1.80 Realization of a full subtractor using three half subtractor.

1.26 Addition of Two *n* Bit Binary Numbers

Methods are as follows:

1. serial addition/adder
2. parallel addition/adder
3. carry look ahead adder

Method 1 is the slowest and method 3 is the fastest with simplest hardware complexity.

For two speed of addition is more compared to serial adder and also more hardware complexity than serial adder.

Serial addition:

1. one *n* bit right shift Augend register (Fig. 1.81)
2. one *n* bit right shift Addend register

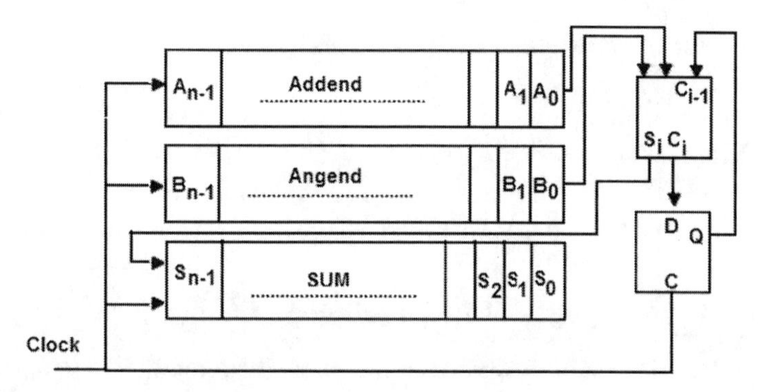

Figure 1.81 Shift *n*-bit Augend register.

3. one *n* bit ight shift sum register

Registers are some cells where binary bits are stored and can be shifted by clockwise shifting right.

4. one D-type flip-flop
5. one clock generator
6. only one full adder

So complete *n* bit addition between two binary numbers requires *n* clock pulses. If the time period of clock pulse = *T* sec1 bit addition then the time required for *n* bit serial addition = *nT* sec and after *n* clock pulse, the sum register will contain the *n* bit sum and the carry output of the full adder contain the final carry.

This Augend register can store n bit binary number and it can shift right 1 bit, after arrival of one clock pulse.

Adding two 4 bits binary number

$3 \quad 0 \leftarrow \quad A = 0\ 0\ 1\ 1 \rightarrow A$ 4 bit binary number

$+4 \qquad\qquad B = 0\ 1\ 0\ 0 \rightarrow B$ 4 bit binary number

----

7

Initially we set output Q of D flip-flop = 0

Table 1.23 Truth table of a shift n bit augend register

Clock	A_i	B_i	Q	C_{i-1}	S_i	C_i	S
0	1	0	0	0	1	0	0
1	1	0	0	0	1	0	1
2	0	1	0	0	1	0	1
3	0	0	0	0	0	0	1
4	X	X	X	X	X	X	0

After fourth clock pulse, may be do not care, the sum register = 0 (Table 1.23).

Main disadvantage:

Additional time $\propto n$

Hence, as n increases, the time required to complete the addition also increases and here it is speed will be quite low as n increases.

1.26.1 Advantages of Serial Adder

Less hardware complexity as number of full adder is always equal to 1 being independent of n.

1.26.2 Disadvantages of Serial Adder

It is speed decreases with increase in the number of bits. So, increase the speed parallel full adder is used.

1.27 *n* Bit Parallel Full Adder

1. This requires n number of full adder (Fig. 1.82).

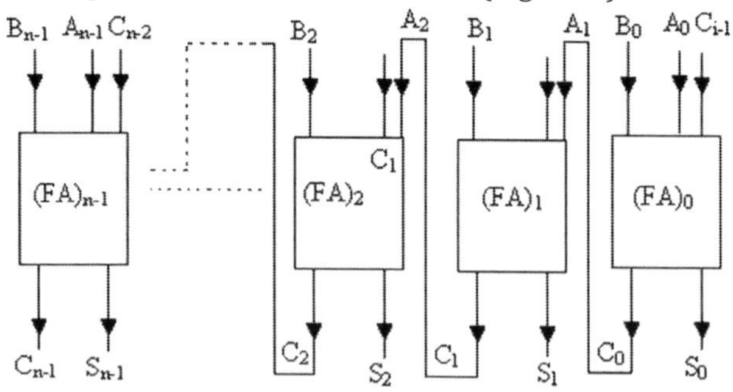

Figure 1.82 Parallel full adder.

As there is number initial carry in first LSB bit so for the first full adder it can be replaced by a half adder.

So $A_{n-1} \ldots A_3\, A_2\, A_1\, A_0 = n$ bit Augend

$B_{n-1} \ldots B_3\, B_2\, B_1\, B_0 = n$ bit Addend

$S_n S_{n-1} \ldots S_3\, S_2\, S_1\, S_0 = (n+1)$ bit Sum

Here the inputs (A_0, B_0) (A_1, B_1) (A_2, B_2) ... (A_{n-1}, B_{n-1}) are applied simultaneously to the inputs of all the full adder. Since the carry has to develop and it takes time so though the parallel adder is much faster operation compared to serial adder which is the advantage of this parallel adder but parallel adder differ from the following disadvantages:

1. The carry propagation delay much less than the bit delay of the serial adder which depends upon the number of stages of full adder present that is the value of *n*.
2. It is hardware complexity is more than the serial adder as it requires n full adder for adding 2*n* bit binary number.

1.28 Combinational and Sequential Circuit

Digital circuits are classified as:

1. Combinational circuit (Fig. 1.83).
2. Sequential circuit (Fig. 1.84).

Distinction between two combinational circuits is one whose outputs depends only on the present value of the input(s) being independent of the of the past history of the input(s). The sequential circuit is on the other hand is a circuit whose output(s) not only depend on the present value of the input but also depend on the past history of the input(s).

So we can say that combinational circuit should not contain any memory element (device which can store some earlier information) where as the sequential circuit contains at least one memory element.

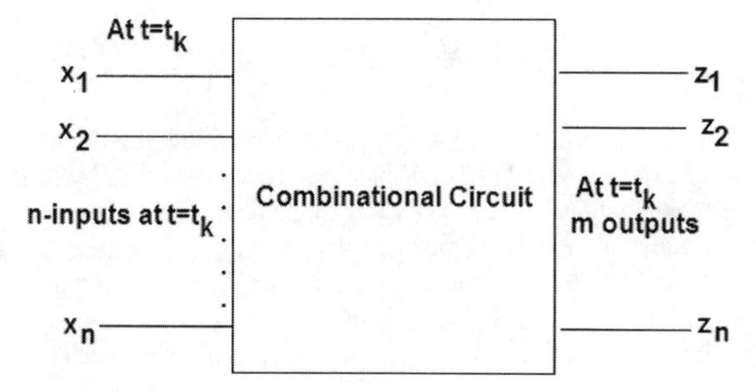

Figure 1.83 Block diagram of a combinational circuit.

The input–output relationship:

$$Z_i\,(k) = f_j\,\{x_1\,(t_k), x_2\,(t_k), \ldots x_n\,(t_k)\} \tag{A}$$
$$i = 1, 2, \ldots m$$
$$j = 1, 2, \ldots n$$

as input and output may not be always same.

S_j determines the nature of the function.

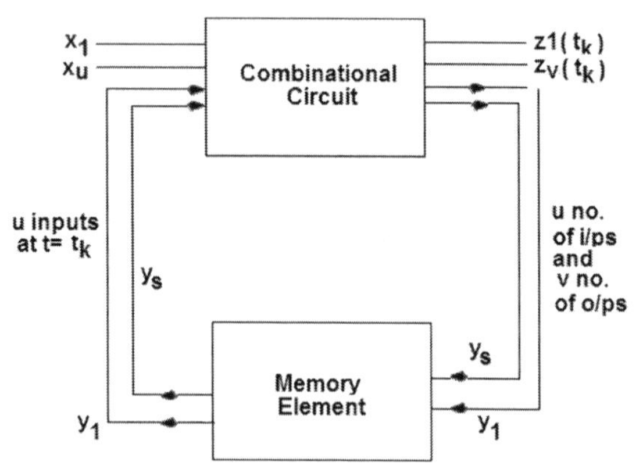

Figure 1.84 Block diagram of a sequential circuit.

The behavior of a combinational circuit is described by the output function as given by Eq. (A) whereas in sequential circuit some output is fed back to the input. In sequential circuit we get two types of outputs. $Z_i\,\{t_k\}$ is called the output function and the other $y_i\,\{t_k\}$ is called the state function.

$$Z_i\,(t_k) = g_i\,\{x_1\{t_k\} \ldots x_u\{t_k\}; y_1\{t_k\} \ldots y_s\{t_k\}\} \tag{B}$$

where i varies from 1 to v. So we have u number of outputs.

And

$$y_j\,(t_k) = h_j\,\{x_1\{t_k\} \ldots x_u\{t_k\}; y_1\{t_k\} \ldots y_s\,\{t_k\}\} \tag{C}$$

where $j = 1, 2 \ldots$ s

Thus we can say that sequential circuit is described by the output function as given by the Eq. (B) and next state function as given by Eq. (C).

For combinational circuit the basic building blocks are logic gates and in sequential circuit the basic building blocks are flip-flops, so, sequential circuit does not require the presence of clock pulses. Examples of combinational circuits are logic gates, multiplexers, demultiplexers, decoders, encoders, and so on. Examples of sequential circuits are flip-flops, registers, counters, memory devices, sequence generator, and so on. Speed of combinational circuit is higher than the sequential circuit.

Classification or gate complexity of digital ICs is classified in one of the four following categories:

1.28.1 Small Scale Integration (SSI)

The small scale integration (SSI) devices have got a gate complexity of less than 10 gates. This includes several gates flip-flops in one package; for example 7400, 7402, 7408 → all have four gates, so SSI.

1.28.2 Medium Scale Integration (MSI)

The medium scale integration (MSI) devices have got gate complexity of 10 to 100 gates (near). This includes multiplexers, demultiplexers, decoders and encoders, and so on.

1.28.3 Large Scale Integration (LSI)

The large scale integration (LSI) devices have a complexity of 100 to 1000 gates. This includes large memory devices, calculator, and microprocessor ICs.

1.28.4 Very Large Scale Integration (VLSI)

The very large scale integration (VLSI) devices have a gate complexity of greater than 1000 gates. (2,35,000 numbers of equivalent transistors may be contained in some CPU,486DX). This includes large memory array, complex microprocessors, and microcomputer chips or data selector.

1.29 Multiplexer Design Procedure and Applications

Multiplexer is a special type of combinational circuit. This includes large memory array, complex microprocessor, and microcomputer chips. Type of multiplexer or a data selector is a very important special combinational circuit (made with MSI devices) with gate complexity of MSI devices. Generally a $(n : 1)$ multiplexer or a data selector contains the following (Fig. 1.85):

1. n number of inputs of the multiplexer
2. only one output of the multiplexer
3. m select/Address lines
4. one strobe/enable input (optional)

1.29.1 General Block Diagram

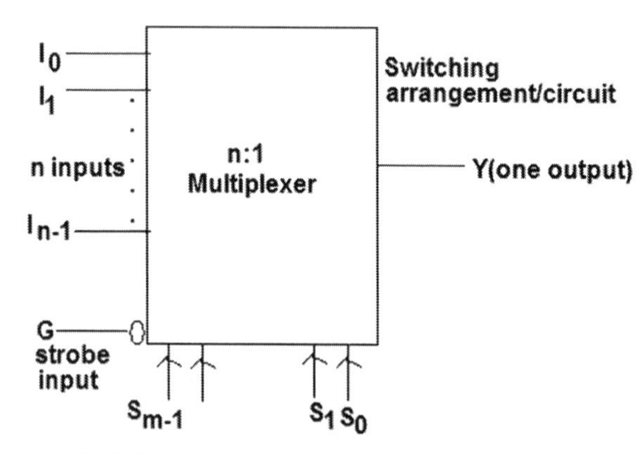

Figure 1.85 Block diagram of a $n : 1$ multiplexer.

All inputs and outputs should be purely logical or binary input. For $n : 1$ multiplexer (Fig. 1.85), each input will be selected at the single output at any instant of time, depends upon the input applied to the select/address inputs. Switching circuit connects a particular input to the output.

Here m and n are related by

$$2^m = n$$

So for a $4 : 1$ multiplexer, the number of inputs is 4 and $m = 2$ $(2^2 = 4)$.

Let us consider, a $4 : 1$ multiplexer having 4 inputs, 2 select lines, one output one strobe inputs (Fig. 1.86). Strobe inputs should be grounded for enabling the multiplexer and can perform its intended function.
Commercial multiplexers will be
$2 : 1, 4 : 1, 8 : 1, 16 : 1, \ldots$
$m = 1, m = 2, m = 3, m = 4$

The corresponding function of the multiplexer can be verified using the data given in Table 1.24

Table 1.24 Truth table of 4 : 1 multiplexer

G	S_1	S_0	Y
0	0	0	I_0
0	0	1	I_1
0	1	0	I_2
0	1	1	I_3
1	X	X	0

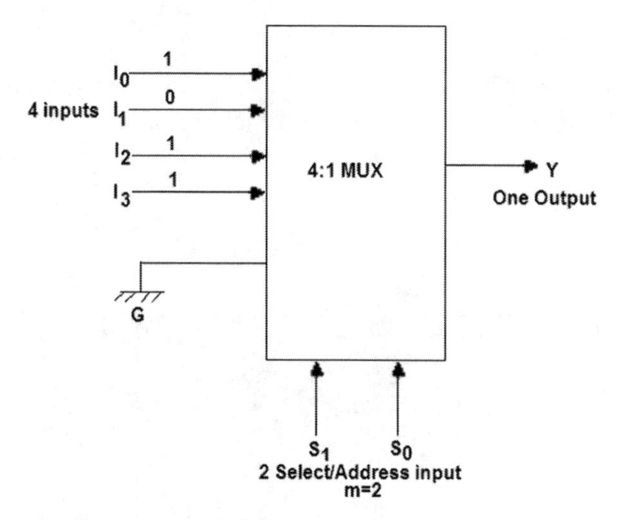

Figure 1.86 Block diagram of a 4 : 1 multiplexer.

Multiplexers are also considered as many to one device (cable TV channels) that is out of many inputs, we select only one input at a time at the output, G = 0 always for enabling the multiplexer.

If G = 1, whatever be the value of S_1 and $S_{0,}$ the output is always 0.

By observing the truth table of 4:1 multiplexers, the logical expression for the output of the multiplexer can be written as:

$$Y = \bar{G}\bar{S}_1\bar{S}_0 I_0 + \bar{G}\bar{S}_1 S_0 I_1 + \bar{G} S_1 \bar{S}_0 I_2 + \bar{G} S_1 S_0 I_3 \quad (1.35)$$

1st case when $G = S_1 = S_0 = 0, I_0$

$$\text{output} = \bar{0}\bar{0}\bar{0} I_0 + \bar{0}\bar{0}0 I_1 + \bar{0}0\bar{0} I_2 + \bar{0}00 I_3$$

$$= I_0 + 0 + 0 + 0 = I_0$$

$$\text{If, } I_0 = 1 \text{ then output} = I_0 = 1$$

$$G = S_1 = 0, I_0 = 1$$

$$\text{So output } Y = 0 + I_1 + 0 + 0 = I_1$$

$$\text{But if } G = 1, Y = 0 + 0 + 0 + 0 = 0$$

$$S_1 = X$$

$$S_0 = X$$

That is when the strobe input is connected to the logical high voltage or logical = 1, the multiplexer is not enabled that is disabled and cannot perform. So, the G input has to be connected to the ground always to enable the multiplexer. So

$$Y = \bar{G}\bar{S}_1\bar{S}_0 I_0 + \bar{G}\bar{S}_1 S_0 I_1 + \bar{G} S_1 \bar{S}_0 I_2 + \bar{G} S_1 S_0 I_3$$

Realization by logic gates

Use of a strobe input of a multiplexer:

1. Strobe input when grounded enables the multiplexer that is G. The strobe input is used for enable or disable a multiplexer Fig. 1.87.

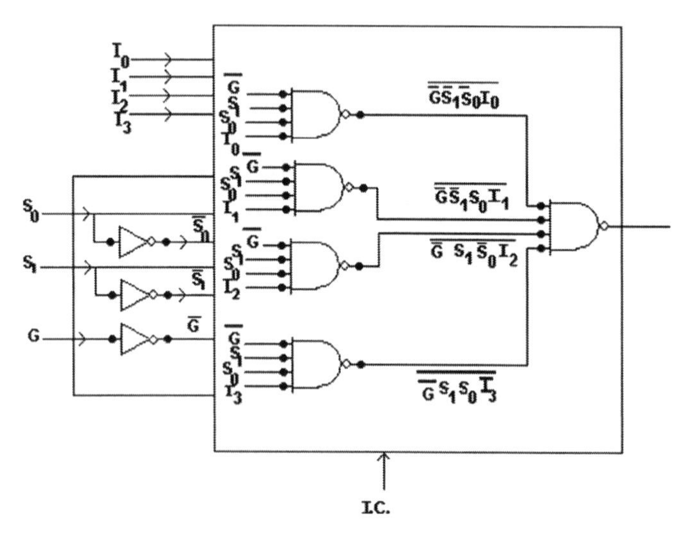

Figure 1.87 Logic circuit of a multiplexer with strobe input.

2. Strobe input can be realized for cascading two or more lower order multiplexers. This is called realization of multiplexer tree.

Advantages of designing with multiplexer:

1. The given logic function or the truth table need number to be minimized that is the minimization of the logic function or truth table is not required.
2. Logic design is simplified.
3. It reduces the IC package-count as only 1 IC instead of 4 ICs are required for FA.
4. The layout of the circuit is simpler.
5. It requires less number of wiring or interconnection.
6. Cost of realization is minimum.
7. Reliability of designing with MUX is more.

If the logical function or truth table is given then express the logical function or truth table either in min term or in max term form.

(1) If the logical function or truth table is in min term form then connect those input of the multiplexer to logical 1 which are present in the min term expression and connect the other inputs to logical 0 which are number present in the min term expression. This is to be followed if the multiplexer output is equal to same as input.

(2) If the logical function or truth table is expressed in max term for just do the reverse.

Note:

1. If the logical function or TT is expressed in max term then just do the reverse.
2. If the output of MUX is complement of the input selected then reverse the operation as stated in (1) and (2) above.
   That is followed
   (a) for min term expression
   (b) for max term expression
3. The inputs for the LF or TT are to be applied to the select or the address input of the MUX.
4. Connect the strobe input G to logical 0 or ground to enable the multiplexer.
5. Apply suitable supply voltage $V_{CC} = 5\,V$ and ground to the respective pins of the multiplexer IC.

Realize two input XOR gates by using a multiplexer:
Choose single 4:1 MUX, as m should be 2 for 2 inputs. Realization of a two input XOR gate using a 4:1 multiplexer is shown in Fig. 1.88

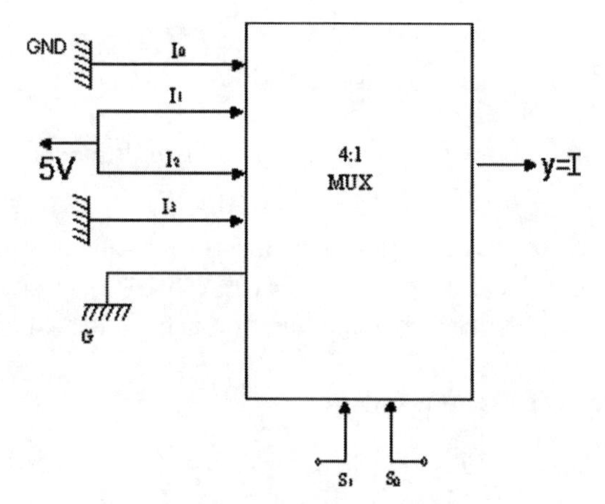

Figure 1.88 Realization of a two input XOR gate using a 4 : 1 multiplexer.

Truth table

A	B	Y
0	0	0
0	1	1
1	0	1
1	1	0

$Y = \sum m\ (1, 2)$
If $Y_{out} = I_{input}$ (here)
If $Y = \overline{I\ \text{Input}}$

Hence, we are to see whether output = Input or output = $\overline{\text{Input}}$. Realization of a two variable function using a 4:1 multiplexer is demonstrated in Fig. 1.89.

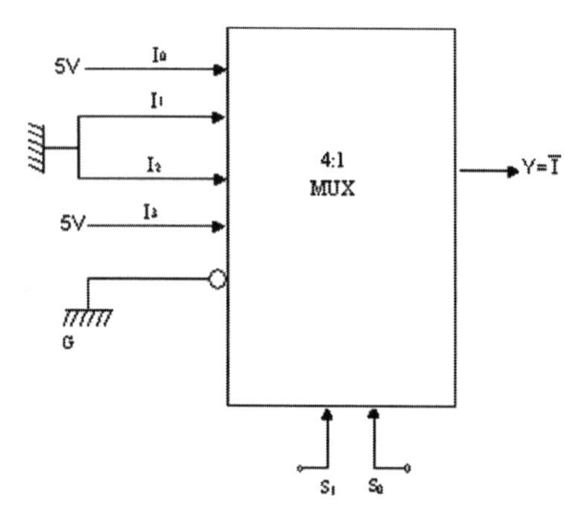

Figure 1.89 Realization of a two variable function using a 4 : 1 multiplexer.

Realize the following logic function by using multiplexers:

$$f(A, B, C) = \prod M\ (0, 1, 3, 7)$$

three select/address inputs are required for realizing three input MUX and so we have to choose $(2^3) = (8 : 1)$ MUX.

Realization of a three variable function using a 8:1 multiplexer is demonstrated in Fig. 1.90.

If we are to realize this function

$$f(A, B, C) = \prod M\ (0, 1, 3, 7)$$

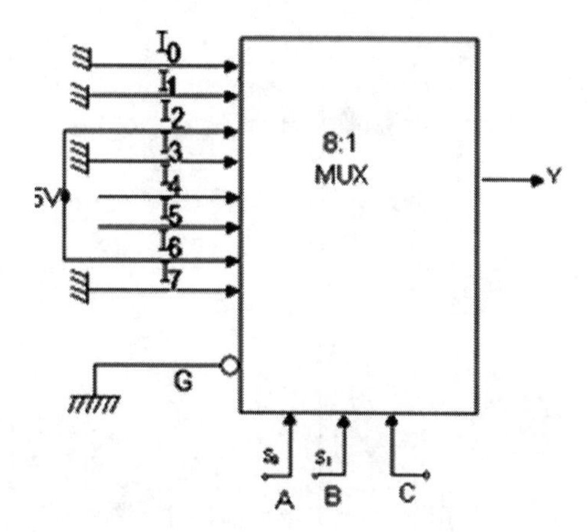

Figure 1.90 Realization of a three variable function using a 8 : 1 multiplexer.

Realize full adder by using 8 : 1 multiplexer.

$$S = \sum m\ (1, 2, 4, 7)$$

$$C_i = \sum m\ (3, 5, 6, 7)$$

So we require 2, 8 : 1 multiplexers.

To realize full adder circuit we require four 7400 ICs and to realize full adder circuit by MUX we require only two multiplexers (Fig. 1.91).

Commercial available multiplexers ICs 74157 = Quad 2 : 1 multiplexers have Output = Input and only 1 select input.

74157 IC has (two data I/Ps and one strobe input for each 2 : 1 MUX). 1 select (S_0) line, 1 enable line, 1 V_{CC}, 1 ground, $2 \times 4 = 8$ input lines, $4 \times 1 = 4$ output lines (Fig. 1.92). Hence, 16 pins DIP type IC.

$3 \times 4 = 12$

1 Select

1= Enable

1 V_{CC}

1 gnd.

So, 16 pins. DIP package for 74157.

74158 = quad 2 : 1 Mux→ out = $\overline{\text{Input}}$

74153 = Dual 4 : 1 Mux; Out = Input

74352 = Dual 4 : 1 Mux Output = $\overline{\text{Input}}$

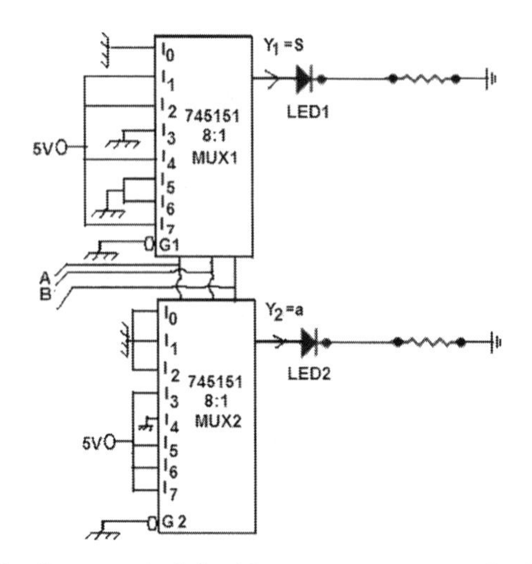

Figure 1.91 Realization of a full adder using two 8 : 1 multiplexers.

So 74153 have 16 pins device (Fig. 1.93). This is getting in lab.

In lab 74151→ Single 8 : 1 MUX with complementary outputs. It is a 16 pin DIP. It has 2 outputs Z and $\overline{Z}$ 74151 is a 16 pin DIP.

74152→ Single 8 : 1 MUX with inverted output→16 pin DIP.

74150→ Single 16 : 1 MUX with inverted 24 pin DIP.

Commercially, maximum order of MUX is 16 : 1.

Realize full adder circuit by using one 74153 and any other gates required.

1.29.2 Advantage

Within 74153 we have 2, 4 : 1 MUX. To realize F/A circuit by using 74153 (Fig. 1.93), we convert 3 input TT of a F/A into 2 input TT as follows: (Fig. 1.94) . The truth table corresponding to the realization of a full adder using two 4:1 multiplexer is given in Table 1.25

Table 1.25 Truth table of the realization of a full adder using two 4:1 multiplexer

A	B	S	C_i
0	0	C_{i-1}	0 from (1)
0	1	C_{i-1}	C_{i-1} from (2)
1	0	$\overline{C_{i-1}}$	C_{i-1} from (3)
1	1	$\overline{C_{i-1}}$	1 from (4)

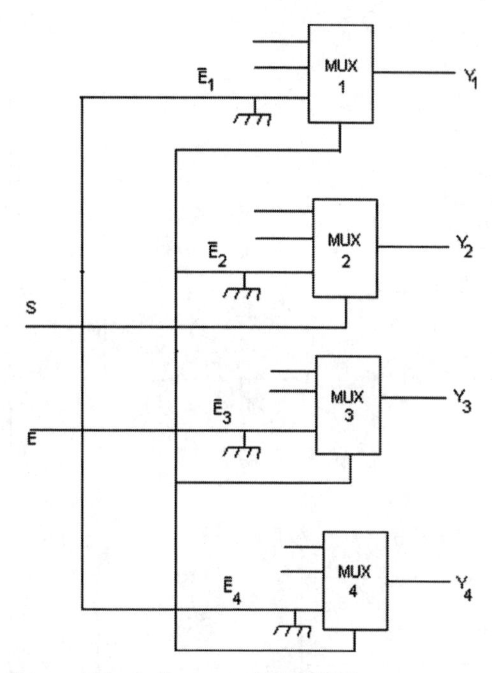

Figure 1.92 Internal block diagram of 74157 IC.

So, the F/A, TT is reduced to 2 input TT, I/P = A and B.

So, only 1(74153) IC is required and 1/6 7404 (as hex-inverter) IC is required.

So, this is much economical than the earlier one. (8 : 1 mux)

Do the same operation for full subtractor. For full subtractor,

$$\beta_i = \sum m(1,2,3,7)$$

A	B	D	β_i
0	0	$\overline{\beta_{i-1}}$	
0	1	$\overline{\beta_{i-1}}$	1
1	0	β_{i-1}	0
1	1	β_{i-1}	β_{i-1}

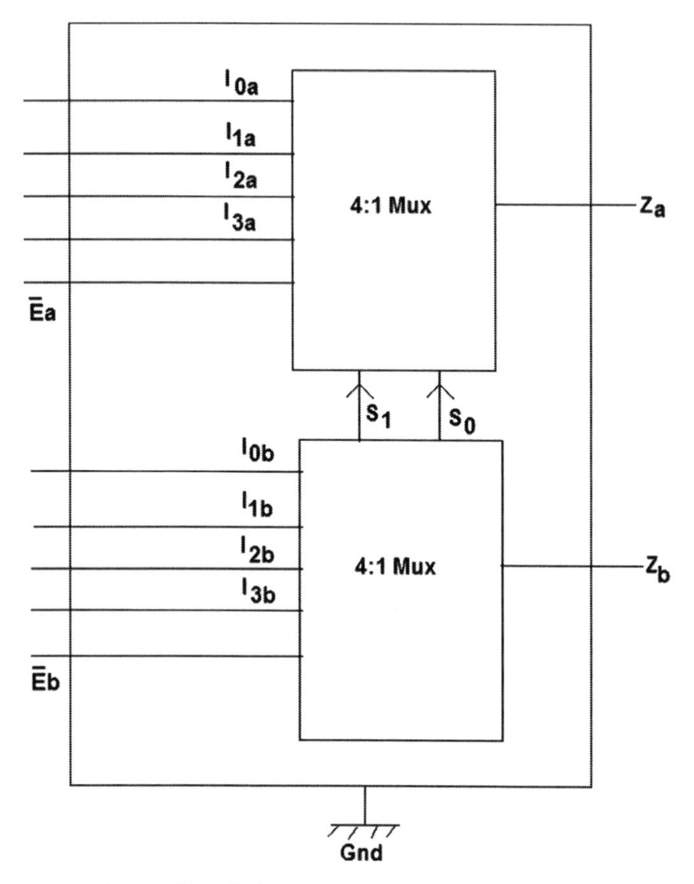

Figure 1.93 Internal block diagram of 74153 IC.

1.29.3 Application of Multiplexer

1. It can be used to realize any given LF and TT without minimizing it.
2. It can be used as a universal logic gate.
3. It can be used for parallel to serial converter.
4. It can be used for the design of the sequence generator.

By cascading lower order multiplexer we can realize higher order multiplexers. This is called multiplexer tree.

Realize 4 : 1 multiplexer by using 2 : 1 multiplexers and other gates (Fig. 1.95) required if any.

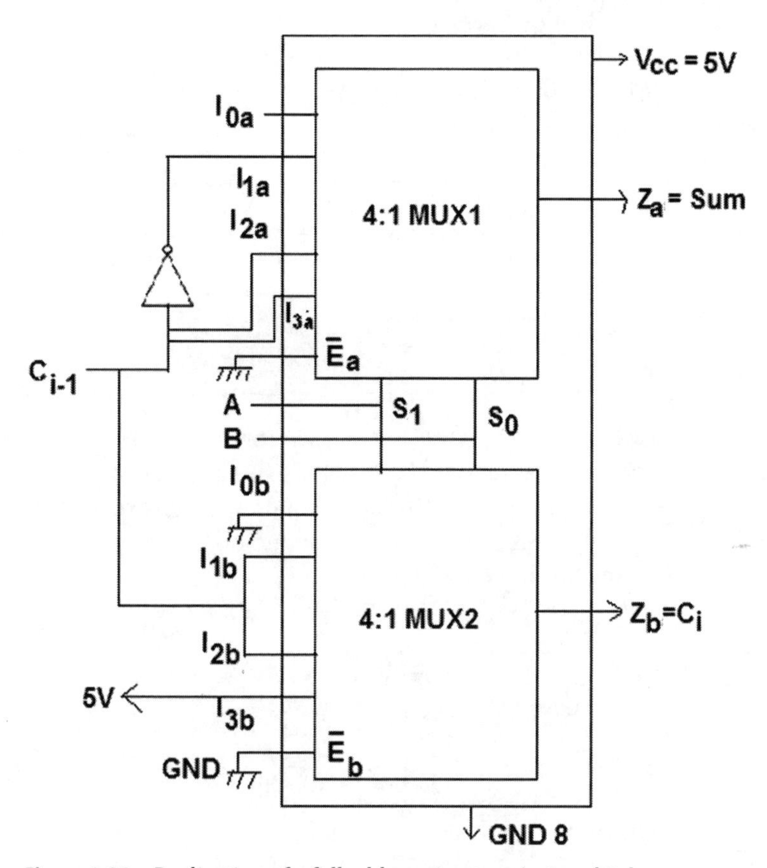

Figure 1.94 Realization of a full adder using two 4 : 1 multiplexers.

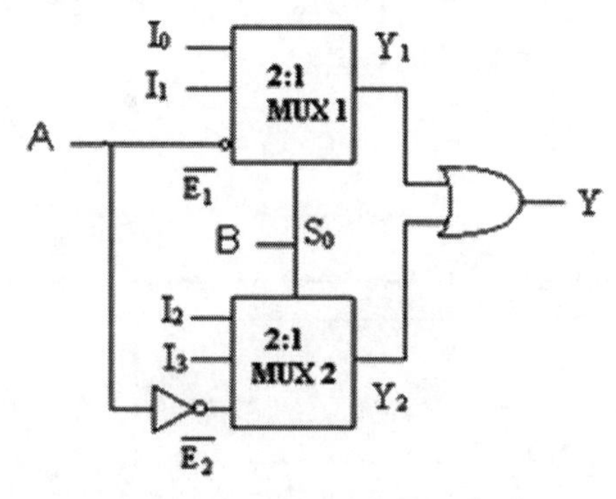

Figure 1.95 Realization of a 4 : 1 multiplexer using two 2 : 1 multiplexers and one OR gate.

I_0, I_1, I_2, I_3 are the 4 inputs of the MUX.

$A, B \rightarrow 2$ select I/P lines.

The corresponding truth table is shown in Table 1.26

Table 1.26 Inputs and outputs of the two multiplexers used for designing a 4 : 1 multiplexer

Select Inputs						
A	B	MUX1	MUX2	Y_1	Y_2	Y
0	0	Enabled $E_1 = 0$	Disabled $E_2 = 1$	I_0	0	I_0
0	1	Enabled	Disabled	I_1	0	I_1
1	0	Disabled	Enabled	0	I_2	I_2
1	1	Disabled	Enabled	0	I_3	I_3

So, this truth table agrees with 4 : 1 MUX. So for realizing the higher order MUX we require not only MUX but also other gates. But if we have to realize higher order MUX by using only lower order MUX then the diagram is (Fig. 1.96). The corresponding functionality can be verified using the Table 1.27.

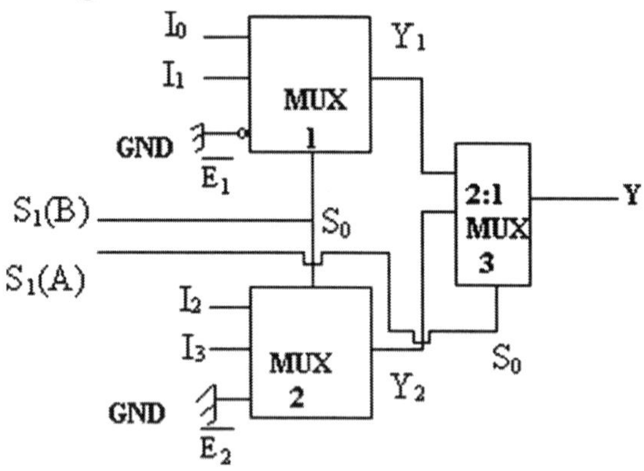

Figure 1.96 Realization of 4 : 1 multiplexer using only two 2 : 1 multiplexers.

Table 1.27 Truth table for realization of higher order mux using lower order mux

A	B	Y_1	Y_2	I_0	I_1	Y
0	0	I_0	I_2	I_0	I_2	I_0
0	1	I_1	I_3	I_1	I_3	I_1
1	0	I_0	I_2	I_0	I_2	I_2
1	1	I_1	I_3	I_1	I_3	I_3

Realizing a 4 : 1 MUX using two 2 : 1 multiplexers and one NOT gate and one OR gate:

This is an application of strobe input as realizing multiplexer tree. To realize 32 : 1 multiplexers from 16 : 1 multiplexers (Fig. 1.97).

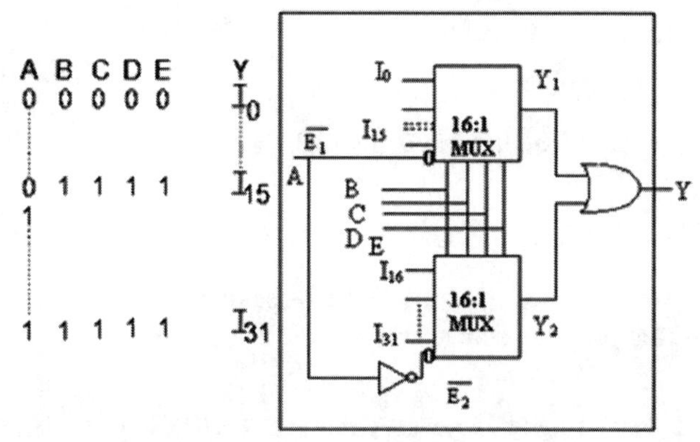

Figure 1.97 Realization of a 32 : 1 multiplexer using only two 16 : 1 multiplexers and one OR gate.

We can design incompletely specified function by using MUX.

$$f(A, B, C) = \sum m\,(0, 1, 2) + \sum d\,(5, 6, 7)$$

Realization of the function using multiplexer (Fig. 1.98).

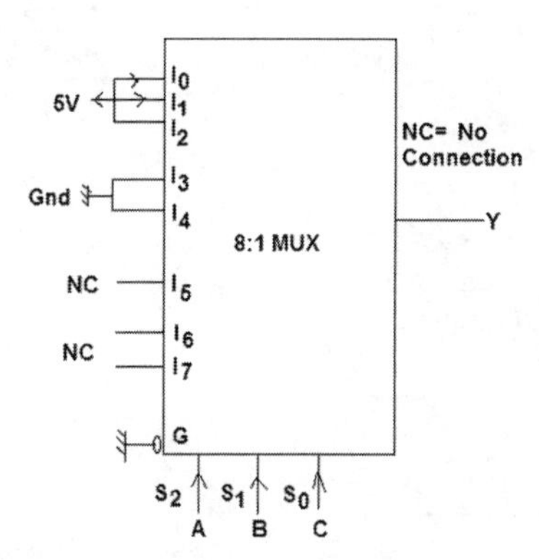

Figure 1.98 Realization of a specific function $f(A, B, C)$ using one 8 : 1 multiplexer.

If any do not care terms are present corresponding to these terms, the I_1 can be kept unconnected and a unconnected input means logical 1 and TTL logic. So, that at the output we get output 1.

for I/P 3 and 4 O/P = 0

and 0, 1, 2 O/P = 1.

For four variable logic function

$$f(A, B, C, D) = \sum m\,(0, 3, 5, 6, 7, 8, 9, 13)$$

Realize it by using 8 : 1 MUX and OR gates if necessary (Convert 4 I/P TT 3 I/P TT)→ take two rows for $f(A, B, C, D)$ at a time from top multiplexers.

Multiplexer can be used as universal logic gate.

1.29.4 Multiplexer as Universal Logic Gate

Multiplexer can also be used as a Universal logic gate. Realization of various logic gates using multiplexer is depicted below.

1.29.4.1 Realizing NOT gate by using 2:1 MUX

The realization of a NOT gate using multiplexer is shown in Fig. 1.99.

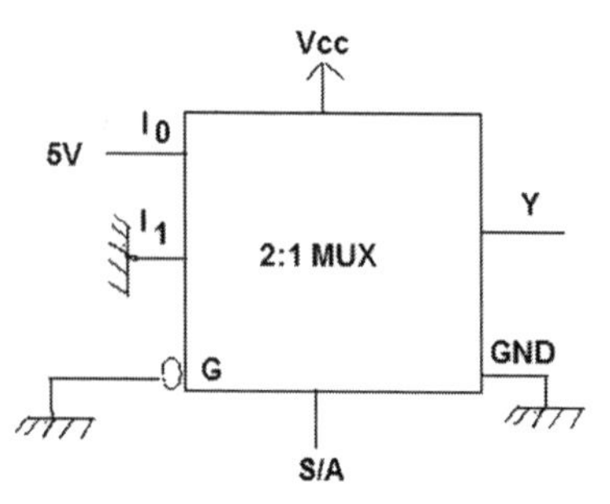

Figure 1.99 NOT gate using a multiplexer.

Truth table of NOT gate

S/A	Y
0	1
1	0

1.29.4.2 Realizing AND gate by using 2 : 1 MUX

The realization of an AND gate using multiplexer is shown in Fig. 1.100.

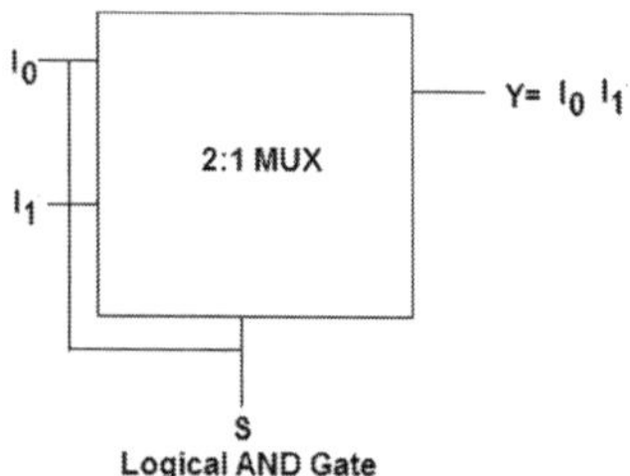

Figure 1.100 AND gate using a multiplexer.

1.29.4.3 Realizing OR gate by using 2 : 1 MUX

The realization of an OR gate using multiplexer is shown in Fig. 1.101.

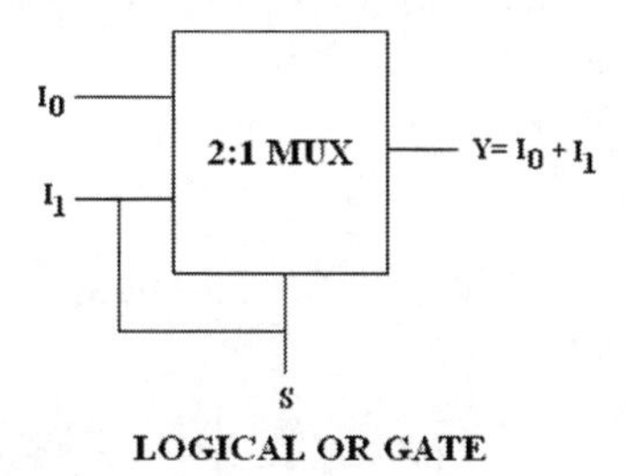

Figure 1.101 OR gate using a multiplexer.

1.30 Demultiplexers and Their Applications

It is also a combinational circuit whose function is just the reverse of multiplexers.

It has got:

(i) one input
(ii) n Outputs and
(iii) m select/address lines.

where $2m = n$.

The block diagram of 1 : n demultiplexer is shown in Fig. 1.102.

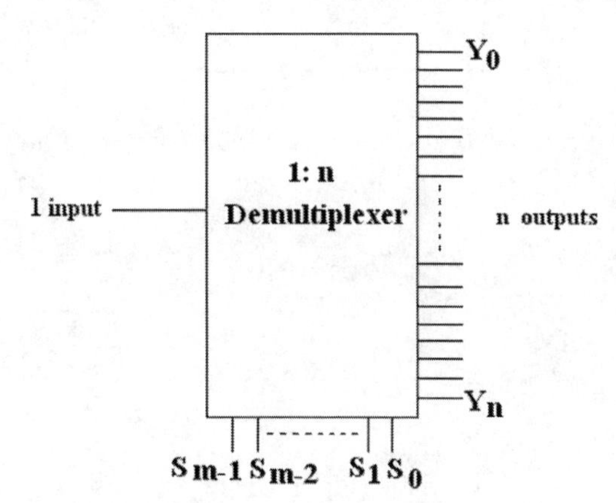

Figure 1.102 Block diagram of a 1 : n demultiplexer.

Single input of Demux can be distributed among any one of the n outputs. At which output will be distributed depends upon the inputs applied to the select/address inputs of the Demux. For this we have to consider the truth table. Let us consider truth table of (1 : 4) Demultiplexer as given in Table 1.28.

The block diagram of a 1 : 4 demultiplexer is shown in Fig. 1.103.

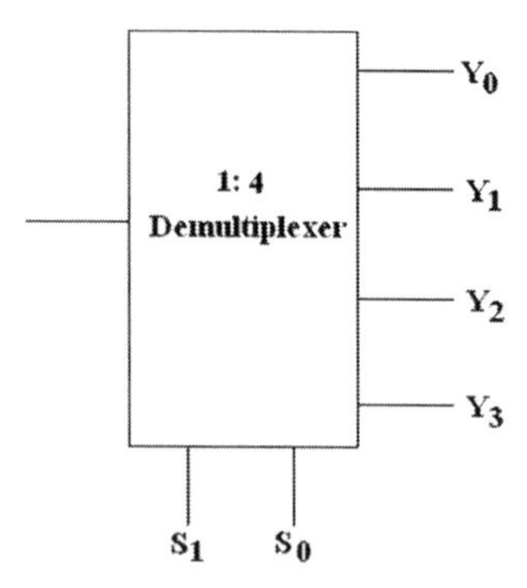

Figure 1.103 Block diagram of a 1 : 4 demultiplexer.

So, demultiplexer can be considered as a one to many devices.

Table 1.28 Truth table of a 4 : 1 demultiplexer

Select inputs			outputs			
I	S_1	S_0	Y_0	Y_1	Y_2	Y_3
1	0	0	$I = 1$	0	0	0
1	0	1	0	$I = 1$	0	0
1	1	0	0	0	$I = 1$	0
1	1	1	0	0	0	$I = 1$

From the truth table, that is by observing the truth table the logical expression for Y_0, Y_1, Y_2, and Y_3 are given by

$$Y_0 = \overline{S_1}\,\overline{S_0}I$$
$$Y_1 = \overline{S_1}S_0I$$
$$Y_2 = S_1\overline{S_0}I$$
$$Y_3 = S_1S_0I$$

Realizing this circuit we have (Fig. 1.104)

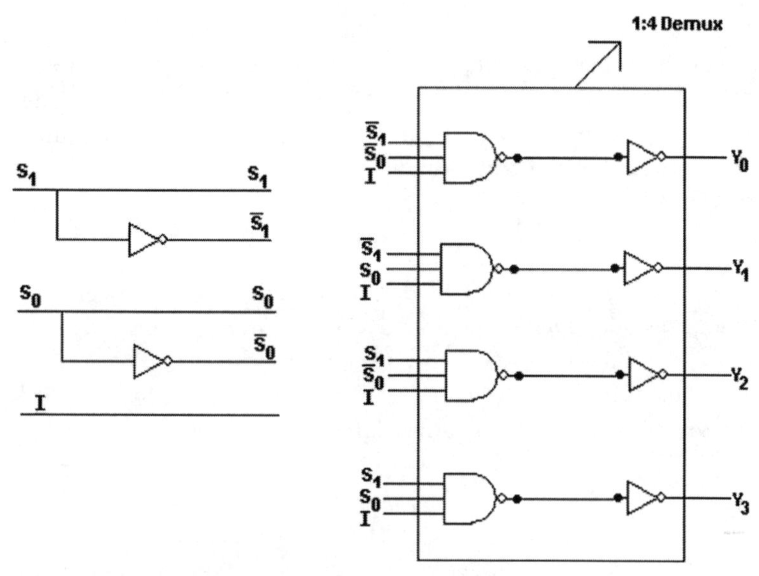

Figure 1.104 Internal circuit of a demultiplexer.

1.30.1 Application of Demultiplexer

(i) It can be used as a one to many devices.
(ii) It can be used as serial in parallel out connection (Fig. 1.105).

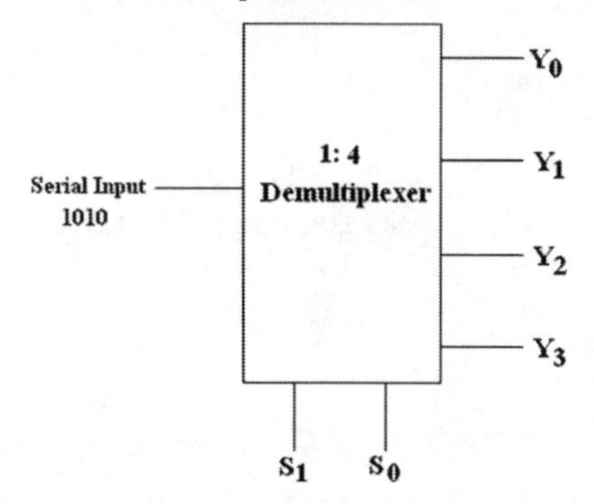

Figure 1.105 Demultiplexer as serial in parallel out connection

1.31 Decoder: Definition and Applications

It is also a special combinational circuit. A demultiplexer will become decoder if the single input of the demultiplexer becomes the enable input of the decoder and the select/address lines for the demultiplexer become the input of the decoder. Output of decoder is same as the output of demultiplexer. The chips of decoder and demultiplexer thus remain same. In general we can say n to m line decoder has got (i) n inputs (ii) m outputs where n and m are related by

$$2^n = m$$

2 : 4, 3 : 8, 4 : 16, and so on decoders are available as IC commercially.

Any decoder higher than 4 : 16 is not commercially available. A decoder may contain 1 or more enable input.

The block diagram of n to m line decoder with active low output is shown in Fig. 1.106 (so negative sign comes in final output).

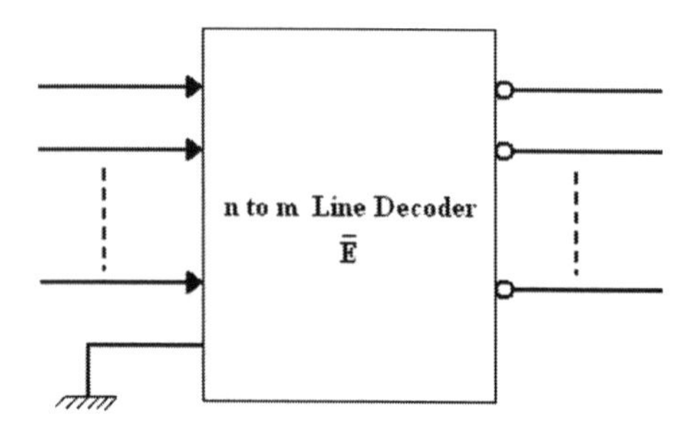

Figure 1.106 Block diagram of a decoder active low outputs.

Let us consider a 2 to 4 line decoder with active low outputs (Fig. 1.107). The truth table is given in Table 1.29.

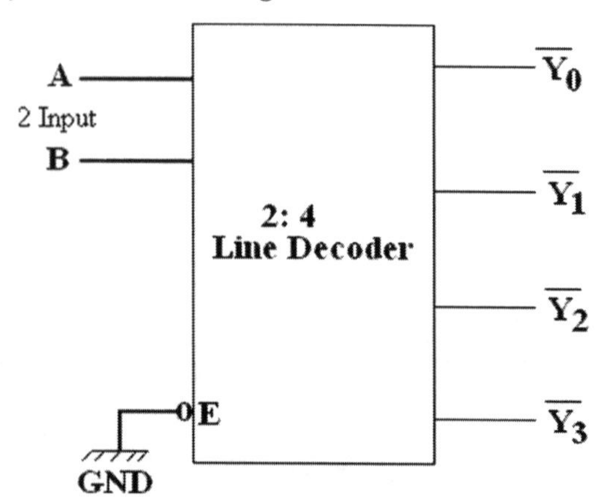

Figure 1.107 Block diagram of a 2 to 4 decoder.

Table 1.29 Truth table of a 2 to 4 line decoder with active low outputs

E	A	B	$\overline{Y_0}$	$\overline{Y_1}$	$\overline{Y_2}$	$\overline{Y_3}$
0	0	0	0	1	1	1
0	0	1	1	0	1	1
0	1	0	1	1	0	1
0	1	1	1	1	1	0
1	X	X	1	1	1	1

By inspecting the truth table, the logical expression for $\overline{Y}_0, \overline{Y}_1, \overline{Y}_2, \overline{Y}_3$ are as follows:

$$\overline{Y_0} = \overline{\overline{E}\,\overline{A}\,\overline{B}}$$

$$\overline{Y_1} = \overline{\overline{E}\,\overline{A}B}$$

$$\overline{Y_2} = \overline{\overline{E}A\overline{B}}$$

$$\overline{Y_3} = \overline{\overline{E}AB}$$

So, the only one output at a time is active low in active low decoder. If we do not know that it is decoder and we only know the output and we are connecting the LEDs in the output:

The diode in LED glows, when the diode is forward biased. So from the output (status of LEDs) we can say what we apply the inputs. This is the function of the decoder. So, decoder is a combinational circuit which produces only one output, as active low or active high and remaining outputs are just the reverse for such value of the inputs provided the decoder is enabled (disabled decoder does not perform its intended function).

Similarly we can have a decoder with active high outputs (Fig. 1.108). Truth table is given in Table 1.30.

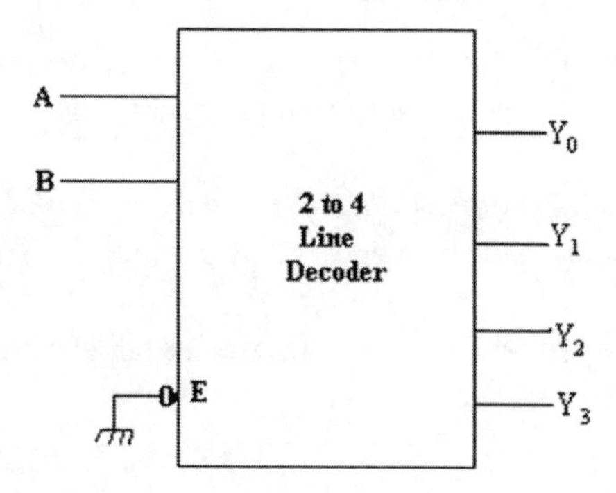

Figure 1.108 Block diagram of a decoder with active high outputs.

$$Y_0 = \overline{A}\,\overline{B}\,\overline{E}$$

$$Y_1 = \overline{A}B\overline{E}$$

$$Y_2 = A\overline{B}\overline{E}$$

$$Y_3 = AB\overline{E}$$

Table 1.30 Truth table of a 2 to 4 line decoder with active high outputs

E	A	B	Y_0	Y_1	Y_2	Y_3
0	0	0	1	0	0	0
0	0	1	0	1	0	0
0	1	0	0	0	1	0
0	1	1	0	0	0	1
1	X	X	0	0	0	0

So, this variety of decoder is possible. But for lab purpose active low output decoders are used, commercially available demultiplexers/ decoder chips is given in Table 1.31.

Table 1.31 Different demultiplexer ICs

IC No	Types of Demux/Decoder	Outputs
74139	Dual 1 : 4 Demultiplexer or 2 : 4 like decoder (has 1 enable input for each decoder)	Inverted output (Active low)
74138	Single 1 : 8 Demux or 3 : 8 line decoder (has got 3 enable inputs for the decoder)	Active low
74154	Single 1 : 16 Demux or 4 : 16 line decoder	Active low

Decoders are more useful than multiplexers.

1.31.1 Applications of a Decoder

1. Decoder can be used to identify the code applied to its input. (what we apply at the input can be detected by observing the output of the decoder)
2. Decoder can be used to convert binary coded decimal that is 4-bit binary coded decimal value into decimal by using BCD to decimal decoder circuit.
3. Decoder can be used to realize the multiple output logical function using some extra gates.

4. Decoder can be used to realize the truth table of any logic gate by using one extra gate.
5. Several lower order decoders can be cascaded to realize higher order decoders. This is called decoder tree.
6. Decoder can be used to drive the chip select signal of different memory devices in a microprocessor-based system.
7. Decoder can be used to drive the chip select signal for different I/O devices in a μP based system.

1.31.2 Application of a Decoder (Example)

Realize a two input XOR gate by decoder and any other gates required (NAND gates are essential at the output of a decoder).

So we have to use a decoder having 2 inputs, 4 outputs (2 to 4 line.) (Fig. 1.109).

Decoder with active low output

$Y_{XOR} = \sum m\ (1, 2)$

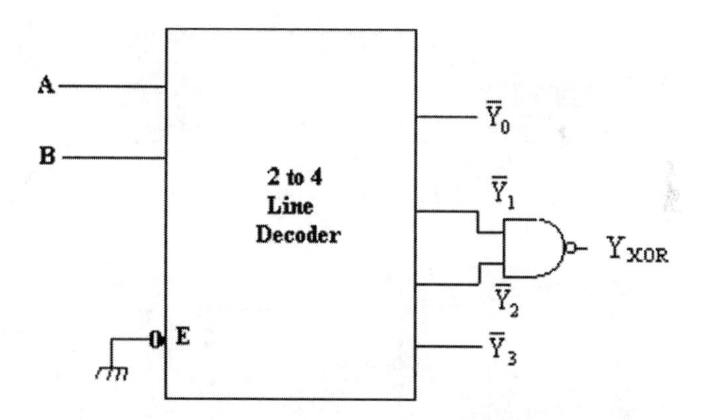

Figure 1.109 Realization of a XOR gate using a 2 to 4 decoder with active low outputs and a NAND gate.

So, it verifies the truth table. XOR is realized by 4 two input NAND gates but here we require only 1 NAND gate along with Decoder to realize the XOR gate.

In case of decode with active high output (Fig. 1.110)

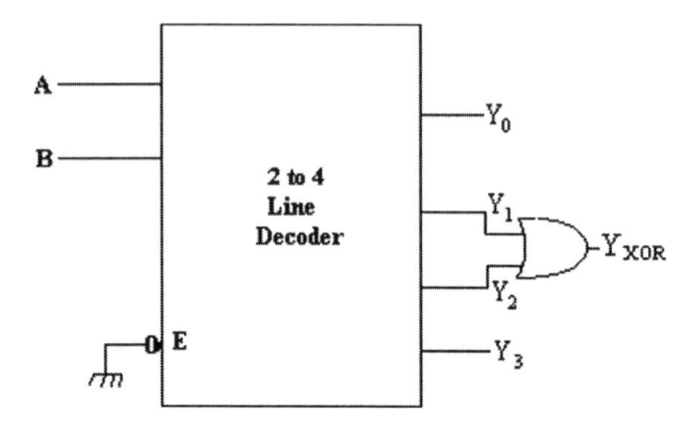

Figure 1.110 Realization of a XOR gate using a 2 to 4 decoder with active high outputs and a NAND gate.

1.31.3 Cascading of Decoders

This can be used as 3 to 8 line decoder. Inputs are A, B, C. (Fig. 1.111). The truth table is given in Table 1.32.

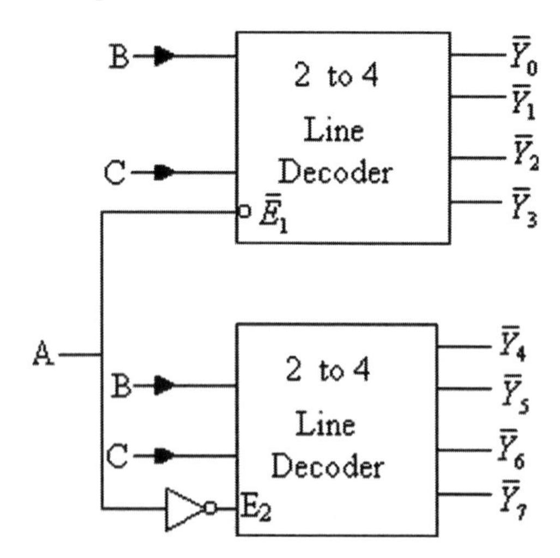

Figure 1.111 A 3 to 8 line decoder using two 2 to 4 decoder.

If any decoder is disabled its output is active high.

Table 1.32 Truth table of a 3 to 8 line decoder

A	B	C	D_1	D_2	$\overline{Y_0}$	$\overline{Y_1}$	$\overline{Y_2}$	$\overline{Y_3}$	$\overline{Y_4}$	$\overline{Y_5}$	$\overline{Y_6}$	$\overline{Y_7}$
0	0	0	E	D	0	1	1	1	1	1	1	1
0	0	1	E	D	1	0	1	1	1	1	1	1
0	1	0	E	D	1	1	0	1	1	1	1	1
0	1	1	E	D	1	1	1	0	1	1	1	1
1	0	0	D	E	1	1	1	1	0	1	1	1
1	0	1	D	E	1	1	1	1	1	0	1	1
1	1	0	D	E	1	1	1	1	1	1	0	1
1	1	1	D	E	1	1	1	1	1	1	1	0

This is the decoder tree. So, we realize 3 to 8 line decoder by using 2 to 4 line decoder and one NOT gate and similarly by using 17 (4 to 16 line decoder). We can realize 8 to 256 line decoder. So, by cascading low order decoder we can realize higher order decoder.

1.32 Seven Segment LED Display

There are two types of seven segment displays:

(a) common anumberde display
(b) common cathode display

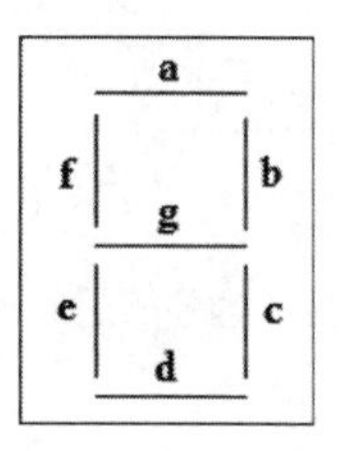

Figure 1.112 Seven segment LED.

A pictorial representation of a seven segment LED is shown in Fig. 1.112.

The dot is used to represent a decimal number. In all the seven segments, there is a LED attached. The dot segment also has a very small LED.

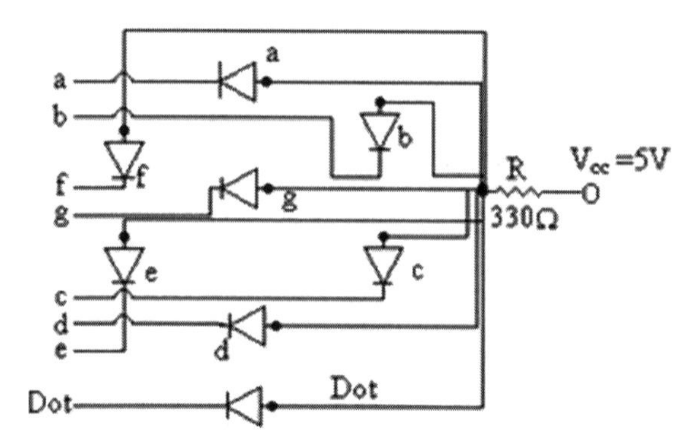

Figure 1.113 Internal circuit of a common anode seven segment LED.

All the anodes are connected to V_{CC} = 5 V and all the cathodes are connected to the inputs. So, there are 8 inputs (Fig. 1.113). The block diagram of a seven segment LED is given in Fig. 1.114. Now we find that input of "a" is 0 V, the diode "a" is forward biased and the LED glows. On the other hand, if the input to "'a" is 5 V then the diode is reverse biased and the LED do not glow and so the segments glows only when the cathode signals are low that is when they are at logical zero.

Logical inputs for displaying decimal 0-9 corresponding to common anode seven segment LED is given in Table 1.33.

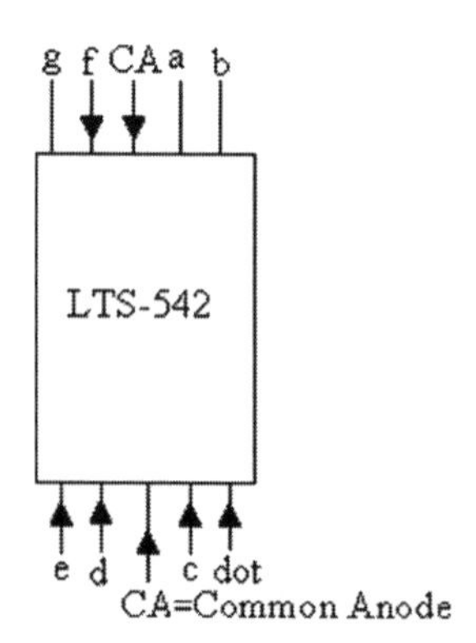

Figure 1.114 Block diagram of a seven segment LED.

R is known as the current limiting resistance and it is of the order of 330 Ω.

The other type of display is the common cathode display device. It is known as cathode display device, all the cathodes of each segment are connected to a common point and the common point is grounded.

Table 1.33 Logical input to the nodes of a common anode seven segment LED for displaying decimal numbers

Decimal	*a*	*b*	*c*	*d*	*e*	*f*	*g*
0	0	0	0	0	0	0	1
1	1	0	0	1	1	1	1
2	0	0	1	0	0	1	0
3	0	0	0	0	1	1	0
4	1	0	0	1	1	0	0
5	0	1	0	0	1	0	0
6	0	1	0	0	0	0	0
7	0	0	0	0	1	1	1
8	0	0	0	0	0	0	0
9	0	0	0	0	1	0	0

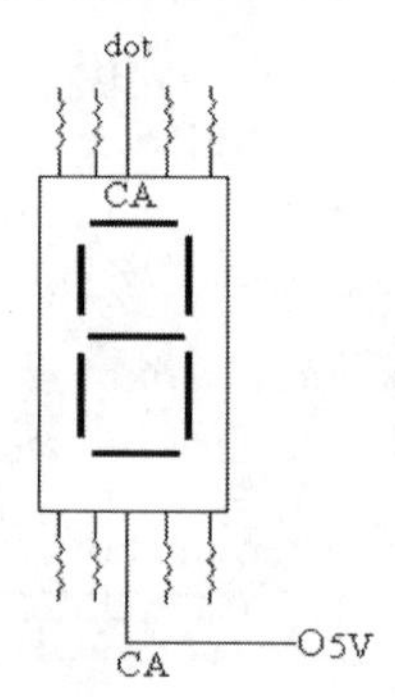

In a common cathode display all the cathodes of all the LEDs in each segment are connected to a common point which is grounded and the inputs are applied to the anodes of each segment to display zero (Fig. 1.115).

Logical inputs for displaying decimal 0–9 corresponding to common cathode seven segment LED is given in Table 1.34.

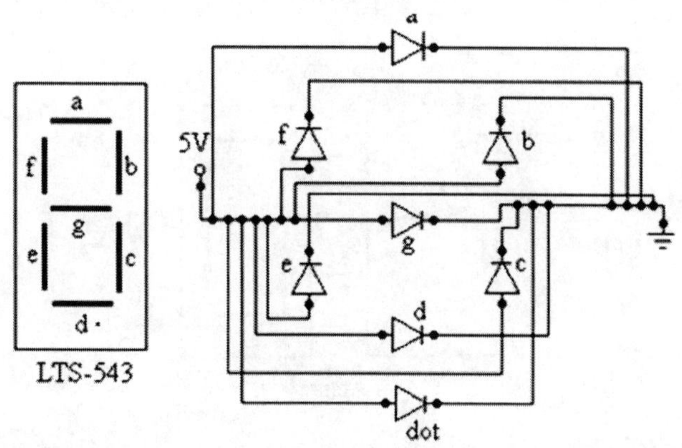

Figure 1.115 Internal circuit of a common cathode seven segment LED.

Table 1.34 Logical input to the nodes of a common cathode seven segment LED for displaying decimal numbers

Decimal	*a*	*b*	*c*	*d*	*e*	*f*	*g*
0	1	1	1	1	1	1	0
1	0	1	1	0	0	0	0
2	1	1	0	1	1	0	1
3	1	1	1	1	0	0	1
4	0	1	1	1	0	1	1
5	1	0	1	1	0	1	1
6	1	0	1	1	1	1	1
7	1	1	1	1	0	0	0
8	1	1	1	1	1	1	1
9	1	1	1	1	0	1	1

In the calculator, we generally do not use LEDs as the power consumption is much higher. Just LCD display is popularly implemented or used. The BCD (binary coded decimal) to decimal decoder driver circuit has been shown in Fig. 1.116. The output is active low.

The verification corresponding data corresponding to common anode display with decoder driver is given in Table 1.35

Table 1.35 Truth table: common anode display with decoder driver

Decimal	*A*	*B*	*C*	*D*	*a*	*b*	*c*	*d*	*e*	*f*	*g*
0	0	0	0	0	0	0	0	0	0	0	1
1	0	0	0	1	1	0	0	1	1	1	1
2	0	0	1	0	0	0	1	0	0	1	0
3	0	0	1	1	0	0	0	0	1	1	0
4	0	1	0	0	1	0	0	1	1	0	0
5	0	1	0	1	0	1	0	0	1	0	0
6	0	1	1	0	1	1	0	0	0	0	0
7	0	1	1	1	0	0	0	0	1	1	1
8	1	0	0	0	0	0	0	0	0	0	0
9	1	0	0	1	0	0	0	0	1	0	0

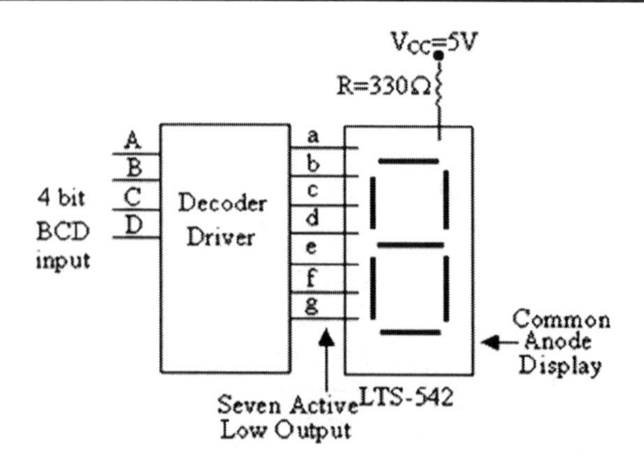

Figure 1.116 Common cathode display circuit with decoder driver.

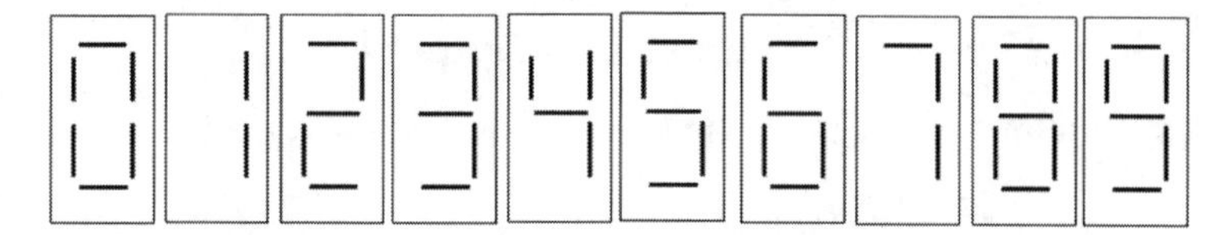

Figure 1.117 All decimal digit display in seven segment LED.

(when cathode is 0 V; and common anode is connected to 5 V).

Decimal	*A*	*B*	*C*	*D*	*a*	*b*	*c*	*d*	*e*	*f*	*g*
10	1	0	1	0	X	X	X	X	X	X	X
11	1	0	1	1	X	X	X	X	X	X	X
12	1	1	0	0	X	X	X	X	X	X	X
13	1	1	0	1	X	X	X	X	X	X	X
14	1	1	1	0	X	X	X	X	X	X	X
15	1	1	1	1	X	X	X	X	X	X	X

So, we can find that when the decimal input is greater than 9 then one display device is not sufficient to represent it and thus all the other conditions becomes do not care from a to g. Logical expression for *a*, *b*, *c*, *d*, *e*, *f*, *g* in the max term form (for the active high output) is given as:

$$a = \prod M\ (0, 2, 3, 5, 7, 8, 9)\ d\ (10, 11, 12, 13, 14, 15)$$
$$b = \prod M\ (0, 1, 2, 3, 4, 7, 8, 9) \prod d\ (10, 11, 12, 13, 14, 15)$$
$$c = \prod M\ (0, 1, 3, 4, 5, 6, 7, 8, 9) \prod d\ (10, 11, 12, 13, 14, 15)$$
$$d = \prod M\ (0, 2, 3, 5, 6, 8) \prod d\ (10, 11, 12, 13, 14, 15)$$
$$e = \prod M\ (0, 2, 6, 8, 9) \prod d\ (10, 11, 12, 13, 14, 15)$$
$$f = \prod M\ (0, 4, 5, 6, 8, 9) \prod d\ (10, 11, 12, 13, 14, 15)$$
$$g = \prod M\ (2, 3, 4, 5, 6, 8, 9) \prod d\ (10, 11, 12, 13, 14, 15)$$

The decoder driver output whose output is active high can drive common cathode seven segment display device. Active low can drive common anode seven segment display device.

$$a = \sum m\ (0, 2, 3, 5, 7, 8, 9) + \sum d\ (10, 11, 12, 13, 14, 15)$$
$$b = \sum m\ (0, 1, 2, 3, 4, 7, 8, 9) + \sum d\ (10, 11, 12, 13, 14, 15)$$
$$c = \sum m\ (0, 1, 3, 4, 5, 6, 7, 8, 9) + \sum d\ (10, 11, 12, 13, 14, 15)$$
$$d = \sum m\ (0, 2, 3, 5, 6, 8) + \sum d\ (10, 11, 12, 13, 14, 15)$$
$$e = \sum m\ (0, 2, 6, 8) + \sum d\ (10, 11, 12, 13, 14, 15)$$
$$f = \sum m\ (0, 4, 5, 6, 8, 9) + \sum d\ (10, 11, 12, 13, 14, 15)$$
$$g = \sum m\ (2, 3, 4, 5, 6, 8, 9) + \sum d\ (10, 11, 12, 13, 14, 15)$$

1.32.1 Decoder for Active Low Output

For minimizing the max term form we redraw the K map

AB \ CD	00	01	11	10
00	0		0	0
01		0	0	
11	X	X	X	X
10	0	0	X	X

So, the possible combination becomes:

$$(1{-}{-}{-})({-}1{-}1)({-}01{-})({-}0{-}0)$$

$$a = A\ (B+D)\ (B+C)(B+D) \tag{1.36}$$

and similarly we can find for other.

$$b = B\ (\overline{C}+\overline{D})\ (C+D) \tag{1.37}$$

$$c = \overline{B}C\overline{D} \tag{1.38}$$

$$d = (\overline{C}+D)\ (B+\overline{C})(\overline{B}+C+\overline{D})\ (B+D) \tag{1.39}$$

$$e = (\overline{C}+D)\ (B+D) \tag{1.40}$$

$$f = \overline{A}\ (C+D)\ (\overline{B}+C)\ (\overline{B}+D) \tag{1.41}$$

$$g = \overline{A}\ (\overline{B}+C)\ (\overline{C}+D)\ (B+\overline{C}) \tag{1.42}$$

These are the expression for decoder whose output is active low.

1.32.2 Decoder for Active High Output

AB \ CD	00	01	11	10
00	1	1	1	
01	1	1	1	1
11	X	X	X	X
10	1	1	X	X

Analyzing the min term expression for C

$$\text{So } C = \overline{C} + D + B + B\overline{C}D$$

Minimizing expressions for a, b, c, d, e, f, g for decoder whose output is active high is therefore given by:

$$a = A + BD + \overline{B}\,\overline{D} + \overline{B}C$$

$$b = B + \overline{C}\,\overline{D} + CD$$

$$c = B + \overline{C} + D$$

$$d = C\overline{D} + \overline{B}C + B\overline{C}D + \overline{B}\,\overline{D}$$

$$e = C\overline{D} + \overline{B}\,\overline{D}$$

$$f = A + \overline{C}\,\overline{D} + B\overline{C} + B\overline{D}$$

$$g = A + B\overline{C} + C\overline{D} + \overline{B}C$$

1.33 Decoder Driver IC and Its Application

The decoder driver chips, which is available directly in the market is 7447 (with active low output).

7447 can be used for common anode seven segments (Fig. 1.118).

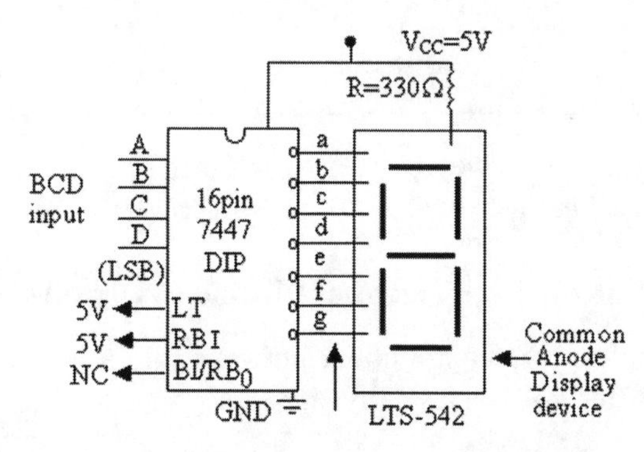

Figure 1.118 Common anode display circuit with decoder driver.

Where LT = lamp test input terminal

RBI = ripple blanking input

BI = blanking input

RBO = ripple blanking output used for cascading of display

7447 (Decoder Driver IC with active low output)

LT	RBI	BI/RBO	BCD	Operation display blank
X	X	used as input 0 as BI input	XXXX	

This mode is used to conserve power in a multiplexed display (irrespective of whatever BCD input). For this we use common anode seven segment display.

The inputs to different pins of 7447 IC and its corresponding outputs is demonstrated in Table 1.36.

Table 1.36 Inputs to different pins of 7447 IC and its corresponding outputs

LT	RBI	BI/RBO This can be used either as input or output.	BCD	Output states operation
0	X	Normally 1 as output	X	It is used to test if all seven segments are functioning i.e. whether all LEDs are glowing.
1	1	Normally output 0 during 0 blanking interval.	Any input	Normal decoding
1	0 This feature is to suppress display of leading zeroes. for example (18 −12) = 06 ↑ This is suppressed here.	RBO output = 1 during normal decoding but RBI output = 0 during zero blanking	Any input 0–9	Normal decoding but 0 is not displayed.
X	X	Used as input 0 as BI input	XXXX	Display blank

7448 (Decoder Drive IC with high output) common cathode display is used in this when BI = 0 the RBO will also be zero.

1.33.1 Multiple Digit Decimal Display (4 Digits)

Say we want to display a number between 0 to 9999; for this we have to cascade 7447 in a suitable manner.

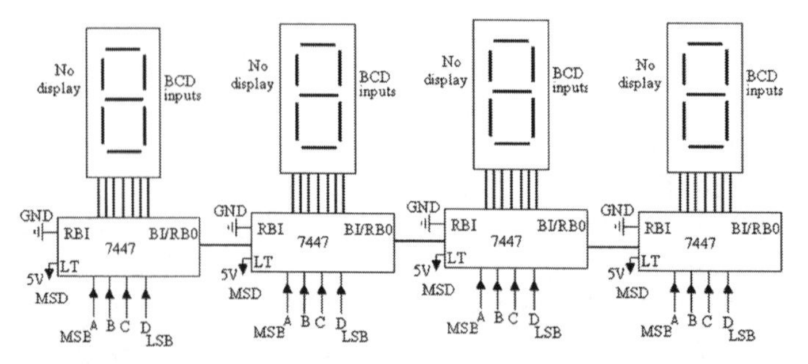

Figure 1.119 Cascading of 7447 ICs.

The LT input is always connected to 5 V.

0000	0000	0000	0000
LT RBI	BI/RBO	Any input	Normal alcidine but is not displayed
1 0	1 at output but becomes 0 during zero blanking interval. Output is also zero during zero blanking interval		

If

0010	0100	0101	1000
2	4	5	8

LT	RBI	BI/RBO	BSD	output
1	1	1 for normal decoding 0 for zero blanking interval.	Any input	Displaying 0 to 9 any number.

For MSD decoder driver RBI is connected to ground just to suppress the display of leading zeroes.

[So, 0013 if applied then 0 and 0 will not be displayed; only 13 is displayed].

1.34 Encoder

Encoder is a combinational circuit whose function is just the reverse of decoder (Fig. 1.120). An encoder may generally have *n* inputs and *m* outputs where $2^m \geq n$. For a normal encoder, out of *n* inputs only one input can be activated at any instant of time corresponding to each activated input, we get a specified *m* bit code at its output.

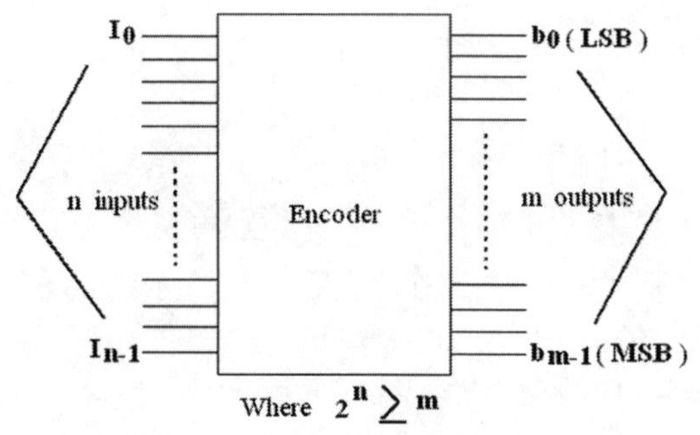

Figure 1.120 Block diagram of an encoder.

Out of n inputs, only one input would be activated at a time. Corresponding to each input, we get only one output.
Example: In a computer keyboard alphabets = J_2 = (26 Lower Case + 26 Upper Case)

Memories = 10 (0 to 9)

And some special characters = 24

So, total = 86 keys or characters.

So, for a keyboard encoder, there may be 86 inputs corresponding to each input, there is seven bit code.

$$2^m \geq n \; n = 86$$
$$2^m \geq 86$$

So, minimum value of m should be 7. So, minimum number of output bit number = 7. This 7 bit code generated for each character in a keyboard is called ASCII (American standard code for information interchange).

Encoder is a device by which any character or symbols or key can be converted into a suitable code and the process of such conversion is called encoding. Its information is just the reverse. Only one input is activated at a time by the encoder, whereas only one output is activated in the case of decoder.
Only one input can be activated. So at a time one switch is closed.

Hence for 9 the $b_3\, b_2\, b_1\, b_0$ gives 1001 and so on.

for 8 the $b_3\, b_2\, b_1\, b_0$ gives 1000.

[Decimal 0, 1,...9 are represented by switches W_0, W_1,...W_9, respectively]
So, it is called the decimal to BCD encoder (Fig. 1.121). By inspection of the Truth table (Table 1.37) the logical expression for the outputs are

$$b_3 = W_8 + W_9 \ldots \tag{1.43}$$

$b_3 = 1$ when switch 8 or 9 is pressed.

$$b_2 = W_4 + W_5 + W_6 + W_7 \ldots \tag{1.44}$$

$$b_1 = W_2 + W_3 + W_6 + W_7 \ldots \tag{1.45}$$

$$b_0 = W_1 + W_3 + W_5 + W_7 + W_9 \ldots \tag{1.46}$$

Let us realize them by OR gates.

The encoder is realized by one 4 input, two three input, and one two input OR gates as shown in Fig. 1.122.

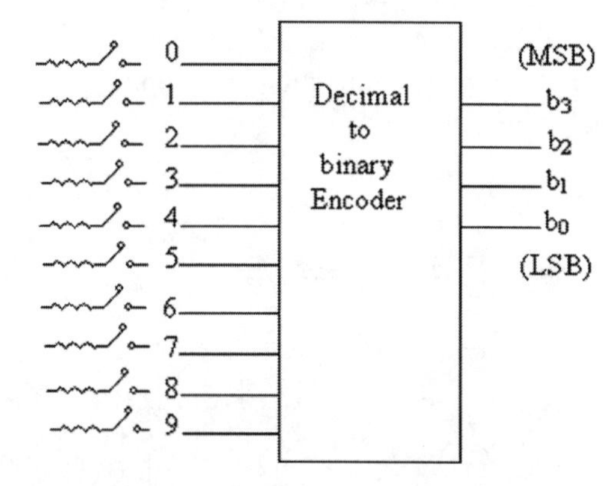

Figure 1.121 Block diagram of a decimal to binary encoder.

Table 1.37 Truth table for decimal to binary encoder

W_0	W_1	W_2	W_3	W_4	W_5	W_6	W_7	W_8	W_9	b_3	b_2	b_1	b_0
1	0	0	0	0	0	0	0	0	0	0	0	0	0
0	1	0	0	0	0	0	0	0	0	0	0	0	1
0	0	1	0	0	0	0	0	0	0	0	0	1	0
0	0	0	1	0	0	0	0	0	0	0	0	1	1
0	0	0	0	1	0	0	0	0	0	0	1	0	0
0	0	0	0	0	1	0	0	0	0	0	1	0	1
0	0	0	0	0	0	1	0	0	0	0	1	1	0
0	0	0	0	0	0	0	1	0	0	0	1	1	1
0	0	0	0	0	0	0	0	1	0	1	0	0	0
0	0	0	0	0	0	0	0	0	1	1	0	0	1

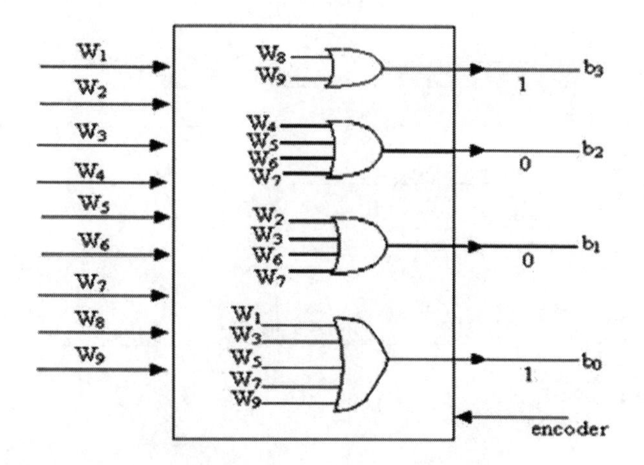

Figure 1.122 Logic circuit for decimal to binary encoder.

The encoder can also be realized by three two input OR gates (Fig. 1.123). There may be some glitch due to unequal propagation in different input lines.

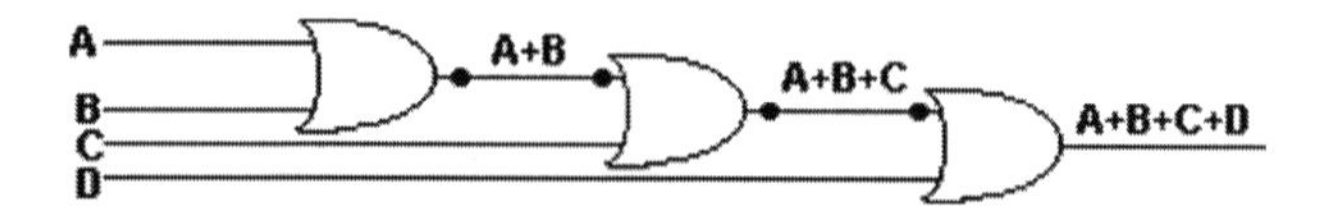

Figure 1.123 The encoder realization using three 2-input OR gates.

When $W_9 = 1$ $(W_0 = W_1 = \ldots = W_8 = 0)$

If $W_5 = 1$, $(W_0 = W_1 = W_2 = W_3 = W_4 = W_6 = W_7 = W_8 = W_9 = 0)$

Output = 0101

Same encoder circuit can be realized by using a series of diodes called the diode-encoding matrix as shown in Fig. 1.124.

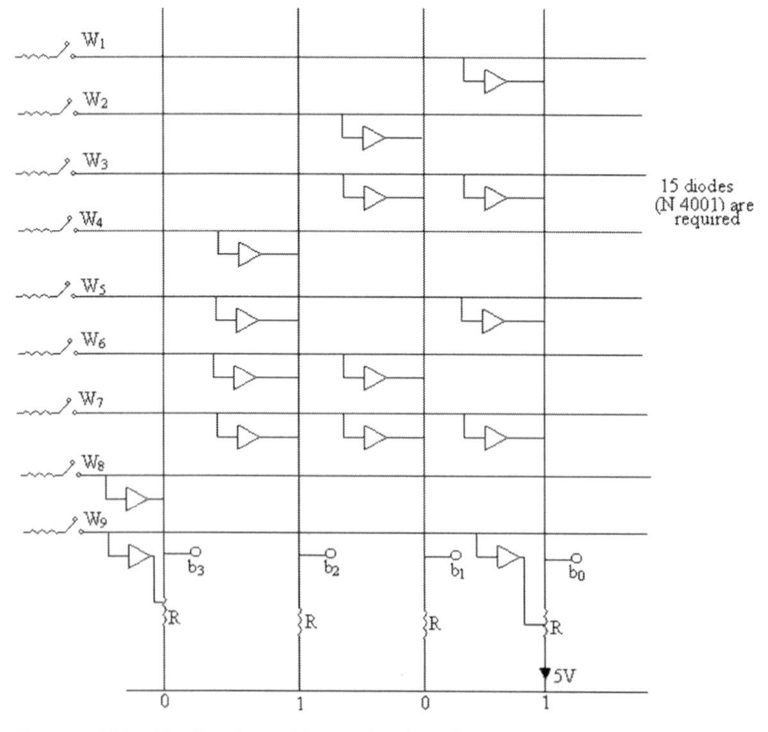

Figure 1.124 Realization of decimal to BCD binary encoder by using diode encoding matrix.

The horizontal and vertical lines can be short circuited by placing the diodes. When none of the nine switches are pressed, the current through the output resistance

b_3	b_2	b_1	b_0
0	0	0	0

If W_5 key is pressed

$I_0 \times \mathrm{R} = I_2 \times \mathrm{R} = 5$ V and so the output is 0101. Here I_0, I_1, I_2 and I_3 represents the current flowing through the nodes b_0, b_1, b_2 and b_3 respectively.

The rows and columns can only be connected via the diodes and when diodes are nonconductive, the output is 0000. The diodes are considered ideal, that is number voltage drop is there across the diode when forward biased.

$$\text{So } b_2 = I_1 R = \frac{5}{R} \times R = 5\,\text{V} \quad \text{(nearly or close to 5 V)}$$

In similar way we can realize an octal to binary encoder (Fig. 1.125).

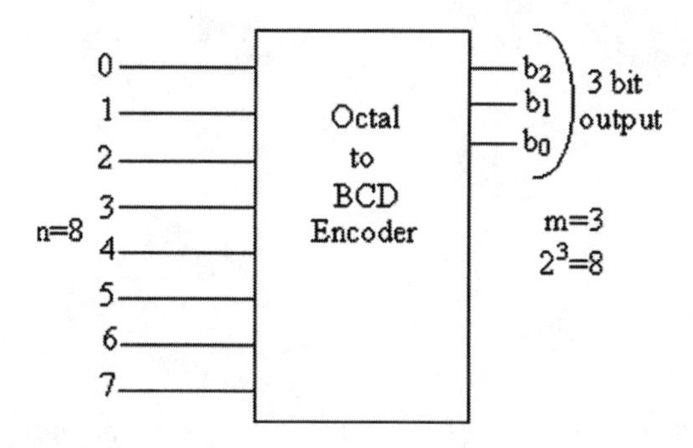

Figure 1.125 Block diagram of an octal to binary encoder.

Example

If 1, 3, and 5 key are pressed simultaneously then the code is produced for the key 5 not for 1 and 3 because if we assume that the highest key number has the highest priority.

Similarly, if 8 and 9 key are pressed, 9 is produced. To realize the priority encoders, we use the inhibit gates.

1.35 Priority Encoder

So far we have discussed normal encoder. If more than one input is activated at a time then the priority encoder will produce the binary code for the highest priority inputs. The block diagram of a priority encoder is shown in Fig. 1.126. For an example If 1,3 & 5 key are pressed simultaneously, then the code is produced for the key 5(not for 1 and 3. Because if we assume that the highest key number has the highest priority) Similarly if 8 and 9 key are pressed, 9 is produced. To realize the priority encoders, we use the inhibit gates.

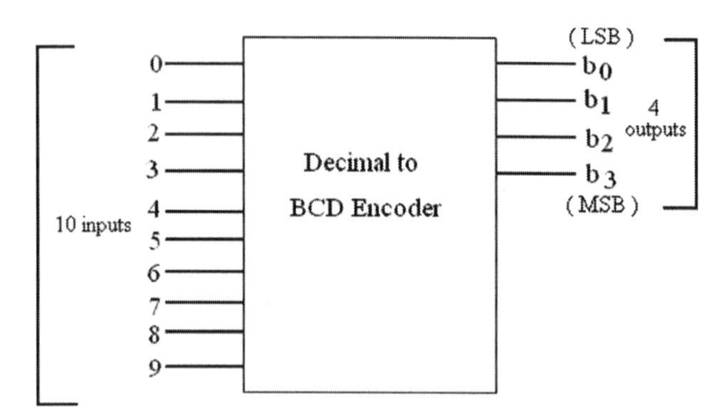

Figure 1.126 Block diagram of a priority encoder.

We have the expression for the output of normal encoder (decimal to binary encoder).

Output for normal encoder

$$b_3 = W_8 + W_9 = 8 + 9 \text{ (When 8 or 9 is activated } b_3 = 1\text{)}$$
$$b_2 = 4 + 5 + 6 + 7$$
$$b_1 = 2 + 3 + 6 + 7$$
$$b_0 = 1 + 3 + 5 + 7 + 9$$

For the priority encoder the priority encoder detects the priority input and produces the binary code for highest priority.

Expression for priority encoder

$$b_3 = 8 + 9 \tag{1.48}$$

(If 8 or 9 or both are pressed then $b_3 = 1$)

For priority encoder $b_2 = 1$, if $W_4 = 1$ and all higher inputs other than 4(which are not present for the expression) for b_2 must be 0.

1. $b_2 = 1$ if $W_4 = 1$ and $W_8 = W_9 = 0$
2. $b_2 = 1$ if $W_5 = 1$ and $W_8 = W_9 = 0$
3. $b_2 = 1$ if $W_6 = 1$ and $W_8 = W_9 = 0$
4. $b_2 = 1$ if $W_7 = 1$ and $W_8 = W_9 = 0$

Combining all these we can write the expression for b_2

$$b_2 = 4\bar{8}\bar{9} + 5\bar{8}\bar{9} + 6\bar{8}\bar{9} + 7\bar{8}\bar{9}$$

If 4 and 5 both are activated then also:

$$b_2 = 1 \tag{1.49}$$

For priority encoder:

$b_1 = 1$, if $W_2 = 1$, input 2 is activated and the other higher priority inputs that are not present in the expression for b_2 must be zero.

1. $b_1 = 1$, if $W_2 = 1$, but $W_4 = W_5 = W_8 = W_9 = 0$
2. $b_1 = 1$, if W_3 1, but $W_4 = W_5 = W_8 = W_9 = 0$
3. $b_1 = 1$, if $W_6 = 1$, but $W_8 = W_9 = 0$
4. $b_1 = 1$, if $W_7 = 1$, but $W_8 = W_9 = 0$

Combining all these, we can write the logical expression for

$$b_1 = 24589 + 34589 + 689 + 789 \tag{3}$$

1. $b_0 = 1$, if $W_1 = 1$ but $W_2 = W_4 = W_6 = W_8 = 0$
2. $b_0 = 1$, if $W_3 = 1$ but $W_4 = W_6 = W_8 = 0$
3. $b_0 = 1$, if $W_5 = 1$, but $W_6 = W_8 = 0$
4. $b_0 = 1$, if $W_7 = 1$, but $W_8 = 0$
5. $b_0 = 1$, if $W_9 = 1$

Combining all these we can write the logical expression for

$$b_0 = \overline{12468} + \overline{3468} + \overline{568} + 78 + 9 \tag{4}$$

Realizing (Fig. 1.127).

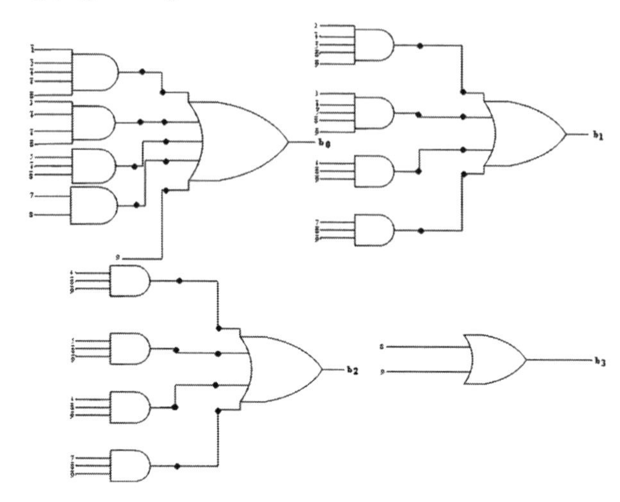

Figure 1.127 Logic circuit realization of the encoder outputs.

1.35.1 Case 1

Let, $W_1 = W_5 = W_7 = 1$ (that is when 1, 5, and 7 inputs are activated simultaneously and $W_2 = W_4 = W_3 = W_6 = W_8 = W_9 = 0$)

So,

$$b_3 = W_8 + W_9 = 0 + 0 = 0$$

$$\begin{aligned} b_1 &= 0\,() + 0\,() + 0\,() + 1\,\bar{0}\,\bar{0} \\ &\uparrow \quad\;\; \uparrow \qquad\;\; \uparrow \\ &2 \qquad 0 \qquad\;\; 6 \\ &= 0 + 0 + 0 + 1 \\ &= 1 \end{aligned}$$

$$\begin{aligned} b_2 = (0)\,() + (1)\,(\bar{0})\,&(\bar{0}) + 0\,() + (1)\,(\bar{0}\,\bar{0}) \\ &\uparrow \\ &0 \\ &= (1 + 1) \\ &= 1 \end{aligned}$$

$$\begin{aligned} b_0 = (1)\,\bar{0}\,\bar{0}\,\bar{0}\,\bar{0} + 0\,() + 1\,\bar{0}\,\bar{0} + 1\,\bar{0} + 0 \\ &= 1 + 1 \\ &= 1 \end{aligned}$$

So, at the output $b_3b_2b_1\mathrm{b}_0$ = 0111 which is the binary code for decimal 7.

74148 → Octal to binary priority encoder with active low inputs and outputs (Fig. 1.128). EI = 0 implies that 74148 will be enabled. The truth table of the octal to binary priority encoder is given in Table 1.38.

Carry outputs (carry of this may act as enable input for the next cascade) are used for cascading two or more 74148.

$$8 + 5 + 1 + 1 + 1 = 16 \text{ pin DIP.}$$

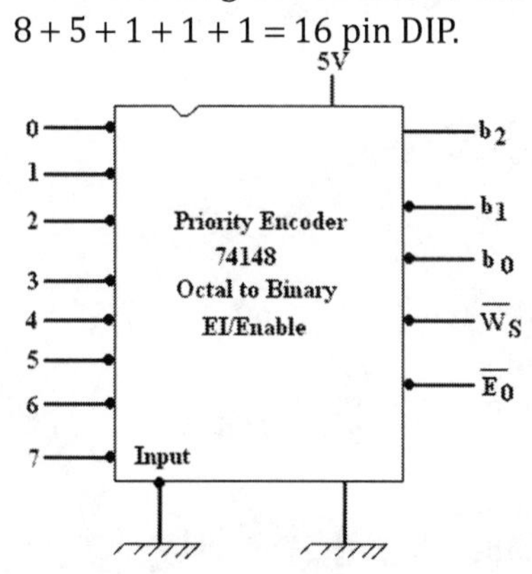

Figure 1.128 Block diagram of an octal to binary priority encoder.

When we want to activate or apply 1 as input then we apply 0 V or ground (for that input). This priority encoder can be used as a normal encoder. If we activate 5, we get at the output (010) as the complement of 101 output is active low output.

Table 1.38 Truth table for octal to binary priority encoder IC 74148

EI	0	1	2	3	4	5	6	7	b_2	b_1	b_0	$\overline{W_S}$	$\overline{E_0}$
1	X	X	X	X	X	X	X	X	1	1	1	1	1
0	0	1	1	1	1	1	1	1	1	1	1	0	1
0	X	0	1	1	1	1	1	1	1	1	0	0	1
0	X	X	0	1	1	1	1	1	1	0	1	0	1
0	X	X	X	0	1	1	1	1	1	0	0	0	1
0	X	X	X	X	0	1	1	1	0	1	1	0	1
0	X	X	X	X	X	0	1	1	0	1	0	0	1
0	X	X	X	X	X	X	0	1	0	0	1	0	1
0	X	X	X	X	X	X	X	0	0	0	0	0	1
0	1	1	1	1	1	1	1	1	1	1	1	1	0

74147 is also used which is decimal to binary priority encoder. If we use a NOT gate in the output of (1) and connect them to the LED and in this we get the actual output, not the complement of it.

Realize octal to binary normal encoder by using diode encoding matrix.

Problems

1. What is Digital Electronics?
2. What are combinational circuits?
3. Define binary logic.
4. Define logic gates.
5. Define Truth Table.
6. What is Min term?
7. What is Max term?
8. Define K Map.
9. Define duality property.
10. State De Morgan's theorem.
11. What are called 'don't care conditions'?
12. Define half adder.
13. Define full adder.
14. Define binary adder.
15. What are decoders?
16. What are encoders?
17. Define priority encoder.
18. Define multiplexer.
19. Define binary decoder.
20. What are the two steps in gray to binary conversion?
21. Write the names of Universal gates.

Chapter 2

Sequential Circuit

2.1 Introduction

In a combinational circuit shown in Fig. 2.1, the input at any instant is totally determined by the presents inputs. This indicates that for every change of input there is immediate change in the output. However, there are some occasions when we wish the output not to change, once set, when there is a change in the input combination. This basically gives the idea of sequential circuits. A digital logic circuit is said to be a sequential logic circuit if it consists of memory elements in addition to the combinational circuit. Moreover, the outputs of the circuit at any instant of time will depend upon the present input as well as the present state of the memory elements. Figure 2.2 represents a sequential circuit that contains a combinational circuit to which memory element are connected to from a feedback path. The function of the memory is to store binary information at any given instant representing the states of the sequential circuit. The sequential circuit generates the binary output determined by the received binary input together with the present state of memory element. To store binary information, we can use a circuit called flip-flop, which is the basic memory element in digital systems.

Foundation of Digital Electronics and Logic Design
Subir Kumar Sarkar, Asish Kumar De, and Souvik Sarkar

ISBN 978-981-4364-58-4 (Hardcover), 978-981-4364-59-1 (eBook)
www.panstanford.com

The flip-flop retains its output between consecutive clock pulses. Flip-flops are employed for the construction of counters and resistor and in some other applications. They are called latches as they can hold or latch in either stable state.

2.2 Definition of Combination and Sequential Circuits

The digital circuits are classified as follows:

1. combinational circuit
2. sequential circuit

2.2.1 Distinction Between Combinational and Sequential Circuits

The combinational circuit is one whose outputs depend only on the present value of the input(s), independent of the past history of the input(s). The sequential circuit is, on the other hand, is a circuit whose output depends not only on the present value of the input but also on the past history of the input.

Hence, we can say that a combinational circuit should not contain any memory element (device that can store some earlier information), whereas the sequential circuit contains at least one memory element.

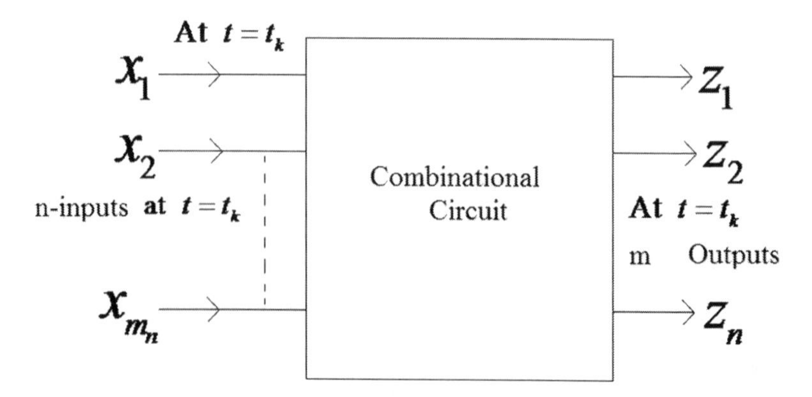

Figure 2.1 A combinational circuit.

2.2.2 The Input–Output Relationship

$$Z_i(k) = f_j\{x_1(t_k), x_2(t_k), \dots x_n(t_k)\} \tag{2.1}$$
$$i = 1, 2, \dots m$$
$$j = 1, 2, \dots n$$

as input and output may not be always same.
S_j determines the nature of the function.

1. The behavior of a combinational circuit is described by the output function as given by Eq. (2.1), whereas in a sequential circuit, some output is fed back to the input.
2. In a sequential circuit, we get two types of outputs. $Z_i\{t_k\}$ is called the output function and $y_i\{t_k\}$ is called the state function.

$$Z_i(t_k) = g_i\{x_1\{t_k\} \dots x_u\{t_k\}; y_1\{t_k\} \dots y_s\{t_k\}, \tag{2.2}$$

where i varies from 1 to v. So we have u number of outputs, and

$$y_j(t_k) = h_j\{x_1\{t_k\} \dots x_u\{t_k\}; y_1\{t_k\} \dots y_s\{t_k\},$$
$$\text{where } j = 1,2, \dots s \tag{2.3}$$

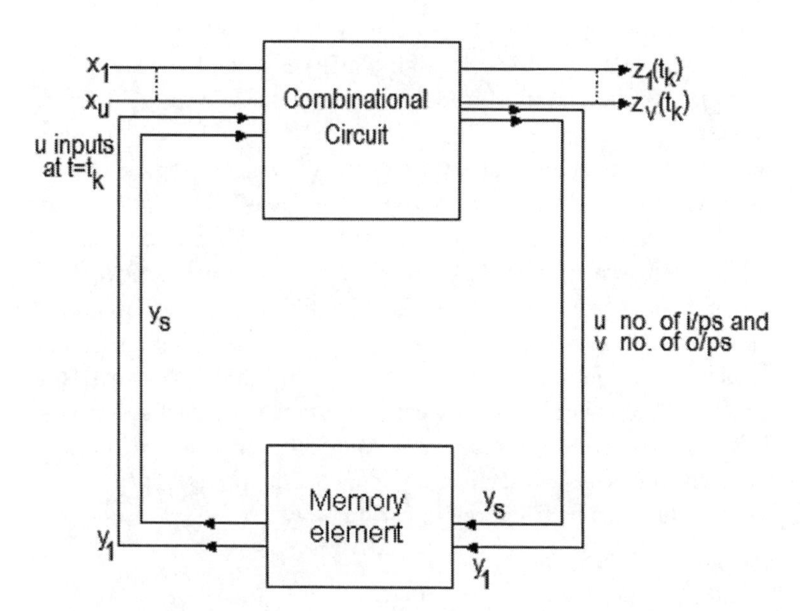

Figure 2.2 Input–output relationships.

3. Thus, we can say that sequential circuit is described by the output function as given by the Eq. (2.2) and next state function as given by Eq. (2.3).
4. For a combinational circuit, the basic building blocks are logic gates. In a sequential circuit, the basic building blocks are flip-flops.
5. Hence, a sequential circuit requires the presence of clock pulses.
6. Examples of combinational circuits are logic gates, multiplexers, demultiplexers, decoders, and encoders. Examples of sequential circuits are flip-flops, registers, counters, memory devices, and sequence generator.
7. The speed of a combinational circuit is higher than that of a sequential circuit.

2.3 Flip-Flop

A flip-flop is a basic memory element in digital circuits and can be used to store 1 bit of information. According to the data inputs, the state of a flip-flop can change only when a clock pulse is present.

It has two stable states: 1 state and 0 state. It can be constructed by cross coupling two NOT circuit in the manner depicted in Fig. 2.3.

In Fig. 2.3, the output of transistor Q_1 is connected to the input of transistor Q_2 and vice versa. This circuit is called a flip-flop. It can stay in one of two stable states. The flip-flop has two stable states and, hence, is called a bistable multivibrator as it can store 1 bit of information ($Q = 0$ or $Q = 1$), it is called a 1 bit memory or 1 bit storage cell. It is also called a latch because it locks the information between consecutive clock pulses. A logic gate has truth tables, but a flip-flop has an "excitation table," which provides information about what its (flip-flop) input should be if the outputs are specified before and after the clock pulses. Some flip-flops have preset and clear inputs for some specific purposes. "Clear" input is to clear or reset the output of the flip-flop ($Q = 0$) where "preset" input is used to set the output of the flip-flop ($Q = 1$).

A flip-flop may be triggered in various ways:

1. level triggering
2. master–slave or pulse triggering
3. positive and negative triggering

In the level triggering scheme, the output of the flip-flop responds to input(s) according to its truth table as long as the clock is present. In the master–slave (pulse) triggering, both negative and positive edges are needed for triggering. In the positive and negative triggering either positive clock pulse or negative clock pulse is required.

2.4 Different Types of Flip-Flops and Their Application

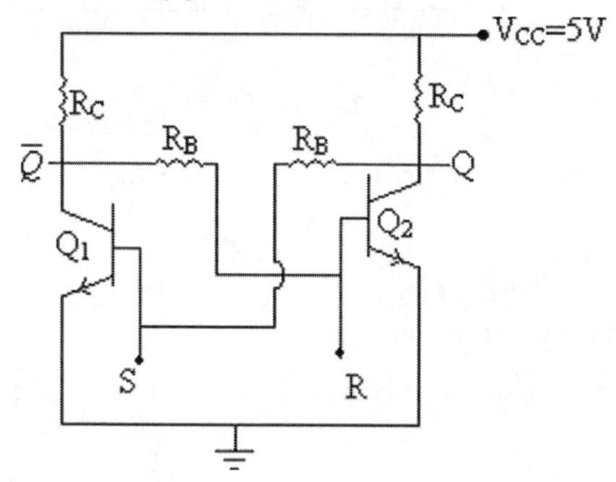

Figure 2.3 S–R Flip-flop.

2.4.1 S–R Flip-Flop

1. There are two additional inputs S and R.
2. Feedback path of resistance R_B is fed to the base of Transisitor Q_2

S and R are two inputs to the transistors Q_1 and Q_2 respectively, which are used to change the outputs of the flip-flops (Q and Q'). When $\overline{Q_1}$ is ON, Q_2 is OFF and vice versa.

Here due to the presence of R_B (base resistances) and ($R_B = 10R_C$) the collector swing is (V_{CC}–$V_{CE,\,SAT}$) = 5 V–0.4/0.2 = 4.6 to 4.8 V, which is nearly equal to logical 1.

However, the actual collector swing is smaller than this due to load and voltage divider action for the resistance R_B and the actual collector swing is of the order of 3.8 to 4.2 V and for transistor–transistor logic (TTL) 3.5 V is logic 1. The truth table of SR flip-flop is given in Table 2.1.

Table 2.1 Characteristics table of S–R flip-flop

S	R	Q_{n+1}	
0	0	No Change	If zero earlier will remain zero and same for 1
0	1	0	Flip-flop is reset
1	0	1	Flip-flop is set
1	1		Output ambiguous as both $Q_1$and Q_2 cannot be simultaneously operated

S = Set
R = Reset
Therefore, this is also called set–reset flip-flop.

Basically there are four different types of flip-flops:

1. S–R Flip-flop.
2. J–K Flip-flop.
3. D–type or Delay flip-flop.
4. T–type or Toggle flip-flop.

Again, depending on the presence of clock input, flip-flops can be of two types

1. Asynchronous flip-flop or unclocked flip-flop.
2. Synchronous flip-flop or clocked flip-flop.

Table 2.2 is shown for comparison.

Table 2.2 Comparison between synchronous and asynchronous flip-flop

Asynchronous flip-flop or unclocked flip-flop	Synchronous flip-flop or clocked flip-flop
The output of asynchronous or unclocked flip-flop is independent of any clock pulse or in such flip-flop no clock pulse is used. The output of such flip-flop depends only on the data inputs.	The output of synchronous flip-flop or clocked flip-flop changes state in synchronization with a clock pulse when data inputs are applied. Changes in the output will take place with the specific instant of time.

The clock pulse is a train of rectangular pulse in time domain where the wave form may be periodic or aperiodic.

For simplicity, we always consider periodic clock pulse.

Following are the main applications of a flip-flop:

1. It can be used as one bit to multiple bit memory elements. (Cascade of two flip-flops is a 2 bit memory element and for n cascaded flip-flops, it is n element memory.)

2. It is used in the architecture section of a computer as
   (a) an accumulator as (where data is permanently stored) in (conjunction with) an adder to store and update data when clocked,
   (b) a status indicator,
   (c) a storage buffer for a computer input/output system.
3. It can be used for alphanumeric display (to display alphabets or numerals, which can be stored in the flip-flop output and displayed accordingly).
4. It can be used for various types of registers such as shift register, e.g., SI-SO, SI-PO, PI-SO, PI-PO, and bidirectional shift register (Here S = Serial, P = Parallel, I = Input, O = Output).
5. It can be used in different types of counter applications.
6. It can be used as a simple divided by N circuit or frequency divider (the f_r at the output of flip-flop may change from the input; input f_r is divided, depending upon the structure of the flip-flop).
7. It is used for the design of sequence generator.
8. It is used for error detection and data conversion. If overflow flip-flop gives 1 at the output, the circuit can be used for overflow detection.

2.4.1.1 Unclocked/asynchronous S–R flip-flop

Figure 2.4 shows an Unclocked/asynchronous S–R flip-flop.

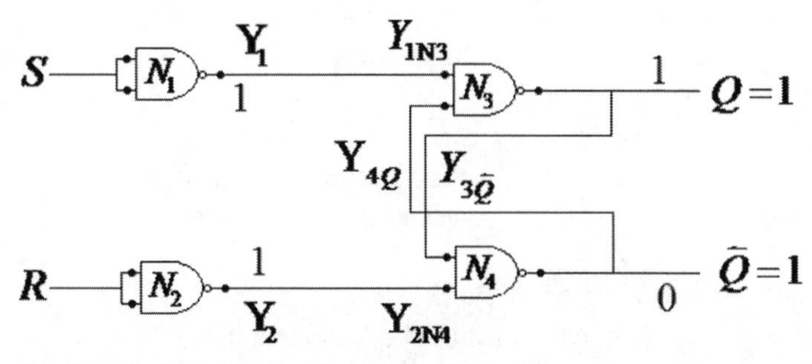

Figure 2.4 Unclocked/asynchronous S–R flip-flop.

It is unclocked as no clock impulse is required. However, it requires four, two-input NAND gates, i.e. one 7400 IC.

The corresponding truth table is given in Table 2.3.

Table 2.3 Truth table of S–R flip-flop

S	R	Q_{n+1}	
0	0	Q_n	No change
0	1	0	Flip-flops is reset
1	0	1	Flip-flops is set
1	1		Not allowed, which is evident from logic diagram.

2.4.1.2 Synchronous or clocked S–R flip-flop

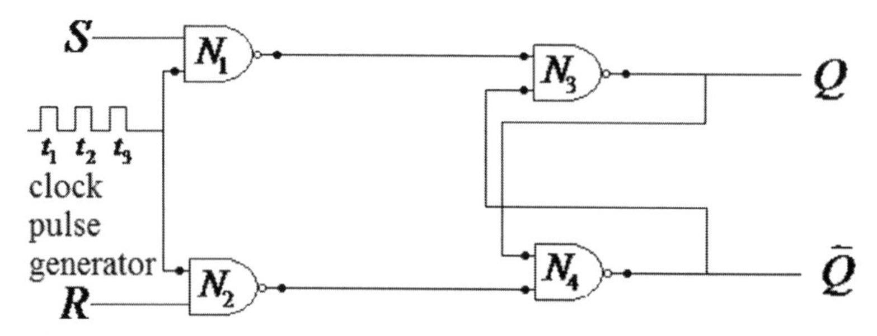

Figure 2.5 Synchronous or clocked S–R flip-flop.

The output of the flip-flop can change only when the clock pulse is 5V, that is, the output changes only during some specific instant of time. For the clocked S–R flip-flop shown in Fig. 2.5, the output of S–R flip-flop changes during the interval over which the clock pulse is high so that the output of the S–R flip-flop may change state only during time intervals t_1, t_2, t_3, and so on.

As shown in figure for the clock pulse, i.e. at instants t_1, t_2, t_3 if S = R = 1 then Q_{n+1} and $\overline{Q}_{n+1}$ may change from Q_n and $\overline{Q}_n$.

2.4.1.3 Advantages of clocked S–R flip-flop

1. It is easy to change the output at certain instant of time.
2. The output will not change when the clock pulse is 0. When the output changes state with the positive edge of the pulse, it is positive-triggered clock and when the output changes state with the negative edge of the pulse, it is negative clock. This is edge-triggered clock.

2.4.2 Jack–Kibby Flip-Flop

Jack–Kibby (J–K) flip-flop is a bistable multivibrator (0 and 1). J and K are the trigger inputs. Figure 2.6 shows the block diagram of a clocked J–K flip-flop. The corresponding truth table is given in Table 2.4.

Table 2.4 Truth table of clocked J–K flip-flop

Clock	J	K	Q_{n+1}	
1	0	0	Q_n	Previous state=present state
2	0	1	0	Always independent of previous state
3	1	0	1	Always independent of previous state
4	1	1	$\bar{Q}_n$	Present state=complement of the previous state

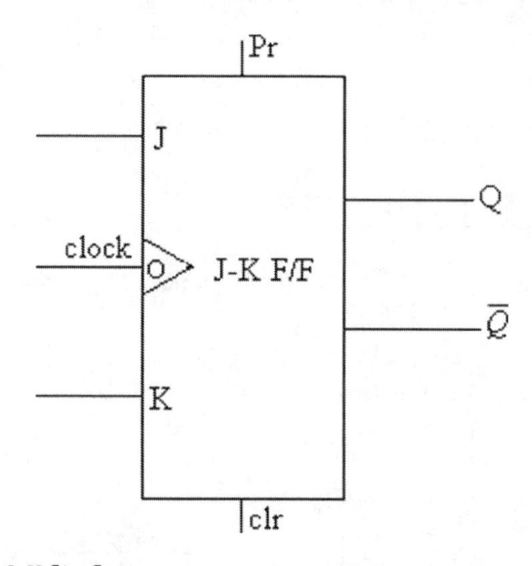

Figure 2.6 J–K flip-flop.

2.4.2.1 To realize J–K flip-flop from S–R flip-flop

This is called flip-flop conversion.

We write the state transition table (ST table) from S–R to J–K flip-flop conversion as shown in Table 2.5. The simplification of the expression is shown in the Fig. 2.7a

Table 2.5 S–R to J–K flip-flop conversion

Dec	J	K	Q_n	S	R	Q_{n+1}
0	0	0	0	0	X	0
1	0	0	1	X	0	1
2	0	1	0	0	X	0
3	0	1	1	0	1	0
4	1	0	0	1	0	1
5	1	0	1	X	0	1
6	1	1	0	1	0	1
7	1	1	1	0	1	0

Truth table of S–R flip-flop

S	R	Q_{n+1}
0	0	Q_n
0	1	0 (Reset)
1	0	1 (Set)
1	1	Not allowed

$$S = F(J, K, Q_n) = \sum m\,(4, 6) + \sum d\,(1, 5)$$
$$R = F(J, K, Q_n) = \sum m\,(3, 7) + \sum d\,(0, 2)$$

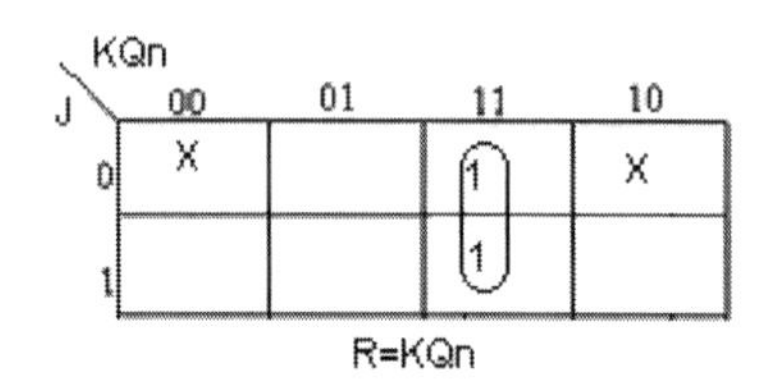

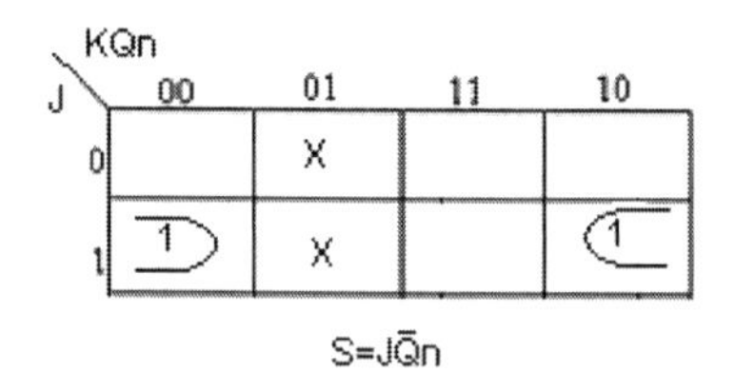

Figure 2.7a Simplified Expression.

Using S–R flip-flop we can get J–K flip-flop and the circuit is shown in the Fig. 2.7b. And the verification table is shown in the Table 2.5a

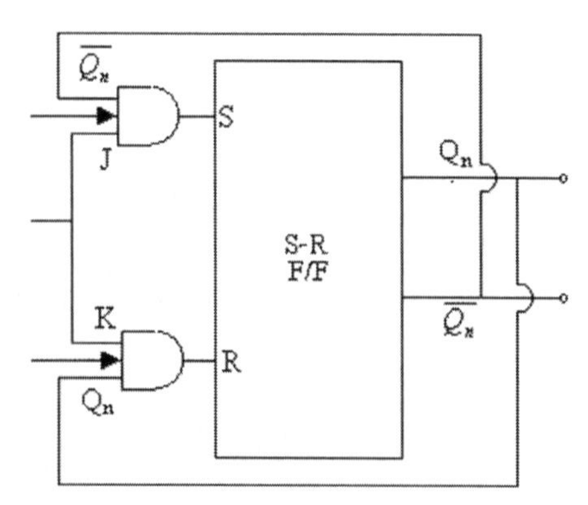

Figure 2.7b S–R flip-flop.

Table 2.5a Truth table verification

J	K	Q_n	S	R	Q_{n+1}
0	0	0	0	0	0
0	0	1	0	0	1

So $Q_n = Q_{n+1}$.

J	K	Q_n	S	R	Q_{n+1}
0	1	0	0	0	0
0	1	1	0	1	0
1	0	0	1	0	1
1	0	1	0	0	1

For J, K=1,0, Q_{n+1}=1

J	K	Q_n	S	R	Q_{n+1}
1	1	0	1	0	1
1	1	1	0	1	0

J,K=0,1, $Q_{n+1} = 0$ and for J, K = 1,1, $Q_{n+1} = \overline{Q}_n$.

2.4.3 Clocked J–K Flip-Flop

Figures 2.8a and 2.8b represent a clocked J–K flip-flop. In Fig. 2.8a, lumped delay line is used and preset and clear terminals are added.

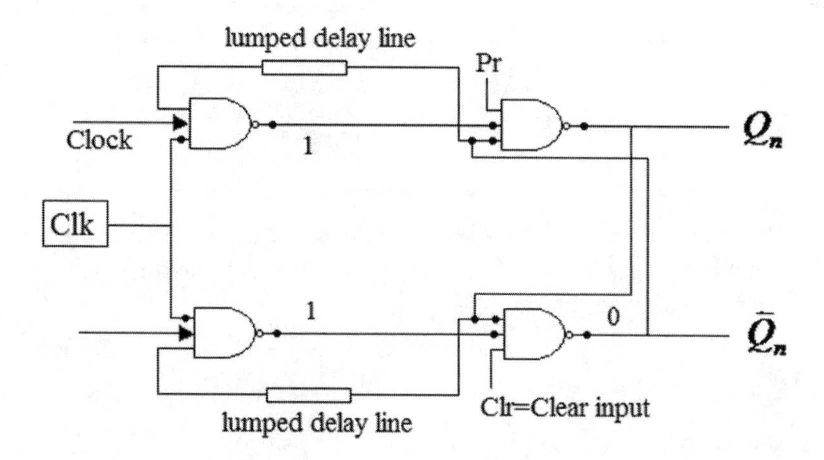

Figure 2.8a Clocked J–K flip-flop.

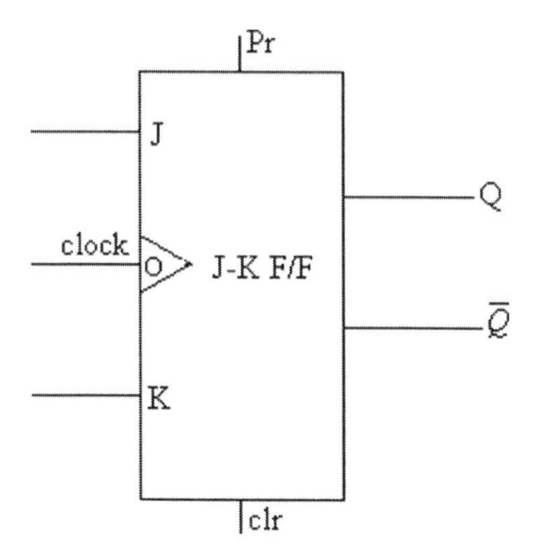

Figure 2.8b Clocked J–K flip-flop.

1. *Pr* = preset input, *Clr* = clear input. The function of Pr and Clr line is shown in Table 2.6 of flip-flop as by applying suitable value at these inputs with *CLK* = 0 that is with no clock pulse we can set or reset a flip-flop. All flip-flops will generally contain at least 1 or 2 asynchronous inputs. The funcwtion of *Pr* and *Clr* line is shown in the Table 2.6.

Table 2.6 Function of *Pr* and *Clr* line

Clk	*Pr*	*Clr*	Q_{n+1}	
0	0	1	1	Setting a flip-flop output
0	1	0	0	To clear the output of flip-flop
1	1	1		Flip-flop is enabled to change state (BLANK) changes in data inputs and clock pulse.

The J and K inputs are called synchronous inputs because by suitable changes in J and K and in synchronism with the clock pulse, the output can be changed.

2.4.4 D–Flip-Flop

The block diagram, truth table and the output waveform of D-Flip-flop are shown in Fig. 2.9, Table 2.7 and Fig. 2.10, respectively.

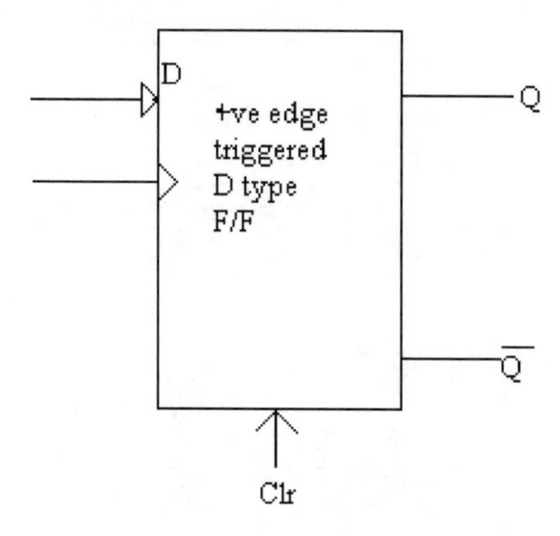

Figure 2.9 D–Flip-flop.

Table 2.7 Truth table of D–flip-flop

D	Q_{n+1}
0	0
1	1

The data at the input of the D– F/F will be delayed at its output by one clock period. We are assuming that the F/F is positive edge triggered.

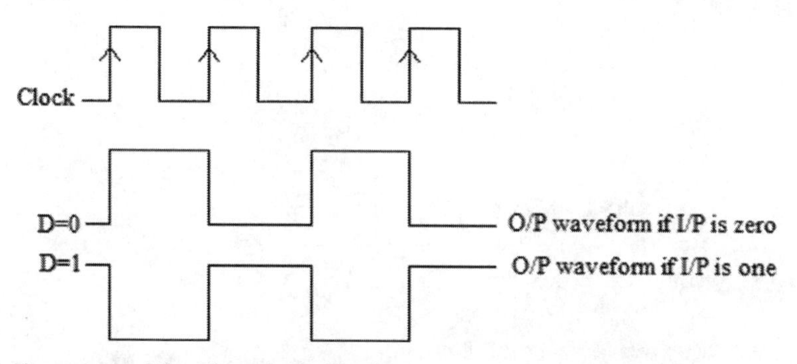

Figure 2.10 Wave form for F–flip-flop.

2.4.5 T–Flip-Flop

The block diagram, truth table and the output waveform of D–Flip-flop are shown in Fig. 2.11, Table 2.8, and Fig. 2.12, respectively.

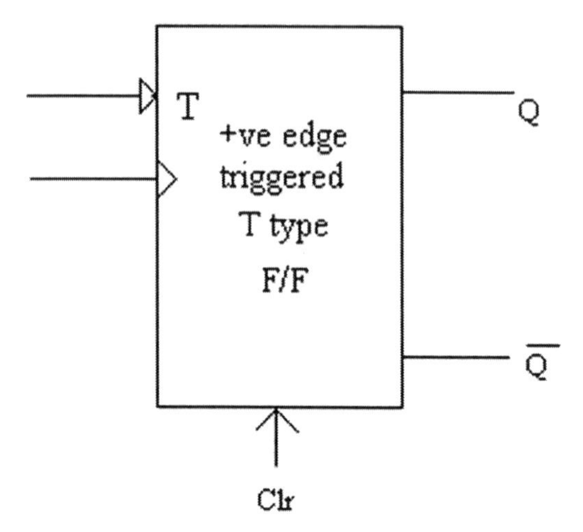

Figure 2.11 T–Flip-flop.

Table 2.8 Truth table of T–flip-flop

T	Q_{n+1}	
0	Q	⇒ present state = previous state
1	$\bar{Q}$	

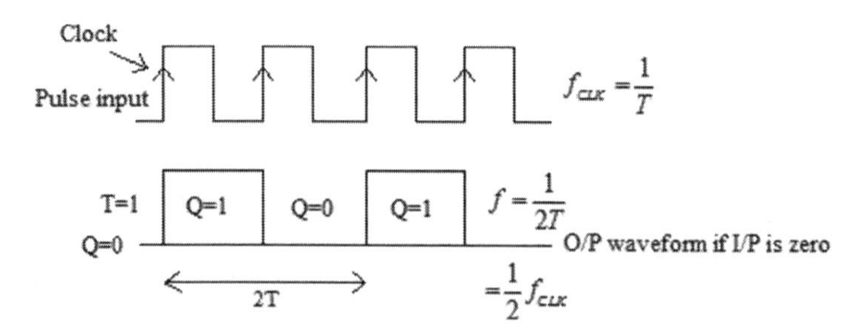

Figure 2.12 Wave form for T–flip-flop.

2.4.5.1 Realization of T flip-flop from D and J–K flip-flop

The realization of T–Flip-flop from D–Flip-flop is shown in the Fig. 2.13 and realization of T–Flip-flop from J–K Flip-flop is shown by the Fig. 2.14 and Tables 2.9 and 2.10.

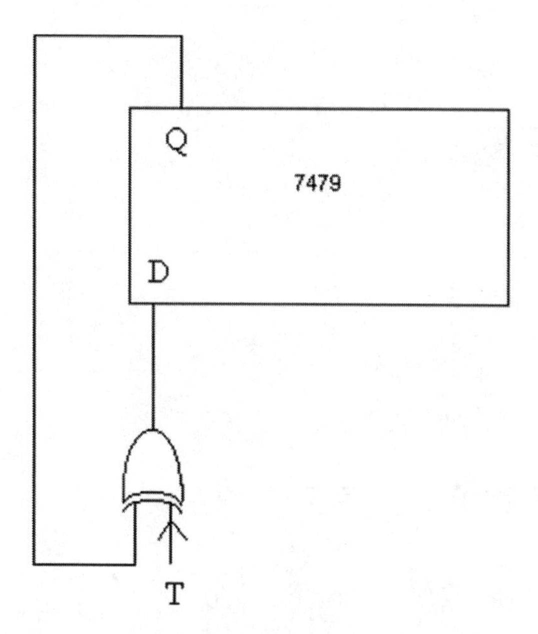

Figure 2.13 Implementation of T–flip-flop using D–flip-flop.

$$D = T \oplus Q_n$$

At $T = 1$, the output of the flip-flop will vary from 0 to 1.

Table 2.9 T–Flip-flop from J–K flip-flop

T	Q_n	J	K	Q_{n+1}
0	0	0	X	0
0	1	X	0	1
1	0	1	X	1
1	1	X	1	0

Table 2.10 Truth table of J–K flip-flop

J	K	Q_{n+1}
0	0	Q_n
0	1	0
1	0	1
1	1	$\bar{Q}_n$

$$\mathrm{J} = f_1\,(T, Q_n)$$
$$= \sum m\,(2) + \sum d\,(1, 3)$$
$$\mathrm{K} = f_2\,(T, Q_n)$$
$$= \sum m\,(3) + \sum d\,(0, 2)$$

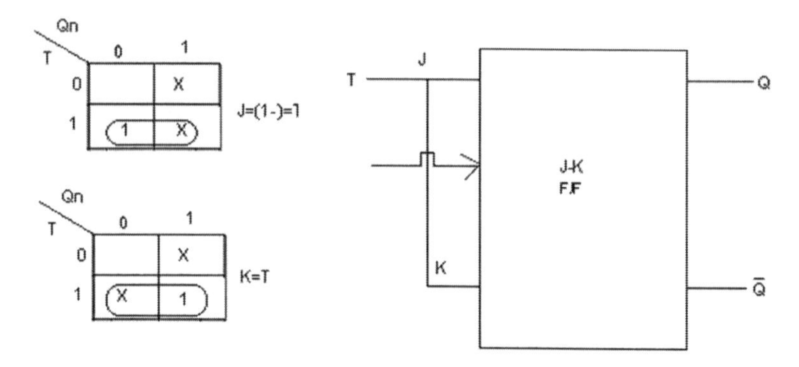

Figure 2.14 Implementation of T–flip-flop using J–K flip-flop.

2.5 Flip-Flop Used as a Divider Circuit

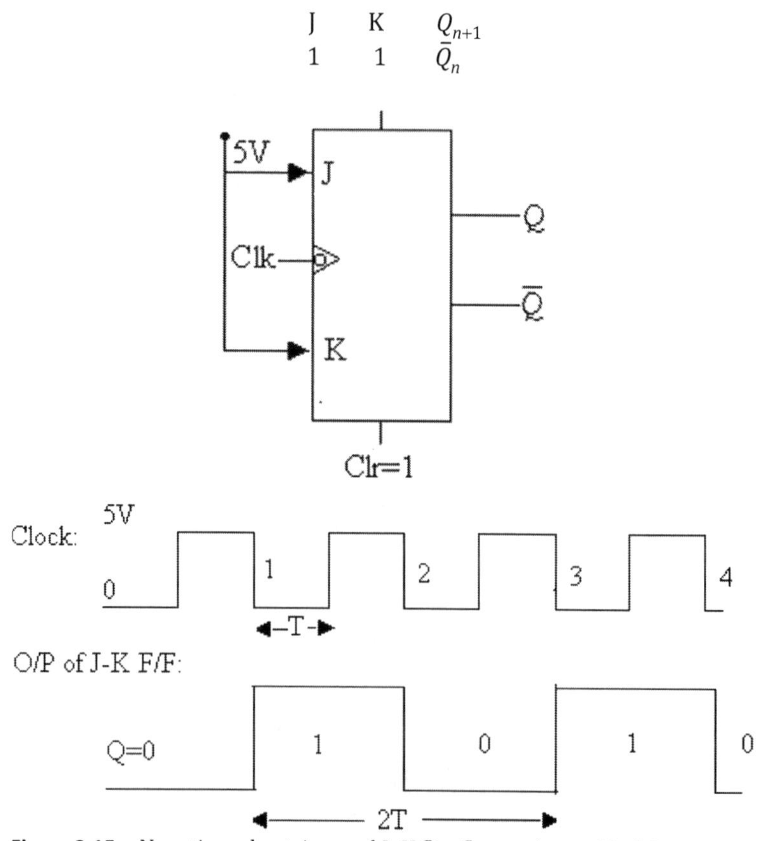

Figure 2.15 Negative edge triggered J–K flip-flop acting as Mod 2 counter.

Let us consider that the J–K F/F considered here is negative edge trigger flip-flop as shown in the Fig. 2.15. The output changes state at the level 1,2,... before applying clock pulse $Q = 0$ and $\bar{Q} = 1$.
So the frequency of the output waveform is

$$f_Q = 1/(2T)$$
$$= (1/2)(1/T)$$

So it behaves as a divider circuit as f_r is divided by a factor of 2.

2.5.1 Conclusion

A simple J–K flip-flop with J = K = 1 as inputs can be used as a divided by 2(÷2) circuit.

or Mod 2 Counter $(0 \rightarrow 1 \rightarrow 0)$

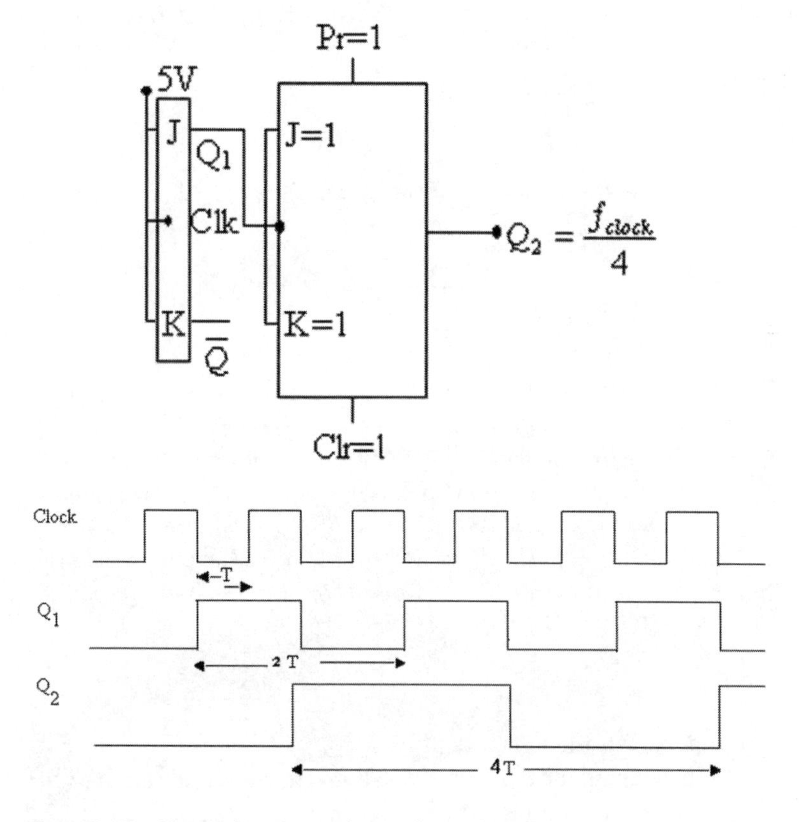

Figure 2.16 Circuit diagram and waveform of J-K flip-flop acting as divide-by-four circuit.

As it switches from 0 to 1 after first clock pulse and switches from 1 to 0 after second clock pulse. Figure 2.16 represents a divide-by-four circuit.

The J–K flip-flop has got another disadvantage that is due to racing problem.

2.6 Racing Problem

For the J–K flip-flop let J = K = 1 and say $Q_n = 0$. If we apply clock pulse whose duration is equal to t_p (pulse duration) that is during t_p interval clock pulse = 1 then the output of J–K flip-flop changes from 0 to 1 after Δt as can be seen in the Fig. 2.17. (Since the flip-flops are made of gates which have their respective propagation delay) where Δt = propagation delay of J–K flip-flop. But as $\Delta t << t_p$ after Δt we get J = K = 1. $Q_n = 1$.

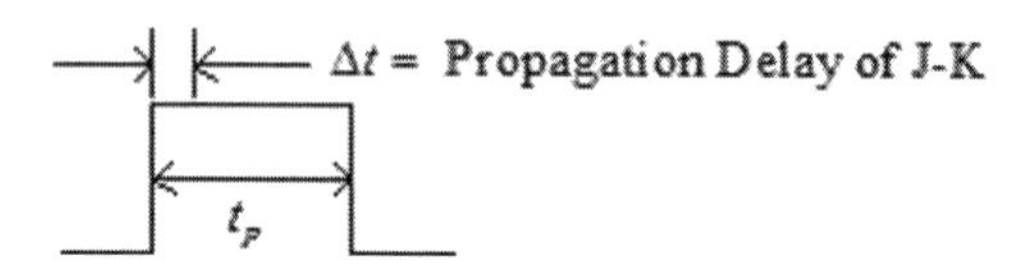

Figure 2.17 Clock pulse.

Since after Δt, the clock pulse is still high Q_n will again change state from 1 to 0. So, during t_p where $t_p > \Delta t$, the output of J–K flip-flop oscillates back and forth between 0 and 1 and thus at the end of the clock pulse, the output of the J–K flip-flop will be undetermined. This phenomena of fluctuating output of J–K flip-flop when $t_p > \Delta t$ is called racing. So, present racing problem we have to satisfy the following condition called race around condition that is:

$$t_p < \Delta t < T \tag{2.4}$$

where

t_p = pulse duration of clock pulse
Δt = propagation delay of J–K flip-flop (time interval required for change in output to the change in input)
T = time period of the clock pulse.

Equation (2.4) can be satisfied; that is the racing problem can be avoided by using

(a) a lumped delay line in the feedback path from the output to the input. The Δt_1 is the additional delay due to the lumped delay line $(\Delta t + \Delta t_1) > t_p$

Then clock pulse becomes 0 and this can be done by the R-C circuit in feedback path.

(b) the most widely used IC solution is the use of master–slave clocking.

There are three methods of clock pulse:

1. DC or edge triggered clock
2. AC coupled clock
3. master–slave clock

1. DC or edge triggered clock:

The DC or edge triggered clocks are generally used for TTL devices and the output of a flip-flop change state due to transition of voltage corresponding to clock input.

(a) A positive edge triggered clock is one where output changes state when the clock pulse changes from 0 to 1 or low to high state i.e. during the positive transition as shown in Fig. 2.18.

(b) A negative edge triggered clock is one when output changes state when the clock pulse changes from 1 to 0 or high to low, that is during negative transition of clock pulse as shown in Fig. 2.19.

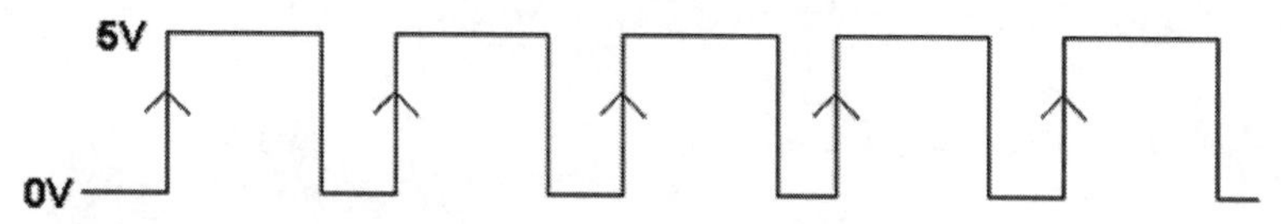

Figure 2.18 Positive clock transition.

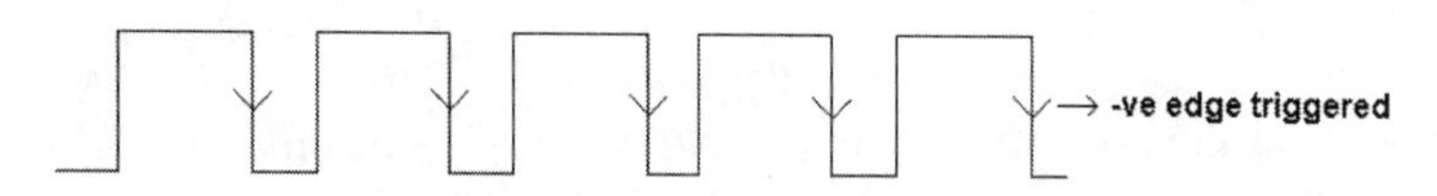

Figure 2.19 Negative clock transition.

A particular flip-flop can be either negative or positive edge triggered not both simultaneously. However in some flip-flops both can be present. Generally, any of these two types is used for a flip-flop. This edge triggered clock or this technique provides high speed transition independent of the clock rise and fall time (assumed to be equal to 0). But if rise or fall time increases beyond 150 ns then noise immunity of the device will decrease. Noise immunity is defined as the amount of noise which can be withstood by the flip-flop without giving any error.

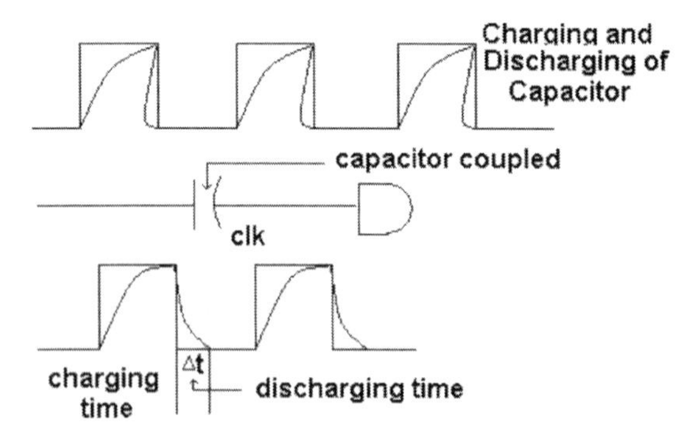

Figure 2.20 Capacitive coupled clock.

2. An AC coupled clock is used in diode transistor logic (DTL) device and this clock is capacitively coupled (as shown in the Fig. 2.20) internally to the latching mechanism in a flip-flop. For AC coupled clock the change in.

If the rise time or fall time of the AC coupled clock is greater than 200 ns, the operation of flip-flop with AC coupled clock, will be either interrupt or impossible (disadvantages).

2.7 Master–Slave Clock

It is used in the master–slave J–K flip-flop. This uses two latch of flip-flop in serial. One is called the master and the other is called the slave latch (Fig. 2.21).

When master is enabled, slave is disabled and vice versa. So, in this technique, when master is enabled, slave is disabled i.e. when clock = 1, master is enabled, slave is disabled.

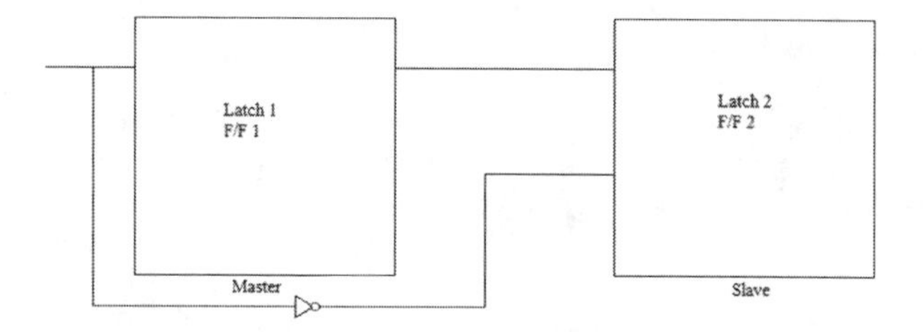

Figure 2.21 Master–slave clock.

When clock = 0, master is disabled, slave is enabled.

This isolates the output of slave from input of master and this prevents racing problem.

To prevent racing problem

$$t_p < \Delta t < T$$
$$\Delta t > t_p$$
$$\Delta t < T$$

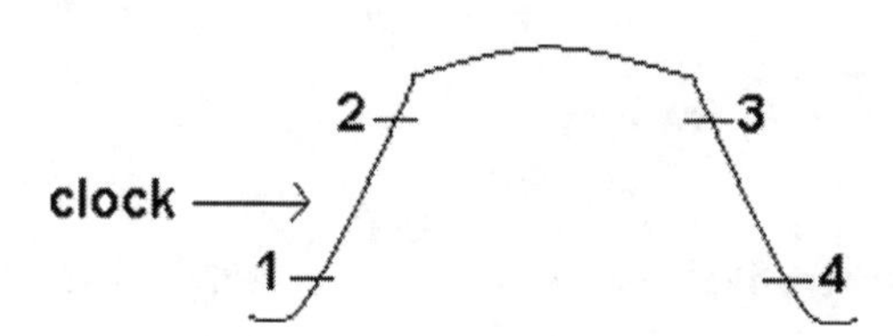

Figure 2.22 Clocking action of master–slave clock.

The clocking action of master–slave clock has got four steps (Fig. 2.22).

At step (1), the slave is isolated from the master.

At step (2), voltage rises and data is enabled at the input of the master and this implies master enabled and slave disabled.

At step (3), data is disabled at the input of the master but enabled at the input of slave which implies that master is disabled and slave is enabled.

At step (4), data transfer from the output of the master to the output of the slave. This prevents racing problem as the output of 1 cannot affect the input of the other.

2.7.1 Input Circuit of a Positive Edge Triggered

As there is no clock input as we use a synchronous flip-flop only during the positive spike the output changes and so, it is called the positive edge triggered.

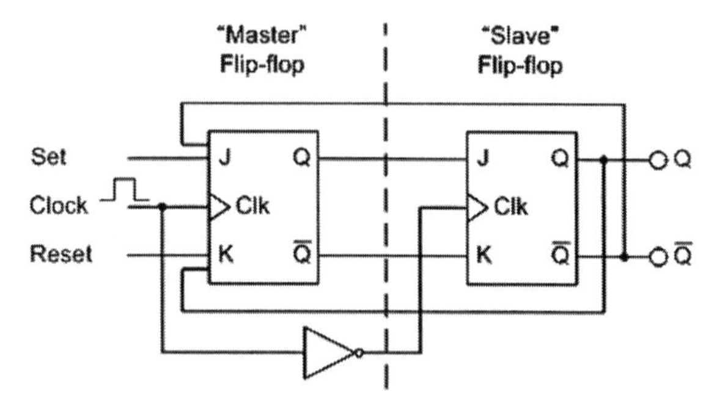

Figure 2.23 Positive edge triggered.

By differentiating clock pulse positive edge triggered pulse is produced.

$$D = T\ (+)\ Q_n$$

2.7.2 Operation of J–K Master–Slave Flip-Flop

According to Fig. 2.24, master is a J–K flip-flop and slave is an S–R flip-flop. The J–K flip-flop receives clock signal directly, whereas the slave which is an S–R flip-flop that receives the inverted clock. With

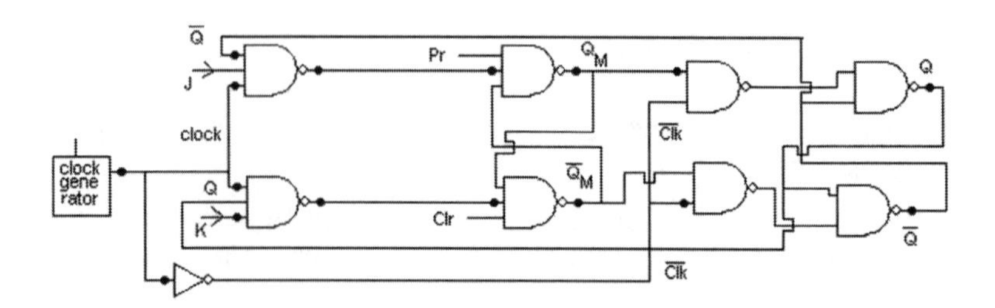

Figure 2.24 J–K Master–slave F/F.

$Clk = 1$ and $\overline{Clk} = 0$ master is enabled and slave is disabled. So, the master then satisfies the truth table for the J–K flip-flop. The data applied via J and K inputs to the master during $Clk = 1$ will appear as an output of the master but the output of the slave does not change state, (as S–R flip-flop do not change output when there is no clock pulse). When the clock = 0 that is during this time, the master is disabled and slave is enabled and the data is transferred from the output of the master to the output of the slave during which $Clk = 0$ that is $\overline{Clk} = 1$.

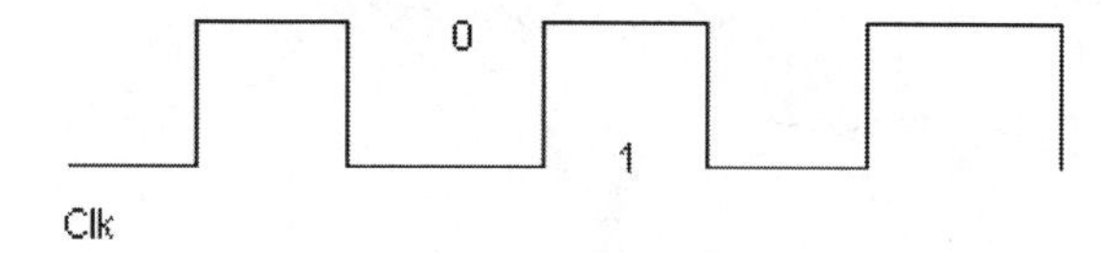

Figure 2.25 Clock pulse.

During $Clk = 0$, any change in the J–K inputs will not change the final output. Only one thing is to be taken care of is that during the interval over which $Clk = 1$, the synchronous data inputs J and K remain constant, otherwise we may get erroneous results.

Suppose at Δt_1, if J = K = 1 then $Q = 1$ and $\bar{Q} = 0$, and after the interval Δt_2, J = 0, K = 1 then $Q = 0$ and $\bar{Q} = 1$.

So, this prevents the race around condition, as the output and input are isolated.

Advantages

1. We can apply J and K = 1 simultaneously.
2. It does not suffer from racing.

2.8 Counters

Counter is a sequential device which can count the number of clock pulses applied at its input or which can remember the number of clock pulse applied at its input. On the second basis, counters can also be considered as a memory device. A single flip-flop can produce a count 0 and 1 that is maximum count = 2. Suitable cascade of n-flip-flops can count upto maximum of 2^n count.

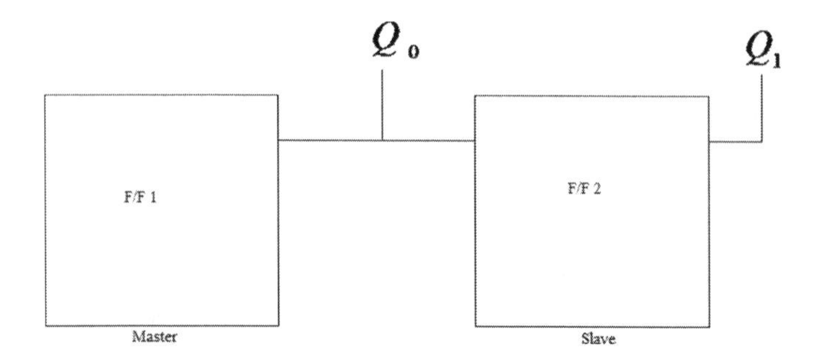

Figure 2.26 Counter.

When $n = 2$ (Fig. 2.26)

Table 2.11

Decimal	Q_1	Q_0	Clk
0	0	0	0
1	0	1	1
2	1	0	2
3	1	1	3
and repeat sequence	0	0	4

So, a counter is a suitable connection or cascade of several flip-flops which always passes through some specified sequence of states with the arrival of each clock pulse. First clock pulse count sequence

$$0 \xrightarrow{(1)} 1 \xrightarrow{(2)} 2 \xrightarrow{(3)} 3$$
$$\xleftarrow[(4)]{\hspace{4cm}}$$

Such a counter is a binary counter.

Counter is of two types:

1. binary counter.
2. nonbinary counter.

1. Binary Counter

A binary counter is one which passes through all possible sequence of states for a given number of flip-flop cascaded to realize a counter.

Binary Counter (BC) $\rightarrow$ State sequence or state transition diagram with three flip-flops.

$$0 \rightarrow 1 \rightarrow 2 \rightarrow 3 \rightarrow 4 \rightarrow 5 \rightarrow 6 \rightarrow 7$$
$$\longleftarrow 8 \longrightarrow$$

2. Nonbinary Counter

Nonbinary counter on the other hand does not pass through all possible sequence of states that is, a counter in which some possible states are skipped of is called a nonbinary counter.

Example are

$$0 \rightarrow 1 \rightarrow 2$$
$$\longleftarrow$$

Because by means of 2 sequence we can reduce the counter to maximum 3. The count 3 is skipped off.

The counter can be so logically designed that it will never arrive at the count 3.

Nonbinary Counter (NBC) with Three Flip-Flops

$$1 \rightarrow 3 \rightarrow 5$$
$$\longleftarrow$$

The counter can start from any intermediate states, not necessarily 0. The states or counts 2,4,6 and 7 are skipped or missing (i.e. some possible states are missing). So, it becomes a nonbinary counter.

2.8.1 Modulus of a Counter

Counter is said to have modulus K if after starting from some initial state it returns to the same initial state after K number of clock pulses.

$$0 \rightarrow 1 \rightarrow 2 \rightarrow 3 \qquad \text{Modulus} = 4$$
$$\longleftarrow$$

(as it has taken 4 clock pulses)

The minimum number of flip-flop N which is required to design a sequential counter of modulus K is obtained from the inequality.

$$2^{N-1} \leq K \leq 2^N \tag{2.5}$$

For Mod 8 counter that for $K = 8$.

$$N = 3$$
$$2^2 \leq 8 \leq 2^3$$
$$\text{Hence, } N = 3$$

2.8.1.1 Mod 3 nonbinary counter

This is not always applicable because if there is a counter which passes through the following states as stated in the following. Let us consider a mod 3 non binary counter.

$$0 \rightarrow 2 \rightarrow 4$$
$$\longleftarrow$$

For $K = 3$ if we use Eq. (2.5) to find minimum value of N, the value of N is 2. But we cannot realize 4 with two flip-flops. For this we have to consider the highest count and accordingly decide upon the number of flip-flop.

For a nonsequential counter, the minimum number of flip-flop required to design such counter depends upon the maximum counter. Since, there are three clock pulses, in this case N is suitably chosen as 3. So, three flip-flops are required in this case counter may be synchronous and asynchronous. Their design steps are given in tabular form.

2.8.2 Design of Counter

In a nonbinary counter, some count states are skipped off that is unused.

Synchronous counter	Asynchronous counter or ripple counter
1. In this counter the clock pulse is applied simultaneously to the clock input of all the flip-flop.	1. Clock pulse is applied to the clock input of the first or LSB flip-flop and all other flip-flop get clock input from the output of the preceding flip-flop.
2. The design is having more hardware complexity.	2. It is simpler and easier to design, i.e. having less hardware complexity.
3. Its speed is greater, i.e. synchronous counter can accept much higher clock frequency. (i.e. it changes state with much more rapidly than of asynchronous counter)	3. Its speed is less or least and hence the clock frequency it can accept is not very large.
4. Synchronous counter can be designed by using either J–K flip-flop, D–type flip-flop, and so on.	4. Asynchronous counter can be designed only by using J–K flip-flop.
5. To design synchronous counter we have to find the minimized logical expressions for the inputs of each J–K flip-flop for each D–type flip-flop (For this we use the S T table.)	5. To design asynchronous counter we have to find the minimized logical expression for preset and clear inputs of all the J–K flip-flop. These are asynchronous inputs.
6. Preset and clear inputs of all flip-flops are connected to logical 1 to enable all flip-flops with clock pulse.	6. The J–K I/P of all the F/F are connected to logical 1.

2.8.2.1 Lock out condition

If a counter, while passing through its used state, accidentally acquires some unused state and this occurs due to some transmission error so that the counter may not return to the used state from this unused state and the counter remains in unused state forever it is called lock out condition. So, we can design a nonbinary counter by preventing lock out condition or without preventing lock out condition. It is better to design by preventing lock out condition. To prevent lock out condition in a counter, we have to return the counter always from any unused state to the initial state which is one of the used states.

Design procedure for Mod 3 asynchronous counter whose initial count = 0:

Count sequence is

$$0 \rightarrow 1 \rightarrow 2$$
$$\longleftarrow$$

Initial state = 0
Unused state/count = 3

1. Find N, the minimum number of J–K flip-flop which satisfy the following condition:
   $$2^{N-1} \le K \le 2^N \text{ here } K = 3\ N = 2 \text{ (minimum value)}$$
2. Write down the ST table for Mod 3 asynchronous counter which contains outputs of two flip-flops and preset and clear inputs of the two flip–flops.
3. From this S T table, write down the logical expression for preset and clear input of each flip-flop either in minterm form or in maxterm form.
4. Obtain the minimized expression for *Pri* and *Clri* inputs using separate K map for each *Pri* and *Clri.*
5. Realize this minimized expression by using suitable logic gates.
6. Initially, set the counter output to 00 by applying suitable values of preset and clear inputs for each J–K flip-flop.
7. Connect all preset and *Clr* inputs of each flip-flop according to the expression obtained by Step 5.
8. Now connect all J–K inputs of all flip-flops to logical 1.
9. Apply clock pulse to the clock inputs of first flip-flop or LSB and other flip-flop gets clock pulse from the output of the preceding flip-flop.

Preventing the lock out condition ST table or Mod 3 asynchronous counter

Table 2.12

Clk	Count	Q_1	Q_0	Pr_1	Clr_1	Pr_0	Clr_0
0	0	0	0	1	1	1	1
1	1	0	1	1	1	1	1
2	2	1	0	1	1	1	1
3	3	1	1	1	0	1	0

When output changes from one used state to other used state *Pr* and *Clr* = 1.

Table 2.13

Clk	*Pr*	*Clr*	Q_{n+1}
0	0	1	1
0	1	0	0

By observing the ST table

$$Pr_1 = \sum m\ (0, 1, 2, 3),\ Clr_1 = \sum m\ (0, 1, 2),$$

$$Pr_0 = \sum m\ (0, 1, 2, 3),\ Clr_0 = \sum m\ (0, 1, 2)$$

(If all the cells of the K map are filled, the logical expression is 1.)

$$Clr_1 = Clr_0 = \bar{Q}_0 + \bar{Q}_1$$

$$= \overline{Q_1 Q_0}$$

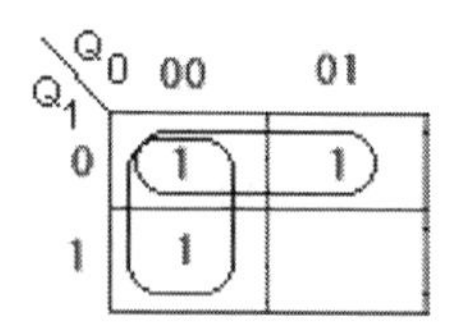

From the Fig. 2.27 at first, let $Clr_0 = Clr_1 = 0$, so that $Q_1 = Q_0 = 0$. Now, put $Pr_1 = Pr_0 = 1$

Now apply J_1, K_1, J_0, K_0 for Mod 4 asynchronous counter.

For Mod 4 asynchronous counter IS = initial stage

$$(1)\ (2)\ (3) \rightarrow \text{clock pulse}$$

$$(\text{IS})\ 0 \rightarrow 1 \rightarrow 2 \rightarrow 3$$

$$\longleftarrow$$

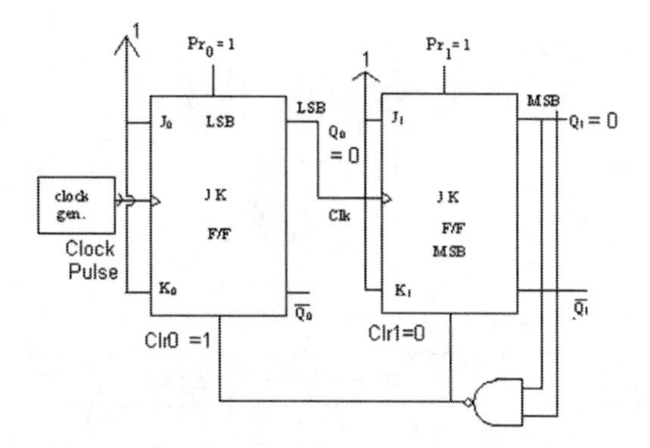

Figure 2.27 Asynchronous circuit design.

Table 2.14 S T table for Mod 4 with initial count = 0

Clk	Count	Q_1	Q_0	Pr_1	Clr_1	Pr_0	Clr_0
0	0	0	0	1	1	1	1
1	1	0	1	1	1	1	1
2	2	1	0	1	1	1	1
3	3	1	1	1	1	1	1
4	0	0	0	1	1	1	1

As this is also used state now.

$Pr_1 = Pr_0 = Clr_1 = Clr_0 = \sum m\ (0, 1, 2, 3)$

$Pr_1 = Pr_0 = Clr_1 = Clr_0 = 1$

Design of Mod 5 counter (Preventing lock out)

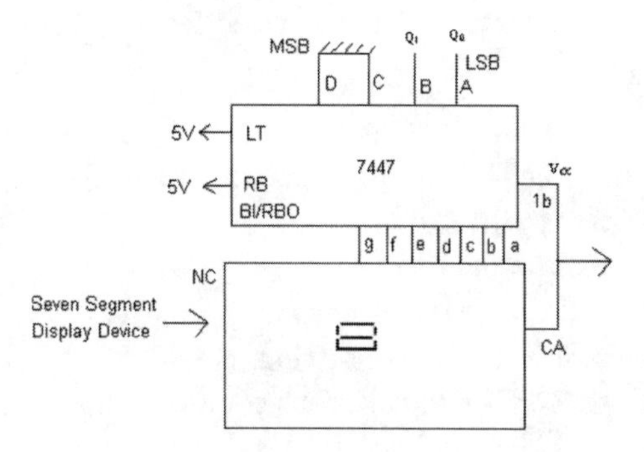

Figure 2.28 Mod 5 counter without lock out.

Initial count = 2

$$\begin{array}{ccccc} & (1) & (2) & (3) & (4) \\ 2 \longrightarrow & 3 \longrightarrow & 4 \longrightarrow & 5 \longrightarrow & 6 \end{array}$$

$\longleftarrow$

(5) $\rightarrow$ 5th clock

Since K = 5, So $N = 3$.

Table 2.15 S T table for Mod 5 counter

Clk	Count	Q_2	Q_1	Q_0	Decimal	Pr_2	Clr_2	Pr_1	Clr_1	Pr_0	Clr_0
0	2	0	1	0	2	1	1	1	1	1	1
1	3	0	1	1	3	1	1	1	1	1	1
2	4	1	0	0	4	1	1	1	1	1	1
3	5	1	0	1	5	1	1	1	1	1	1
4	6	1	1	0	6	1	1	1	1	1	1
5	7	1	1	1		1	0	0	1	1	0
6	8	0	0	0		1	0	0	1	1	0
7	9	0	0	1		1	0	0	1	1	0

$Pr_2 = Clr_1 = Pr_0 = \sum m\ (0, 1, 2, 3, 4, 5, 6, 7)$

$Clr_2 = Pr_1 = Clr_0 = \sum m\ (2, 3, 4, 5, 6)$

$$Clr_2 = \mathrm{Pr}_1 = Clr_0 = Q_2\,\overline{Q_1} + \overline{Q_2}\,Q_1 + Q_1\,\overline{Q_0}$$
$$= Q_2 \oplus Q_1 + Q_1\,\overline{Q_0}$$

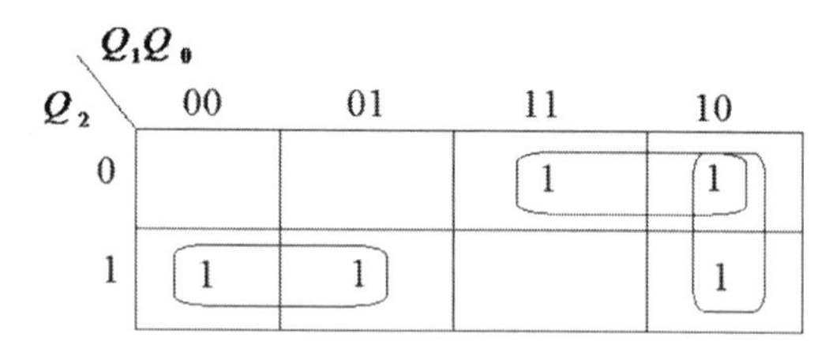

Figure 2.29a K Map simplification.

so to design three J–K flip-flops has been used and it is shown in the Fig. 2.29(a), 2.29(b) 2.29(c)

For $Q_0 = 0 \Rightarrow Clr_0 = 0,\ Pr_0 = 1$

$Q_1 = 1 \Rightarrow Pr_1 = 0,\ Clr_1 = 1$ For initial counter

$Q_2 = 0 \Rightarrow Pr_2 = 1,\ Clr_2 = 0$ setting.

With this, the Mod 5 asynchronous counter will have a sequence

$$2 \rightarrow 3 \rightarrow 4 \rightarrow 5 \rightarrow 6$$

$\longleftarrow$

$f_{\text{clock}} = 1/\text{T}$ and $f_{Q_2} = 1/5\text{T} = (1/5)\,f_{\text{clock}}$ for Mod 5.

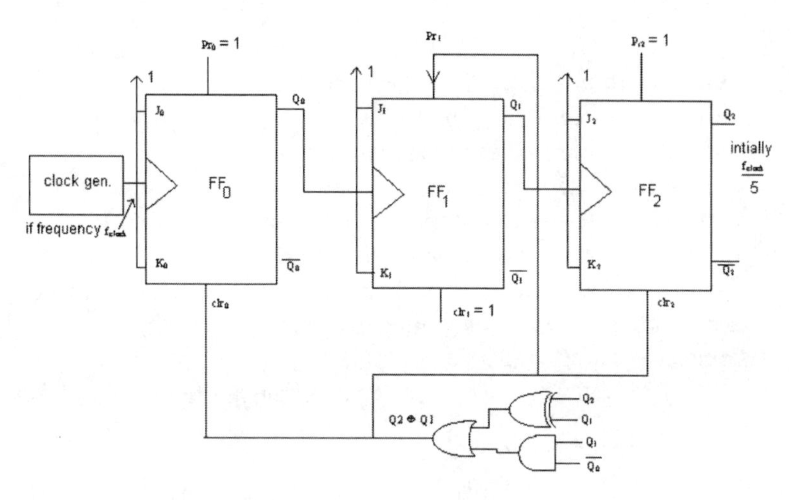

Figure 2.29b Three J–K based design of lock out free Mod 5 counter.

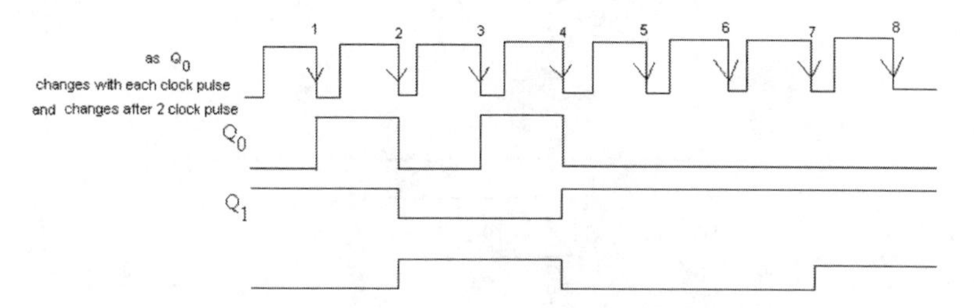

Figure 2.29c Time period of the waveform at the output of Q_2.

Clock	Q_2	Q_1	Q_0	Count
0	0	1	0	2
1	0	1	1	3
2	1	0	0	4
3	1	0	1	5
4	1	1	0	6
5	0	1	0	7
6	0	1	1	8
7	1	0	0	9

For Mod 7, the output f_{Q_2} is divided by 7.

2.8.2.2 Design procedure for synchronous counter

1. Find N the minimum number of J–K F/Fs using J–K F/F whose Initial count = 0 and prevent lock out.

2. Write down ST table for Mod 3 synchronous counter which contains outputs of two flip-flops J and K.
3. Write logical expression for J_i and K_i.
4. Obtain the minimized expression for J_is and K_is.
5. Realize minimized expression of Js and Ks.
6. Initially, set the counter output to 00 by applying suitable values of preset and clear inputs for each J–K flip-flop.
7. Connect J's and K's input of each flip-flop by using the minimized expression for Js and Ks obtained by step 5.
8. Connect *Pr* and *Clr* inputs to logical 1.
9. Apply clock pulse to the clock input of all the flip-flops.

Table 2.16 MOD 3 Synchronous Counter using J–K flip-flop

Clock	Count	Q_1	Q_0	J_1	K_1	J_0	K_0
0	0	0	0	0	X	1	X
1	1	0	1	1	X	X	1
2	2	1	0	X	1	0	X
3	3	1	1	X	1	X	1

J	K	Q_{n+1}
0	0	Q_n
0	1	0
1	0	1
1	1	$\bar{Q}_n$

Q_n		Q_{n+1}	J	K
0	→	0	0	X
0	→	1	1	X
1	→	1	X	0
1	→	0	X	1

$$J_1 = \sum m\,(1) + \sum d\,(2, 3)$$
$$K_1 = \sum m\,(2, 3) + \sum d\,(0, 1)$$
$$J_0 = \sum m\,(0) + \sum d\,(1, 3)$$
$$K_0 = \sum m\,(1, 3) + \sum d\,(0, 2)$$

This will behave as Mod 3 synchronous counter. The design and implementation is shown in the Figs. 2.30(a) and 2.30(b) respectively.

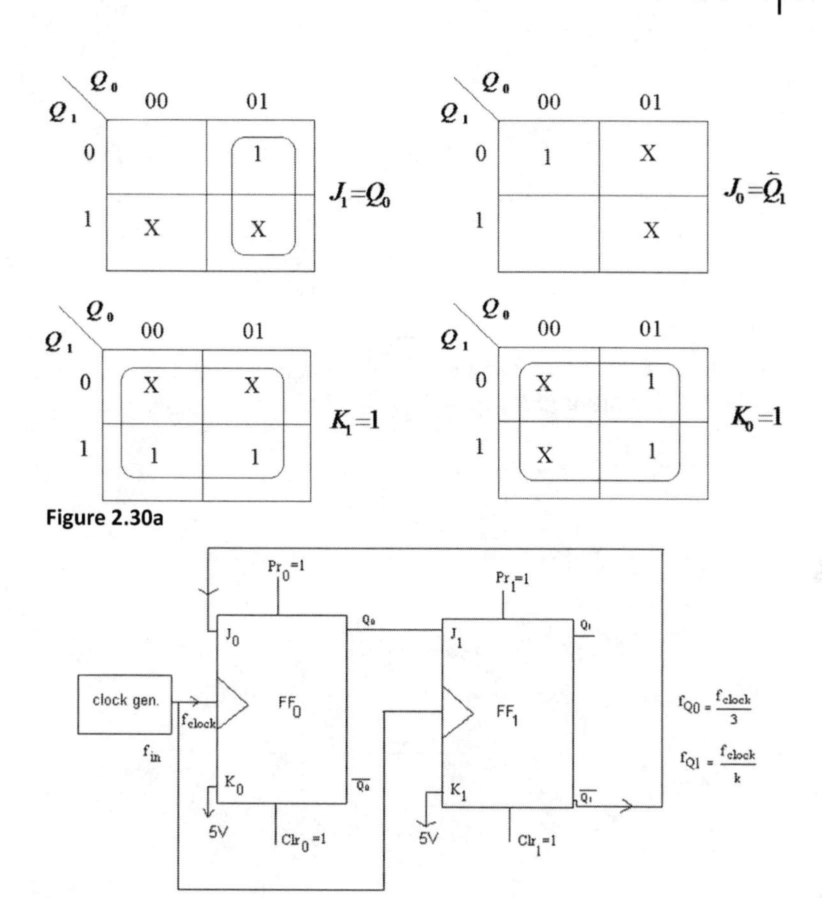

Figure 2.30a

Figure 2.30b MOD 3 Synchronous counter using J–K flip-flop.

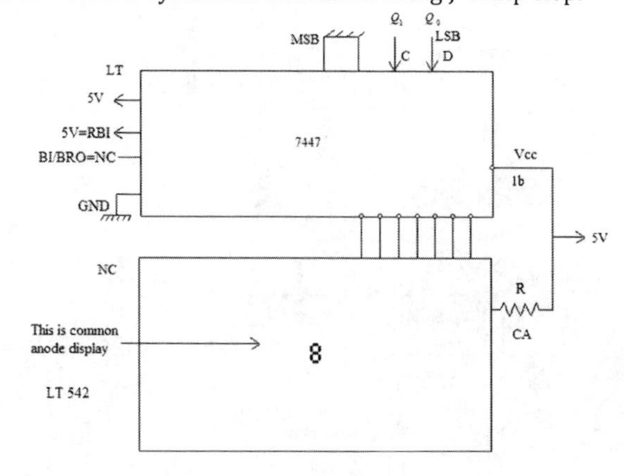

Figure 2.30c IC Implementation of the circuit.

Initial count setting for MOD 5 asynchronous counter using D flip-flop

$$Q_0=0, \quad Pr_0=1, Clr_0=0$$
$$Q_1=0, \quad Pr_1=1 \text{ and } Clr_1=0$$

Then put $Pr_0=1=Pr_1=Clr_0=Clr_1$

D	Q_{n+1}
0	0
1	1

As we are not interested in previous lockout condition, we put do not care condition for inputs D_2, D_1 and D_0 for the unused states.

$$D_2=\sum m\ (3, 4, 5)+\sum d\ (0, 1, 7)$$
$$D_1=\sum m\ (2, 5, 6)+\sum d\ (0, 1, 7)$$
$$D_0=\sum m\ (2, 4)+\sum d\ (0, 1, 7)$$

D_2

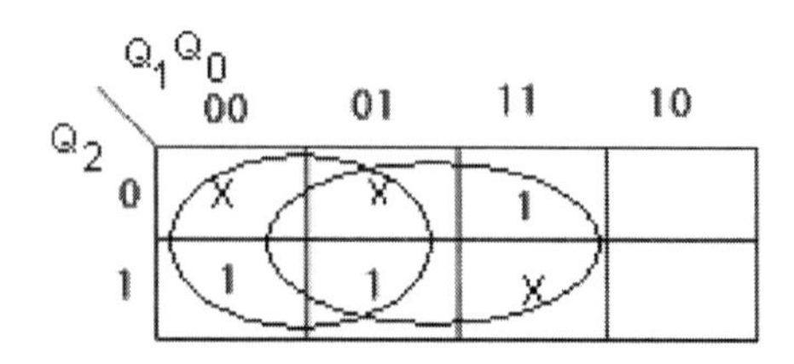

$D_2 = \overline{Q_1} + Q_0$

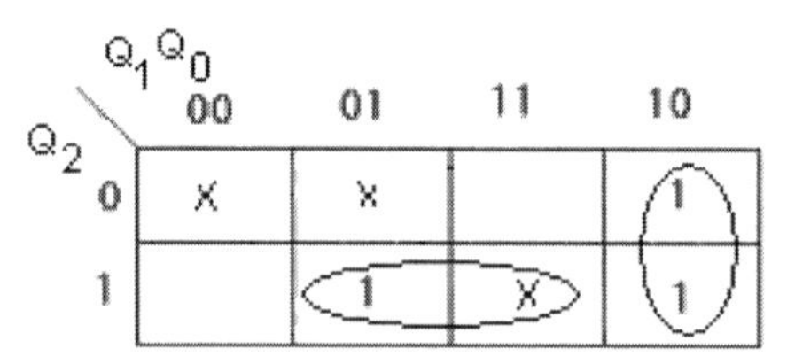

$D_1 = Q_2\, Q_0 + Q_1\, \overline{Q_0}$

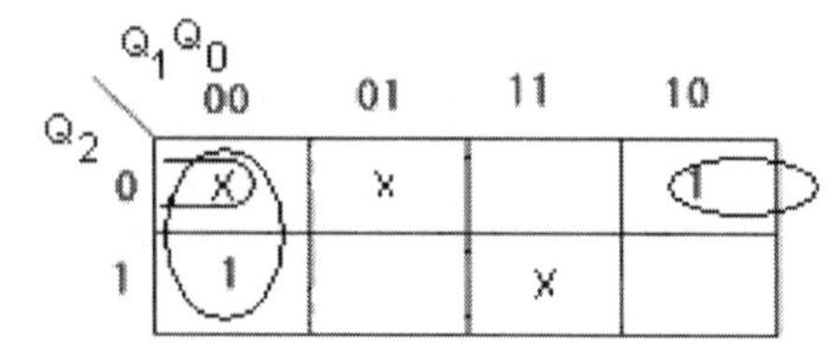

$D_0 = \overline{Q_1}\, \overline{Q_0} + \overline{Q_2}\, \overline{Q_0}$

Figure 2.31(a)

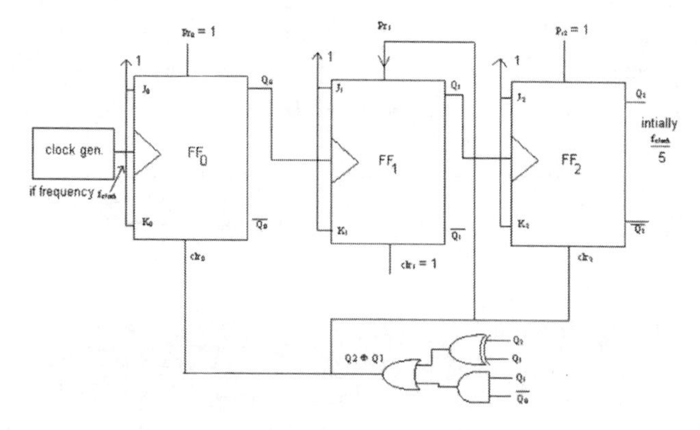

Figure 2.31(b) MOD 3 asynchronous counter using D flip-flop.

2.8.3 Decoding Error in Counter

Decoding errors in a ripple or asynchronous counter:
Two questions before

1. whether any frequency clock frequency can be applied?
2. does it necessarily mean periodic pulse?

Three stage ripple counter whose modulus is 8. (So it is a binary flip-flop.)

Count sequence

$0 \rightarrow 1 \rightarrow 2 \rightarrow 3 \rightarrow 4 \rightarrow 5 \rightarrow 6 \rightarrow 7$

Binary flip-flop as all possible stages are there.

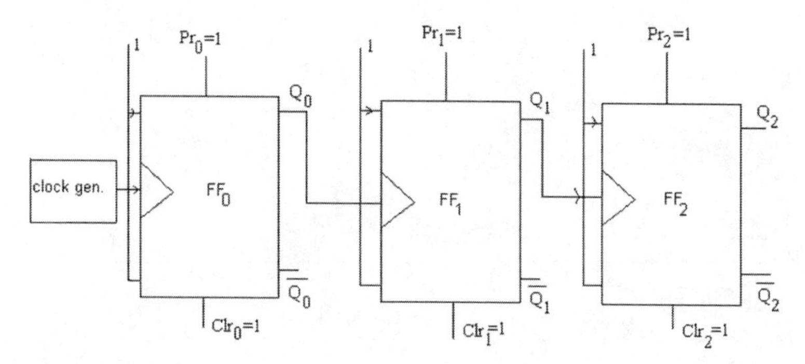

Figure 2.32 Three stage binary counter.

Q_0 changes state when input clock waveform changes from 1 to 0, Q_1 changes state, when Q_0 changes state from 1 to 0 and Q_2 changes state when Q_1 changes state from 1 to 0.

But according to J–K truth table

Clk	J	K	Q_{n+1}
↓ negative edge triggered	1	1	$\bar{Q}_n$

The decoding circuit is shown in the Fig. 2.33.

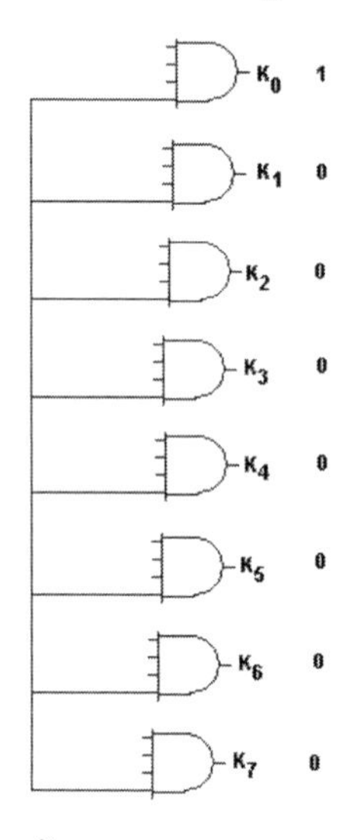

Figure 2.33 Decoding circuit.

Table 2.17

Strobe	K_0	K_1	K_2	K_3	K_4	K_5	K_6	K_7	Count
1	1	0	0	0	0	0	0	0	0
1	0	1	0	0	0	0	0	0	1
1	0	0	1	0	0	0	0	0	2
1	0	0	0	1	0	0	0	0	3
1	0	0	0	0	1	0	0	0	4
1	0	0	0	0	0	1	0	0	5
1	0	0	0	0	0	0	1	0	6
1	0	0	0	0	0	0	0	1	7

$f_{\text{clock}} = 1/T$
$f_{Q_0} = f_{\text{clock}}/2 = 1/2T$
$f_{Q_1} = 1/4T = f_{\text{clock}}/4$
$f_{Q_2} = 1/8T = f_{\text{clock}}/8$

This is an ideal waveform, as output is thought to be got instantaneously, though there is a propagation delay. We assume t_{pd} for all flip-flops are same.

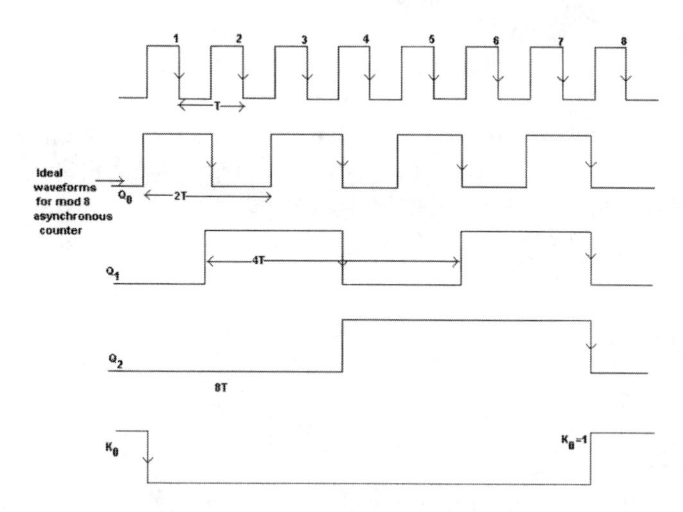

Figure 2.34 Waveform of MOD 8 counter.

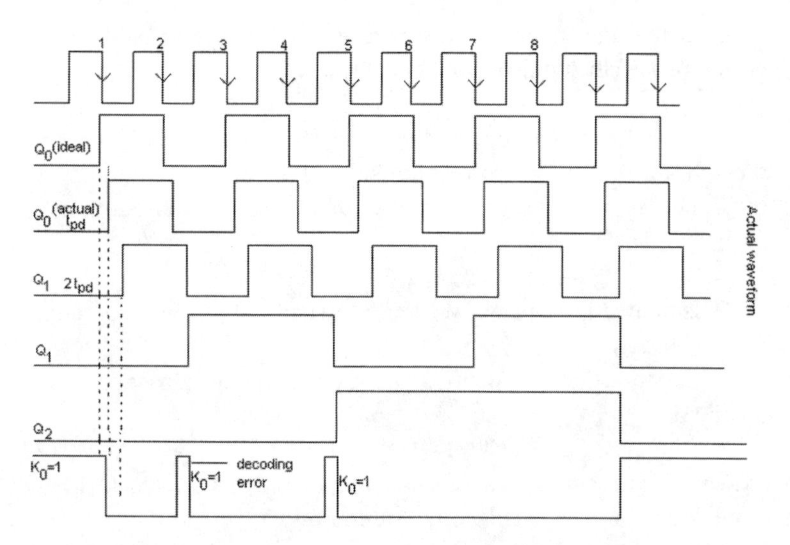

Figure 2.35 Waveform showing decoding error due to propagation delay.

Maximum decoding error = $3t_{pd}$(for Q_2)

Let t_{pd} of each flip-flop be 40 ns.

Output of Q_0 changes state after 40 ns

Output of Q_1 changes state after 80 ns

Output of Q_2 changes state after 120 ns

As 3 t_{pd} is the maximum propagation delay for Q_2.

Let $Q_0\ Q_1\ Q_2 = 111$. So the counter has count = 7

Let us apply a clock pulse whose time period is < 120 ns, say $T = 115$ ns. After 115 ns only Q_0 and Q_1 changes. After 115 ns, the output is 001, and the output will never be reset and after the next clock pulse it becomes 101.

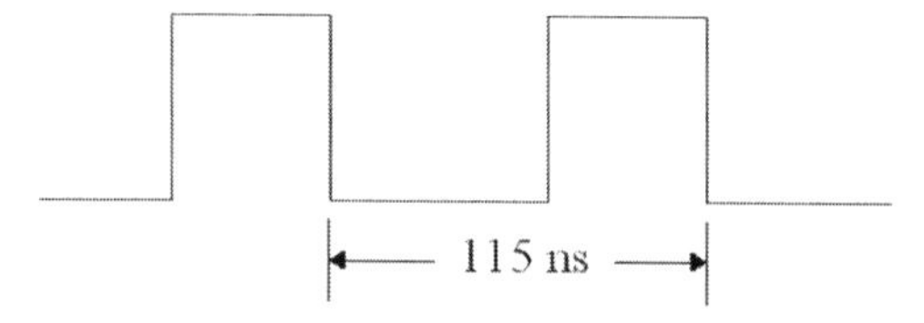

Figure 2.36 Clock pulse.

So after 111 → we get 100 → then 101 → and this is due to the decoding error. So requirement, the time period of the clock pulse for 3 stage asynchronous counter should be

$$T > 3t_{pd}$$

So for N stage asynchronous counter or ripple counter the time period of the clock pulse should be

$$T > Nt_{pd}$$

This is requirement for divider circuit. But to read the count also the counter require a strobe input signal for T_S seconds and hence, the time period of the clock pulse for N stage ripple counter of asynchronous counter should be

$$T \geq Nt_{pd} + T_S$$

(T_S is allowed because if we want to read the strobe time count at same time.)

So the input clock for N stage asynchronous ripple counter should be

$$f \leq 1/(N\, t_{pd} + T_S),$$

where N = number of J–K flip-flop used in N stage asynchronous ripple counter or asynchronous counter and

T_S = Storing time in s to read the count and

t_{pd} = Propagation delay of the flip-flop

For 3 stage ripple counter or asynchronous counter the maximum duration of decoding error interval is $3t_{pd}$.

Limitation:

As N increases f decreases and speed decreases. So, there is a limitation on the frequency, $f_{clock\ max} = 1/(N\ t_{pd} + T_S)$. So, the manufacturer will specify the $f_{clock\ max}$ and thus, we can avoid the decoding error.

Speed of asynchronous counter/ripple counter decreases as N, the number of flip-flop or modulus of asynchronous counter/ripple counter increases.

(The same connection will hold, if the flip-flops are all positive edge triggered, we get $K_0 = 1$.) in the intermediate clock pulses instead of only 0 and 8th clock pulse. This is decoding error, which occurs due to propagation delay.

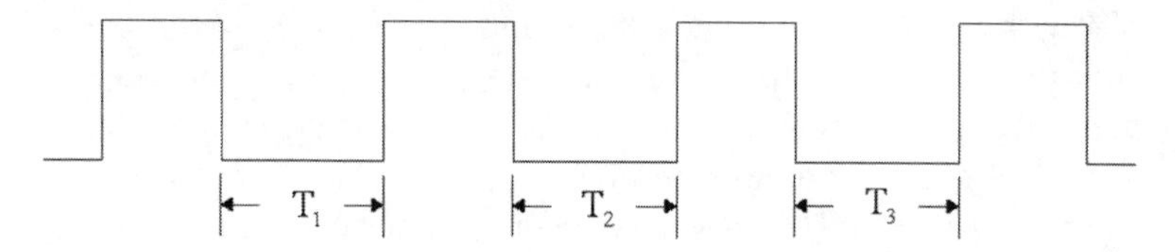

Figure 2.37

It is not necessary in Fig. 2.35 the clock pulse is always periodic. It has only to satisfy the condition:

$$T_{min} \geq (N\ t_{pd} + T_S)$$

Therefore, if this condition is satisfied, nonperiodic clock pulse will do. But we use the periodic clock pulse

1. as output changes after definite interval of time.
2. to draw the timing diagram with more efficiency. The output of one flip-flop ripples through the succeeding flip-flop and provide the clock pulse to succeeding flip-flop. Hence, the name ripple counter.

To improve speed of counter synchronous counter are used:

1. synchronous serial carry counter.
2. synchronous parallel counter.

Three stage Synchronous Serial Carry Counter:

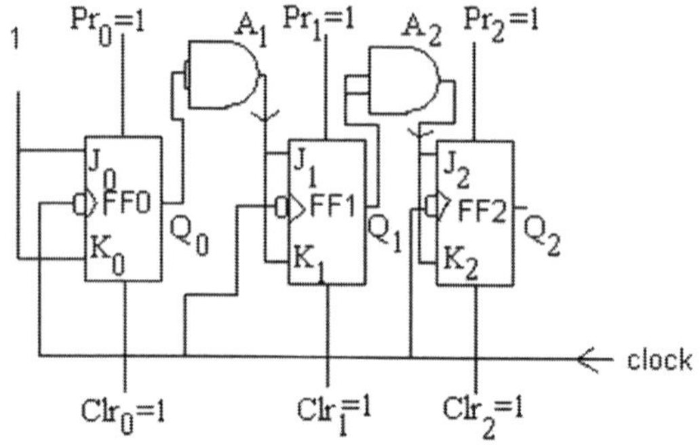

Figure 2.38 Block diagram of three stage synchronous serial carry counter.

The clock pulse is applied simultaneously to the clock input of all the flip-flops.

The waveform (real and actual) for synchronous serial carry counter will almost falsify the same shape as that of asynchronous counter/ripple counter . The output of Q_0 will change state with the arrival of each negative edge of clock pulse. The output of Q will change state when $Q_0 = 1$ when output of AND gate $A_1 = 1$ with each negative edge of the clock. The output Q_2 can change state, when $Q_0 = Q_1 = 1$ that is, $A_1 = 1$ and $Q_1 = 1$ which results in $A_2 = 1$. With each negative edge triggered clock.

There is decoding error, as the time period of the input clock pulse for synchronous serial carry counter (3 stage)

$$= T \geq t_{pd}\,(FF_0) + 2t_{pd}\,(\text{AND gate}) + \underset{\text{for reading count}}{\underset{\downarrow}{T_S}}$$

For N stage synchronous serial counter, the time period should satisfy the following conditions in order to avoid decoding error:

$$T \geq t_{pd}\,(FF_0) + (N{-}1)\,t_{pd}\,(\text{AND}) + T_S$$

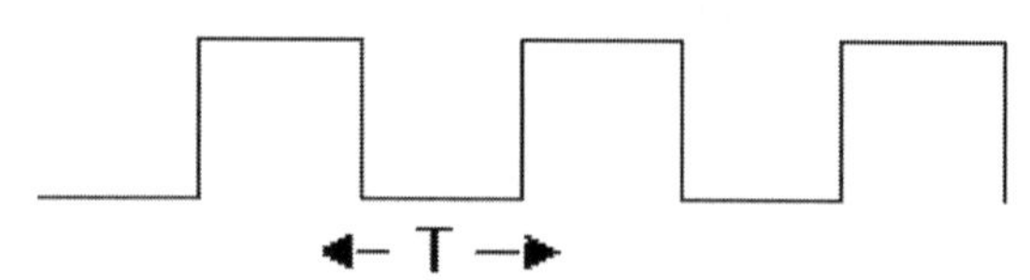

Figure 2.39

Since, the clock pulse is being acted upon simultaneously so clock frequency:

$f \leq 1/\ [t_{pd}\ (FF_0) + (N-1)t_{pd}\ (\text{AND}) + T_S]$

$f_{\text{clock|max AC}} = 1/(Nt_{\text{pd}} + T_S)$

[AC = Asynchronous Counter]

$f_{clock|max\ SSCC} = 1/[t_{pd}\ (FF_0) + (N-1)t_{pd}\ (\text{AND}) + T_S]$ [SSCC = Synchronous Serial Carry Counter]

3. (2) > (1)

t_{pd} of AND gate = 10 ns

Let $t_{\text{pd}} = 50$ ns $= t_{\text{pd}}\ (\text{FF}_0)$

$t_{\text{pd}}\ (\text{AND}) = 10$ ns

$T_S = 40$ ns

For 3 stage counter

$f_{\text{clock|max AC}} = 1/(3 \times 50 + 40) = 1/190$ ns

and

$f_{\text{clock|max SSCC}} = 1/(50 + 2 \times 10 + 40) = 1/110$

So, there is an improvement of speed, if we want to improve the speed, by increasing N, which increases number of AND gates and this limits the f as if reduces to very low value and then use 3 stage synchronous parallel counter.

The clock pulse is connected simultaneously in parallel. The inputs to the AND gates are applied parallely and hence the name synchronous parallel counter. It is the fastest counter. The output of Q_0 changes state with the arrival of each negative edge clock input. The Q_1 changes state with negative edge clock input provided $Q_0 = 1$.

Q_2 changes state with negative edge clock input provided $Q_1 = 1$. So $Q_0 = Q_1 = 1$.

The timing waveform is almost same as that for 3 stage ripple counter or asynchronous counter. Time period of the clock pulse for synchronous parallel counter (for 3 stage) should satisfy the following condition to avoid decoding error.

$T \geq t_{pd}\ (FF_0) + (N - 1)t_{pd}\ (\text{AND}) + T_S$

$t_{\text{pd FF0}}$ (F) = Propagation delay of flip-flop FF0 as a function of fan out F.

The output Q_0 drives the 2 AND gates simultaneously. So the fan out is 2.

So as fan out increases, propagation delay increases as such here is a relation between $t_{\text{pd(FF0)}}$ with F as number of stages increases and $t_{pd\ FF0}$ (F) increases and $t_{pd(FF0)}$ (F) increases and it is the fan out of the synchronous counter which limits frequency or speed of synchronous parallel counter.

$$f_{clock} = \frac{1}{t_{pd}(F) + t_{pd}(AND) + T_s}$$

Table 2.18 Comparison of counter

AC/RC	SSCC	Serial parallel counter
1. It is simpler to design and less complex hardware required.	1. It has the most comples hardware complex circuitry.	1. It has the most complex hardware circuitry.
2. Clock pulse is applied only to the input of FF0 (First flip-flop) and all other flip-flop circuits clock from the output of preceding flip-flop.	2. Clock pulse is applied simultaneously to all the clock input of all flip-flops.	2. Clock pulse is applied simultaneously to all the clock input of all flip-flops.
3. Speed is minimum	3. Speed is > A.C. but < S.P.C.	3. Speed is minimum.
4. Decoding error interval depends upon the number of flip-flops used and propagation delay through it	4. Decoding error interval depends upon the number of AND gates and propagation delay through it.	4. Decoding error interval depends upon the fan out of first flip-flop (FF0)
5. For *N*-stage AC/RC $T \geq N\,t_{pd} + T_S$ $f \leq 1/\,(N\,t_{pd} + T_S)$	5. For *N*-stage SSCC $T \geq t_{pd}\,(FF0) + (N{-}1)\,t_{pd}$ AND $+ T_S$ $f \leq 1/t_{pd}\,(FF0) + (N{-}1)\,t_{pd}$ AND $+ T_S$	5. For *N*-stage SPC $T \geq t_{pdFF0}\,(F) + t_{pd}$ AND $+ T_S$ $f \leq 1/(t_{pdFF0}\,(F) + t_{pd}$ AND $+ T_S)$

Operation of IC7490

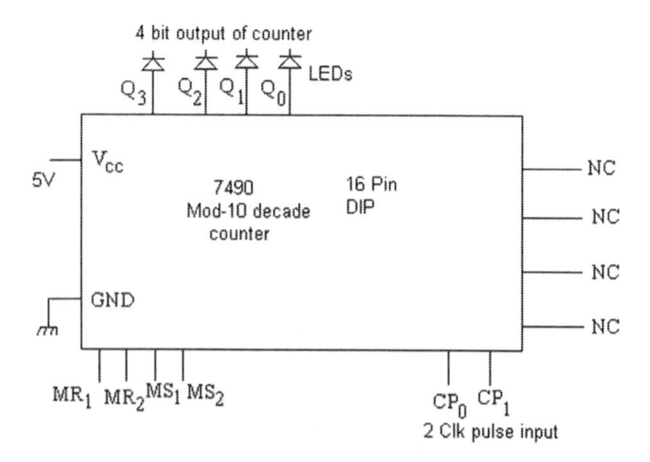

Figure 2.40 Block diagram of IC 7490.

Table 2.19 Truth table for the counter

Master reset input		Master set input					
MR_1	MR_2	MS_1	MS_2	Q_3	Q_2	Q_1	Q_0
1	1	X	0	0	0	0	0
1	1	0	X	0	0	0	0
X	X	1	1	1	0	0	1
0	X	0	X				
0	X	X	0				
X	0	0	X				
X	0	X	0				

This counter is organized as a cascade of two counters.

Mode A:

If Q_0 is connected with CP_1 and external clock pulse is applied at CP_0 then it behaves as a Mod 10 or Decade Counter.

The count sequence is:

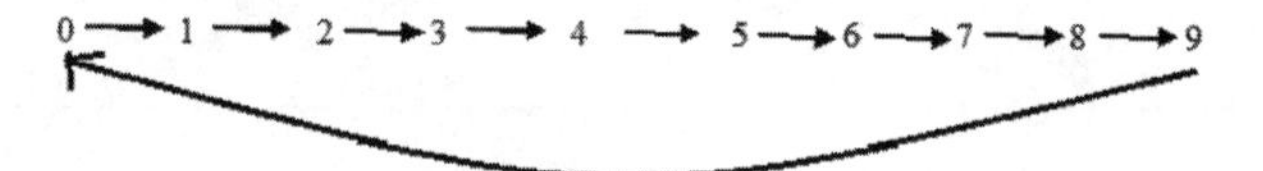

This is a nonbinary counter as out of 16 possible counts, we get only nine counts.

Mode B:

Short circuit Q_3 with CP_0 and apply external clock at CP_1 input. This sequence is called symmetrical biquinary sequence.
The count sequence is:

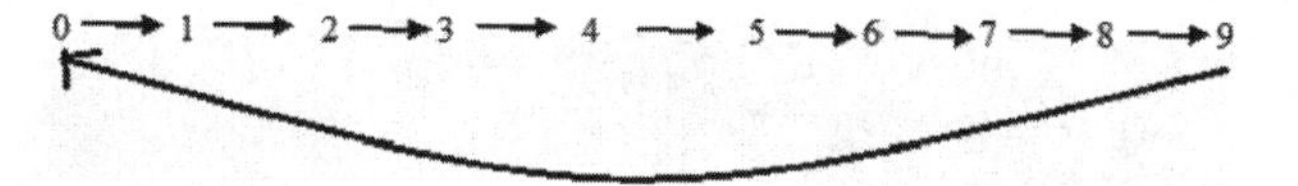

Count sequence:

So enable counter set, if and also check the set and reset conditions before.

All the flip-flops here are negative edge triggered flip-flop.

Biquinary:

$B_i = 2$ quinary = J
It gives two times. The sequence
$0 \rightarrow 2 \rightarrow 4 \rightarrow 6 \rightarrow 8 = 5$ sequence and then
$1 \rightarrow 3 \rightarrow 5 \rightarrow 7 \rightarrow 9 = 5$ sequence (next)

Table 2.20

OUTPUT Mod 5 counter output					
Clk	Q_3	Q_2	Q_1	Q_0	Count
0	0	0	0	0	0
1	0	0	1	0	2
Q_0 changes state when Q_3 changes from 1 to 0					
2	0	1	0	0	4
3	0	1	1	0	6
4	1	0	0	0	8
5	0	0	0	1	1
6	0	0	1	1	3
7	0	1	0	1	5
8	0	1	1	1	7
9	1	0	0	1	9
10	0	0	0	0	0

Mod 5 counter output sequence

$$0 \xrightarrow{(1)} 1 \xrightarrow{(2)} 2 \xrightarrow{(3)} 3$$

$$\xleftarrow{\quad (4) \quad}$$

Mode C:

If we apply *Clk* to CP_1 input the output of Q_3, Q_2, Q_1 gives Mod 5 counter output.

Mode D:

If we apply *Clk* pulse to CP_0 input the output of Q_0 gives Mod 2 counter output.

7493 = Binary asynchronous counter and Mod 16.

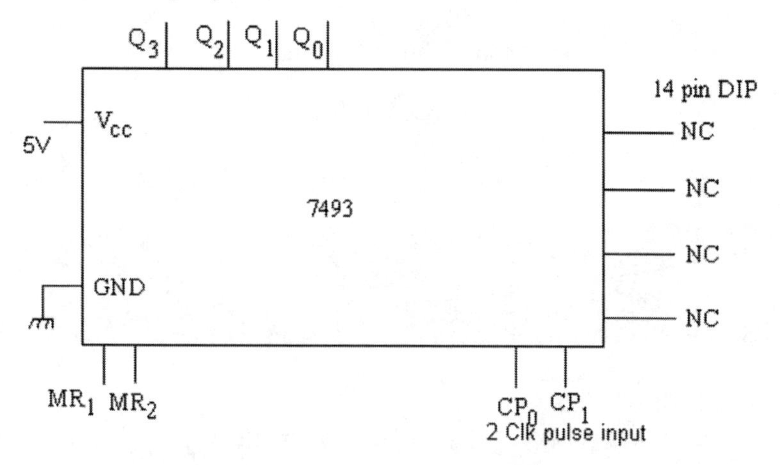

Figure 2.41 Binary asynchronous counter and MOD 16.

This 7493 is arranged as Table 2.21 Counter truth table for 7493

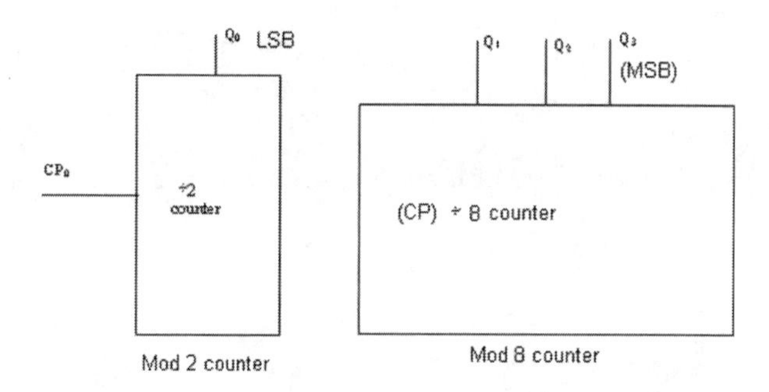

Figure 2.42 IC 7493 arranged as MOD2 and MOD 8 counter.

MR_1	MR_2	Q_3	Q_2	Q_1	Q_0	
1	1	0	0	0	0	Resetting of counter
1	0					Counter enable
0	1					Counter enable
0	0					Hold count (counter stops counting)

Mode A:

If Q_0 is shorted with CP_1 and external *Clk* is applied to CP_0 and if then we get the following Mod 16 count sequence.

MR_1	MR_2
0	1
1	0

0 →(1) 1 →(2) 2 →(3) 3 →(4) 4 →(5) 5 →(6) 6 →(7) 7 →(8) 8 →(9) 9 →(10) 10 →(11) 11 →(12) 12 →(13) 13 →(14) 14 →(15) 1

So it is a binary counter as all possible states are there.

Mode B:

Connect external *Clk* to CP_1 so that the output Q_3, Q_2, Q_1 gives Mod 8 count sequence as follows:

0 →(1) 1 →(2) 2 →(3) 3 →(4) 4 →(5) 5 →(6) 6 →(7) 7

Mode C:

Apply clock pulse to CP_0 input then output Q_0 will give Mod 2 count sequence as

All flip-flops are negative edge triggered. (Connect clock input to another LED and when LED changes from on to off state the output also changes. This negative edge triggered of the counter will be verified.)

Design of Mod 8 up down binary synchronous counter using J–K flip-flop

The up count sequence

Q_2	Q_1	Q_0	
0	0	0	→ 0
0	0	1	→ 1
0	1	0	→ 2
0	1	1	→ 3
1	0	0	→ 4
1	0	1	→ 5
1	1	0	→ 6
1	1	1	→ 7

up count sequence

Up count sequence

0 → 1 → 2 → 3 → 4 → 5 → 6 → 7 [M=Control input=0]

Down count sequence

0 → 1 → 2 → 3 → 4 → 5 → 6 → 7 [M=Control input=1]

The three flip-flops are required.
And so combined OR expression of

$$J_1 = K_1 = \overline{M}Q_0 + \overline{M}Q_0$$

Q_2 changes from 0 to 1 if Q_0 and Q_1 are both 1 for up count sequence and so

$J_2=K_2=\overline{M}Q_1Q_0$ and $M=0$ and Q_2 changes when $Q_0=Q_1=0$ and $M=1$ for down count sequence and so $J_2=K_2=MQ_1Q_0$

Table 2.22

Clk	Count	*M*	Q_2	Q_1	Q_0	Decimal	J_2	K_2	J_1	K_1	J_0	K_0
1	0	0	0	0	0	0	0	X	0	X	1	X
1	1	0	0	0	1	1	0	X	1	X	X	1
1	2	0	0	1	0	2	0	X	X	0	1	X
1	3	0	0	1	1	3	1	X	X	1	X	1
1	4	0	1	0	0	4	X	0	0	X	1	X
1	5	0	1	0	1	5	X	0	1	X	X	1
1	6	0	1	1	0	6	X	0	X	0	1	X
1	7	0	1	1	1	7	X	1	X	1	X	1
1	7	1	1	1	1	15	X	0	X	0	X	1
1	6	1	1	1	0	14	X	0	X	1	1	X
1	5	1	1	0	1	13	X	0	0	X	X	1
1	4	1	1	0	0	12						
1	3	1	0	1	1	11						
1	2	1	0	1	0	10						
1	1	1	0	0	1	9						
1	0	1	0	0	0	8						

So, combining we get

$$J_2=K_2=MQ_1Q_0+\overline{M}Q_1Q_0$$

This is the short cut observation method.

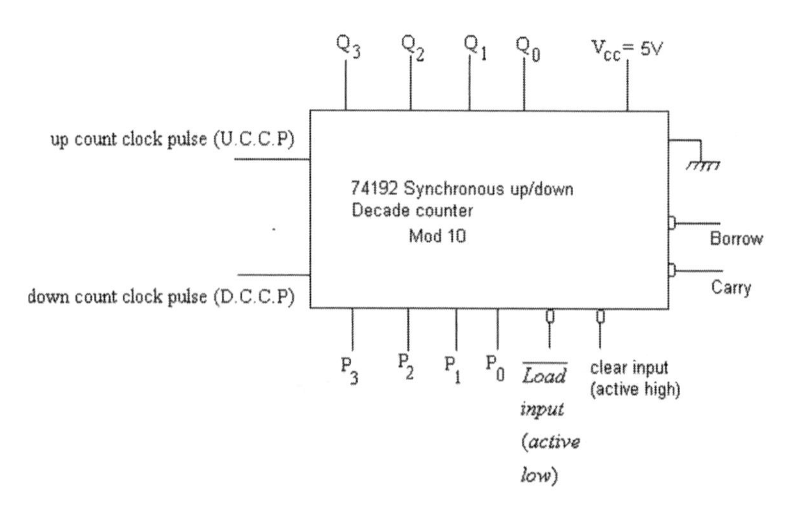

Figure 2.43 Synchronous up/down decade counter.

2.8.3.1 Designing of mod 4 up down counter

IC available for up/down counter is IC 74192 and is shown in Fig. 2.43.

74192 → IC → Synchronous decade or Mod 10 up down counter chip

So it is a 16 pin DIP.

UCCP=Up count clock pulse

DCCP=Down count clock pulse

Table 2.23

R	Load	P_3	P_2	P_1	P_0	UCCP	DCCP	Q_3	Q_2	Q_1	Q_0	
1	1	X	X	X	X	X	X	0	0	0	0	Resetting of counter
0	0					X	X					Parallel load
0	1	X	X	X	X		1					Counter increments i.e. gives upcount sequence
0	1	X	X	X	X	1						Counter decrements and gives down count sequence.

Up count sequence =

$$0 \rightarrow 1 \rightarrow 2 \rightarrow 3 \rightarrow 4 \rightarrow 5 \rightarrow 6 \rightarrow 7 \rightarrow 8 \rightarrow 9$$

Down count sequence =

$$9 \rightarrow 8 \rightarrow 7 \rightarrow 6 \rightarrow 5 \rightarrow 4 \rightarrow 3 \rightarrow 2 \rightarrow 1 \rightarrow 0$$

Modulus is 10, as after 10 clock pulses starting from 0 it returns to 0 in the up clock sequence.

If 12 is applied initially in down count sequence then

$$12 \rightarrow 11 \rightarrow 10 \rightarrow 9 \rightarrow 8 \rightarrow 7 \rightarrow 6 \rightarrow 5 \rightarrow 4 \rightarrow 3 \rightarrow 2 \rightarrow 1 \rightarrow 0$$

$Q_n \rightarrow$	Q_{n+1}	J	K
0	0	0	×
0	1	1	×
1	1	×	0
1	0	×	1

$$f_{\text{clock}} = \frac{1}{t_{\text{pd}}(\text{F}) + t_{\text{pd}}(AND) + T_{\text{s}}}$$

2.8.3.2 Cascading of counter

For cascading it is not necessary that both counters should be synchronous Mod m x n counter or asynchronous counter. It may also be equal to the ÷ m x n circuit.

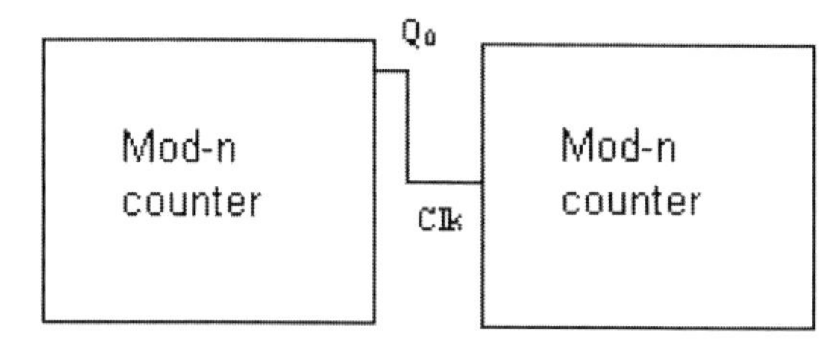

Figure 2.44 Cascading of counter.

For ÷20, we cascade ÷5 and ÷4 counter. Synchronous decade counter chip 74160, 74 indicates the chip is built with TTL logic gates.

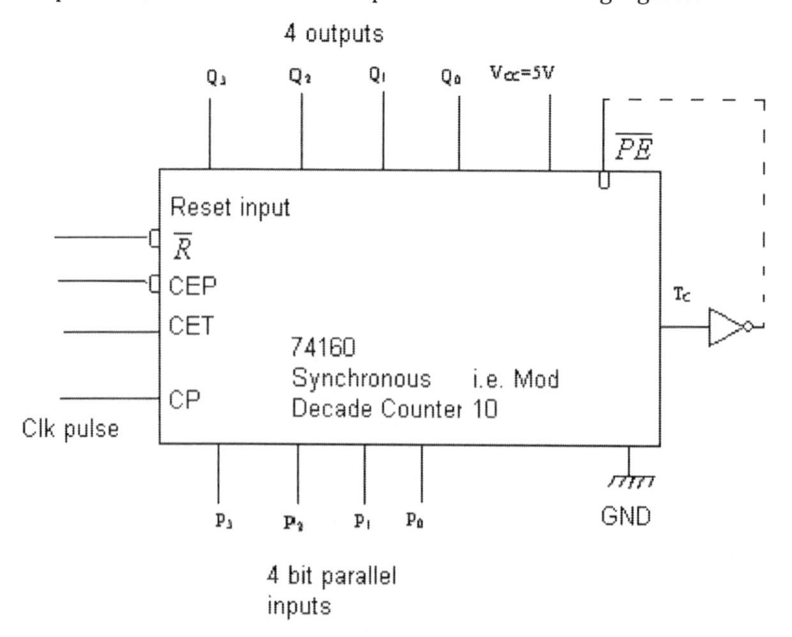

Figure 2.45 Synchronous MOD 10 counter.

$P_3P_2P_1P_0$ are called parallel load inputs.
$Q_3Q_2Q_1Q_0$ are the outputs of the counter.
$\overline{PE}$ → Parallel load input → Active low

This is in synchronism with clock.

$\bar{R}$ → Reset input → Active low
CEP → Count enable parallel input
CET → Count enable trickle input
T_C → Terminal count of output
T_C receives logical 1 output when counter receives the count=9, i.e. 1001
Count sequence of 74160 is

0→1→2→3→4→5→6→7→8→9

TC=1 when count = 9
See schematic diagram for this IC
Mode selects logic of 74160

Table 2.24

$\bar{R}$	$\overline{PE}$	*CEP*	*CET*	P_3	P_2	P_1	P_0	Q_3	Q_2	Q_1	Q_0	
0	X	X	X	X	X	X	X	0	0	0	0	Resetting of Counter
1	0	X	X	P_3	P_2	P_1	P_0	P_3	P_2	P_1	P_0	Parallel loading of counts
1	1	1	1	X	X	X	X					Counter enables, i.e. counter increment with each clock pulse. (active rising edge i.e. positive edge triggered)
1	1	0	1	X	X	X	X					Count hold. Counter stop incrementing.
1	1	1	0	X	X	X	X					Count hold

How by using a single 74160 IC we can realize a modulus ≤ 10.

So, realize a Mod 10 counter by using a single 74160 IC, apply initially $P_3=P_2=P_1=P_0=0$ and $PE=0$ to set the counter at 0 state or 0 count. Now connect *TC* Output via a NOT gate to $\overline{PE}$ input. Now connect *CEP* = *CET* to logical 1 and also connect *R*=logical 1. Now apply clock pulse through clock input of 74160.

Table 2.25

Clk	Q_3	Q_2	Q_1	Q_0	Count	TC	$\overline{PE}$
0	0	0	0	0	0	0	1
1	0	0	0	1	1	0	1
2	0	0	1	0	2	0	1
3	0	0	1	1	3	0	1
4	0	1	0	0	4	0	1
5	0	1	0	1	5	0	1
6	0	1	1	0	6	0	1
7	0	1	1	1	7	0	1
8	1	0	0	0	8	0	1
9	1	0	0	1	9	1	0
10	0	0	0	0	0	0	1

To realize a modulus 6 counter with 74160.

To realize a counter of modulus 6 we require only six possible states. So from 10 possible states we have to skip 4 states that is 10–6=4. Thus, load the counter with 4 = (0100).

Connect four parallel inputs to four switches.

$$(P_3P_2P_1P_0) = (0101)$$

Table 2.26

$\overline{R}$	$\overline{PE}$	CEP	CET	P_3	P_2	P_1	P_0	Q_3	Q_2	Q_1	Q_0
1	0	X	X	0	1	0	0	0	1	0	0
1	1	1	1	X	X	X	X	Counter is enabled			

Clk	Count	TC	$\overline{PE}$
0	4	0	1
1	5	0	1
2	6	0	1
3	7	0	1
4	8	0	1
5	9	1	0
6	4	0	1

Output clock frequency=$f_{\text{clock}}/6$.

So (1) (2) (3) (4) (6)

$4 \rightarrow 5 \rightarrow 6 \rightarrow 7 \rightarrow 8 \rightarrow 9$

(6)

So modulus is six, as after 6th clock pulse it returns to four.

To realize Mod 3 counter we load (10–3=7) initially and thus we can realize any lower order modulus less than 10. This is also called pre settable counter or programmable counter.

By cascading two 74160, we should get 10 × 10=100 Mod counter. We can realize a counter of any modulus, which is ≤ 100 (modulus should always be an integer). So this counter can be used in the frequency synthesizer for generating various frequencies.

2.8.3.3 Designing mod 87 counter

A mod 87 counter has 87 possible states. We have to skip (100–87)=13 possible states. So load the counter initially with 13 in decimal.

Apply $\overline{PE}$ for both the counters = 0.

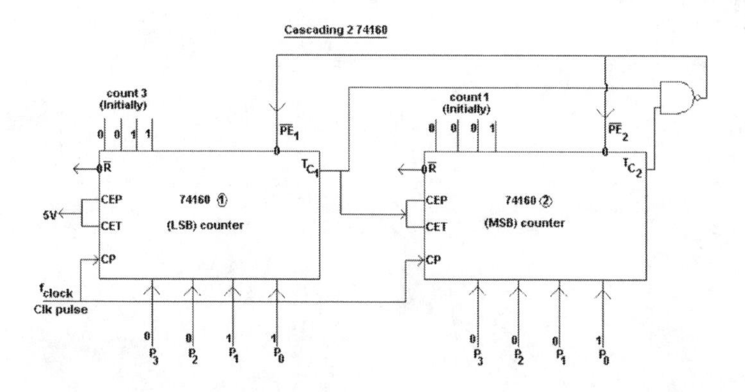

Figure 2.46 Block diagram of MOD 87 counter.

Table 2.27

Clk	Counter 1 count	TC_1	Enabled/Disabled condition of counter 2	Counter 2 count	TC_2	$\overline{PE_1}$	$\overline{PE_2}$	Final Count
0	3	0	D (Unless $TC_1=1$, counter 2 is disabled)	1	0	1	1	13
6 after 6 clock pulse	9	1	E	1	0	1	1	19
7	0	0	D	2	0	1	1	20
17	0	0	D	3	0	1	1	30
27	0	0	D	4	0	1	1	40
37	0	0	D	5	0	1	1	50
47	0	0	D	6	0	1	1	60
57	0	0	D	7	0	1	1	70
67	0	0	D	8	0	1	1	80
77	0	0	D	9	0	0	1	90
86	9	1	E	9	1	1	1	99
87	3	0	D	1	0	1	1	13

Counter returns to 13 after 87 clock pulses. So final output clock frequency $= f_{clock}/87$.

To realize Mod 13 counter by 74160: Mod 13 counter has 13 possible states, so load the two counters = 100–13 = 87 to skip of 87 possible states from 100 possible counts. Seven is the LSB count and 8 is the MSB count.

Table 2.28

Clk	Counter 1 count	TC_1	Counter 2 Enabled/ disabled	Counter 2 count	TC_2	$\overline{PE_1}$	$\overline{PE_2}$	Final count
0	7	0	D	8	0	1	1	87
1	8	0	D	8	0	1	1	88
2	9	1	E	8	0	1	1	89
3	0	0	D	9	1	1	1	90
4	1	0	D	9	1	1	1	91
5	2	0	D	9	1	1	1	92
6	3	0	D	9	1	1	1	93
7	4	0	D	9	1	1	1	94
8	5	0	D	9	1	1	1	95
9	6	0	D	9	1	1	1	96
10	7	0	D	9	1	1	1	97
11	8	0	D	9	1	1	1	98
12	9	1	E	9	1	0	0	99
13	7	0	D	8	0		1	87

(1) (2) (3) (4) (5) (6) (7) (8) (9) (10) (11) (12)
87 → 88 → 89 → 90 → 91 → 92 → 93 → 94 → 95 → 96 → 97 → 98 → 99

For realizing any counter of modulus ≤1000, we cascade three 74160 and in that case we use one, three-input NAND gate.

Problems

1. What is flip-flop?
2. For what conditions, clocked J–K flip-flop can be used as a divide-by-2 circuits when the input signal is applied at clock input?
3. What is basic sequential logic building block in which the output follows the data input as long as the ENABLE input is active?
4. Given that 7483 is a 4-bit parallel adder chip, how do you build a 16-bit parallel adder circuits?
5. Which factor limits the propagation delay of ripple counter?
6. How many flip-flops are required to construct a MOD 10 Johnson counter and a MOD 5 ring counter?
7. A 10 kHz clock signal having a duty cycle of 25% is used to clock a three-bit binary ripple counter. What will be the frequency and duty cycle of the true output of the MSB flip-flop?
8. How many flip-flops are required to produce a divide-by-64 device?
9. How is a J–K flip-flop made to toggle?
10. What is called ones catching?
11. What is the disadvantage of S-R flip-flop?
12. What is the hold condition of JK flip-flop?
13. If both inputs of an S-R flip-flop are LOW, what will happen when the clock goes high?
14. An active-HIGH input S-R latch has a 1 on the S input and a 0 on the R input. What state is the latch in?
15. How many frequency dividers can be produced when a 5-bit Johnson counter is cascade with a 5-bit ring counter?
16. How can the cross-coupled NAND flip-flop be made to have active-HIGH S-R inputs?

17. When T flip-flop does not change its state?
18. Which flip-flop is called transparent?
19. What is the characteristic equation of an R-S flip-flop with active-High inputs?
20. In what condition output does not change in J-K flip-flop?

Chapter 3

Memory

3.1 Computer Memory

Memory is defined as basic unit of a computer where data and instructions are stored. It is organized into locations. Every memory location is called one memory word. The number of bits present in each location is called word length of the memory. It is generally multiplied by 8 bits. The capacity of the memory is define as the total numbered location in the memory. The capacity is the product of memory locations and word length of the memory. Every memory has specific address. The memory is used to store information such as instructions, data, intermediate, and final results.

3.2 Classifications of Memory

A memory is classified according to its function, contents retention, and data access method. In the Fig. 3.1 the performance of different memories is compared according to memory speed, cost per bit, and power dissipation.

Foundation of Digital Electronics and Logic Design
Subir Kumar Sarkar, Asish Kumar De, and Souvik Sarkar

ISBN 978-981-4364-58-4 (Hardcover), 978-981-4364-59-1 (eBook)
www.panstanford.com

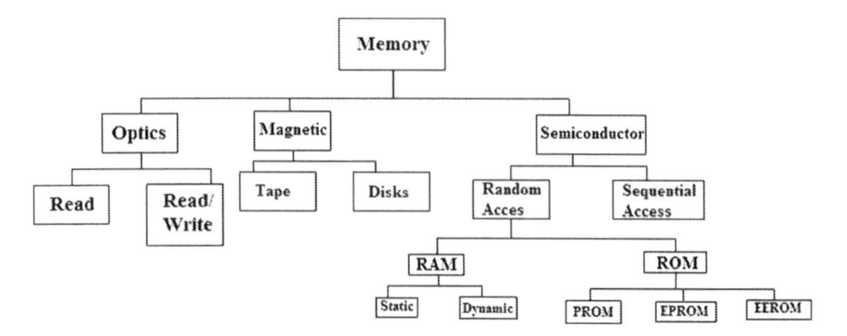

Figure 3.1 Classification of memories.

There are three basic types of memories:

(i) semiconductor memories
(ii) magnetic memories
(iii) optical medium -based memories.

3.2.1 Semiconductor Memories

Semiconductor memories are semiconductor devices in which the basic storage cells are transistor circuits.
They are of two types:

(i) random access memories
(ii) sequential access memories

The random access memories are of two types:

(a) Read Only Memory (ROM)
(b) Read and Write Memory (RAM)

The ROM is again classified into several types:

Programmable ROM—Unlike ROM, it can be programmed by the user only once in its lifetime (by using a special circuit known as PROM programmer).

Erasable Programmable ROM—Erasable Programmable ROM (EPROM) is an application in which data may change from time to time might call for the use of EPROM. The data can be programmed again if desired.

Electrically Erasable ROM—Electrically Erasable ROM (EEROM or Electrically Alterable ROM) is an application in which a portion of data may change for time to time might call for the use of an EEROM.

It can be erased and programmed 10,000 times (both erasing and programming take about 4 to 10 ms. In EEROM, any location can be selectively erased and programmed).

3.2.2 Magnetic Based Memory

Magnetic based memories are of two types:

(i) magnetic tape.
(ii) magnetic disk.

3.2.3 Optical Medium Based Memories

Optical medium based memories are also of two types:

(i) read type
(ii) read/write type

They are similar to magnetic disks.
Note: The bit storage density is the highest for optical type memories and the lowest for the semiconductor memories. There is nearly an order of magnitude difference between the semiconductor and magnetic or the magnetic and optical storage densities. However, semiconductor memories have the following specialties for which they are used as the main memory and the others as the secondary or auxiliary memories:

(i) small size
(ii) low cost
(iii) high reliability
(iv) ease of expansion of memory size
(v) electrical compatibility with the microprocessor

3.2.4 Main or Primary Memory

The main memory is defined as the memory unit that communicates directly with the central processing unit (CPU). It contains only the programs and data currently used by the processor reside inside in main memory. It determines the size and the number of programs that can be stored within the computer as well as the amount of data that can be processed. This is volatile, expensive, and slow read/write (R/W) memory. (Volatile means, as long as the power is there it retains the contents, when power is switched off the content is lost.) It is also random access memory (RAM). The primary memory (PM) is a must for every microprocessor-based system. It is this memory with which the microprocessor directly communicates through its system buses. The size of the PM is limited due to large cost as well as architectural limitation of microprocessor. Some part of primary memory is dedicated for OS process. It is the PM where the operating system (OS) resides, but in some PM area.

3.2.4.1 Classification of primary memory

Primary memory (PM) can be two types. RAM and ROM. The general structure of PM is given in the Fig. 3.2

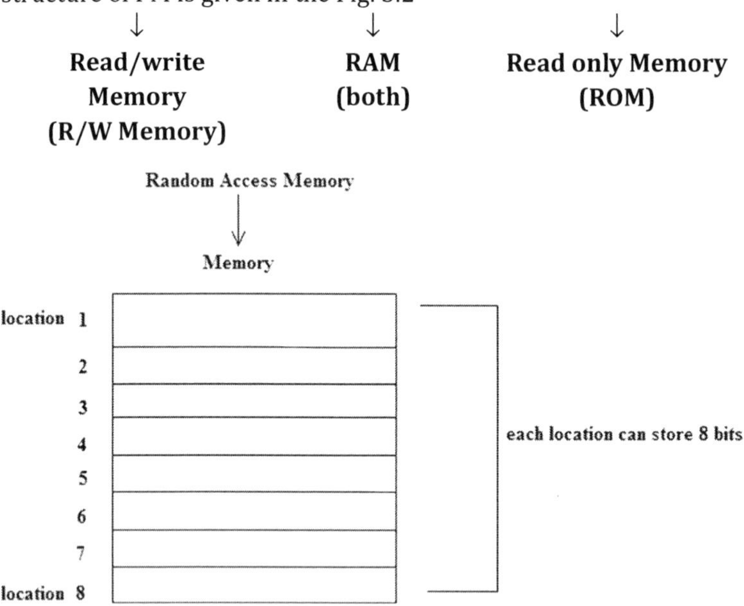

Figure 3.2 The RAM and ROM are the primary memories.

3.2.4.2 Random access memory

The random access memory (RAM) is a volatile storage device in which any memory location can be accessed at random for reading or writing. It retains the stored information as long as the power is not switched off. When power supply is switched off or interrupted the stored information in the RAM is lost.

The RAM may be (i) static or (ii) dynamic

Static RAM retains stored information only as long as the power supply is on. However, dynamic RAM loses its stored information in a very short span of time even though power supply is on. Therefore, dynamic RAM has to be refreshed periodically.

3.2.4.3 Read only memory

Read only memory (ROM) is a onetime programmed, nonvolatile memory device. The user can only read it but cannot write onto ROM. The ROM stores permanent programs and other instructions, information which is needed by the computer to execute user program.

Various types of ROMs are PROM, EPROM, EEROM, and so on.

3.2.5 Secondary or Auxiliary Memory

In a computer, there is not enough storage space in the primary memory until to accommodate all the system programs within for the computer. Further, after installation of a computer as time goes on, there is accumulation of large amount of information, which are not required by the processor at the same time. Hence, it is better to use lower cost storage devices to serve as a backup to store the information that is not currently required by the CPU. These backup storages are called auxiliary or secondary memory (SM).

3.2.5.1 Definition of secondary memory

The SM or auxiliary memory is defined to be the devices that provide backup storage. All information that are not currently used by the CPU are stored in the SM and it is transferred to the primary memory as and when required. Magnetic drams, magnetic disks, and magnetic tapes are the most commonly used secondary memories in the computer system. The SM is much slower than the main memory and it is used for storing system programs and large data which are not currently required by CPU.

3.2.6 Secondary Storage Devices

3.2.6.1 Hard disk

It is a collection of disks called platters. These platters are coated with a material that allows data to be recorded magnetically. The typical rotation speed of the disk is 3,600 revolutions per minute

(RPM). The (R/W) head of drive moves across the disk surface to write or read data on to it. Hard disk is installed inside the computer and can access the data more quickly than floppy disk can. The currently available disk has 80 GB (Giga byte) capacity (it is random access storage device).

It is much larger capacity, less expensive, nonvolatile, and much faster than primary memory. This SM can communicate with PM via microprocessor buses.

Examples of SM are

The hard disk capacity is 1.2 GB (now), 640 MB (earlier), 200 MB (earlier), 2 GB (now), 3 GB (now), and much more.

2 GB hard disk memory cost nearly INR 2000 which is much less than 4 MB primary memory.

4 bits = 1 Nibble

8 bits = 1 byte

1024 bytes = 1 Kbyte

1024 Kbytes = 1 Mbyte

1024 Mbytes = 1 Gbyte

2 Gbytes = $(2 \times 1024)/4 = 512$ times the capacity of 4 MB, PM and at the same time less expensive. As hard discs are nonvolatile, they are used for storing large data, library software, and so on. It is also a R/W memory.

3.2.7 Backup Memory

It is still less expensive, much higher capacity, R/W memory, nonvolatile, which are used for off line backup of large software data, library, programs, and so on.

Example of backup memory magnetic tape, magnetic dish, cartridge tape, digital audio tape (DAT), and video tape. Backup memory of this nature is generally sequential memory, that is, it is not purely RAM.

3.2.7.1 Floppy disk

Unlike hard disk, floppy disks are individually packed disk. The recording medium on a floppy is a Mylar or vinyl plastic disk with magnetic coating on one or both sides. (This plastic disk coated with magnetic material is scaled permanently in a square plastic jacket to protect it from dust and scratches. An elongated slot is cut in the jacket to enable the R/W head of the computer to access information from anywhere on the floppy.) Commonly availbale floppies are of size 3.5 inch diameters and capcity of 150 or 200 MB. A device called the floppy disk drive is required to read from or write onto a floppy disk.

3.2.7.2 Magnetic tapes

Also known as sequential storage device, magnetic tapes are similar to audio or video tapes with the difference that magnetic tapes are coated with magnetic material. Older tapes are 1 inch wide, smaller tapes are also available.

3.2.7.2.1 *Compact disc-read only memory.* A compact disc-read only memory (CDROM) is made up of a reflective layer of aluminum or silver sandwiched between clear polycarbonate coatings. It has a protective layer of acrylic resin over it. Data is stored in digital form within polycarbonate layer. It is 1.2 mm thick and has a diameter of 120 mm. Its storing capacity is 700 MB even more now, which is equivalent to approximately 500 floppy disks.

The CDROM is available in two forms:

1. CD-R: It is called the CD-recordable. It is written once and can be read again and again. Data once written cannot be erased.
2. CD-RW: This is known as erasable-CD. On such a disc, the user can erase previously recorded information and then record new information onto the same physical location.

3.2.8 Cache Memory

The cache is a small, high speed memory between the CPU and the main memory. Its purpose is to increase processor throughput by storing frequently used main memory instructions and data and to prefetch additional instructions (by predicting instruction branching).

3.2.9 Virtual Memory

It is a feature that helps to run a long program in a small physical memory. The OS manages the big program by keeping only part of it in main memory and using the SM for keeping the full program.

Virtual memory describes the following two cases:

1. The main memory space of the processor is not sufficient to run large programs.
2. The physical main memory size is kept small to reduce the cost though the processor has large memory space.

3.2.10 Memory Devices

In most of the digital system, the memory device is one of the most important components. Especially for a microprocessor-based digital system, memory devices are a must.

Bus refers to the hardware lines through which mainly digital signals flows. Address lines is unidirectional. Data lines are bidirectional and can both accept and leave data. The I/O controls the input/output data.

Therefore, memory device is a device in which binary information can be stored. It comes in a variety of organization: capacity (kilobyte or megabyte) and speed (lower and higher speed).

3.3 System Memory and Standard Memory Devices

For RAM, the read access time or write access time for any memory location is constant.

Read/write (R/W) memory is a PM in which we can read from any memory location or also in which we can write onto any memory location (store and read information).

A ROM, is a memory device in which we can only read the contents of any memory locations but cannot write onto any memory location. The difference between R/W and ROM memories is that R/W memory is generally volatile memory unless it is battery backed up, but a ROM is always nonvolatile.

The PM is available as follows:

Primary Memory (PM)	
↓	↓
System Memory Devices	Standard Memory Devices
These are specifically used for a particular microprocessor	These devices can be used with any microprocessor or in any digital system without a μp.

3.3.1 Advantages of System Memory Device

1. Its decoding circuit is on chip and no external decoding circuits are needed for memory decoding.
2. Besides memory it also contains various I/O ports and timer/ counter.
3. Data sheets are provided by the manufacturer of system memory device with 8085 μp.

3.3.2 Disadvantages of System Memory Devices

1. It is not available from variety of sources.
2. It is quite expensive.
3. It is not available in variety of size organization and speed.

3.3.3 Standard Memory Device

The main advantages of semiconductor memories are as follows:

1. It is available from variety of sources.
2. It is less expensive.
3. It is available in different sizes and capacities and can be used anywhere in a digital system with micro processor or without micro processor.

The disadvantages of semiconductor memories are as follows:

1. External hardware circuits are needed for memory decoding.
2. It is relatively or slightly slower than system memory devices due to the requirement of external data decoding circuits.

3.4 Different Semiconductor Memories

Semiconductor memories are of the following types:

1. static bipolar RAM (R/W memory)
2. static MOS RAM (R/W memory)
3. dynamic MOS RAM (DMRAM or DRAM) (it is also R/W memory)

It is just a flip-flop with cross coupled transistors. Static bipolar RAM and Static MOS RAM are shown in the Fig. 3.3 and Fig. 3.4 respectively.

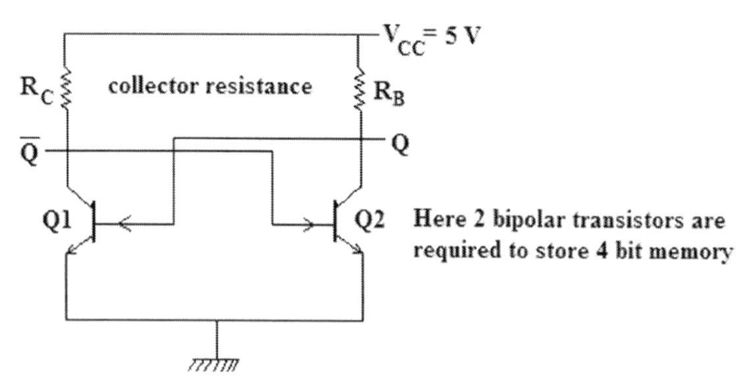

Figure 3.3 1 bit static Bipolar RAM(R/W Memory).

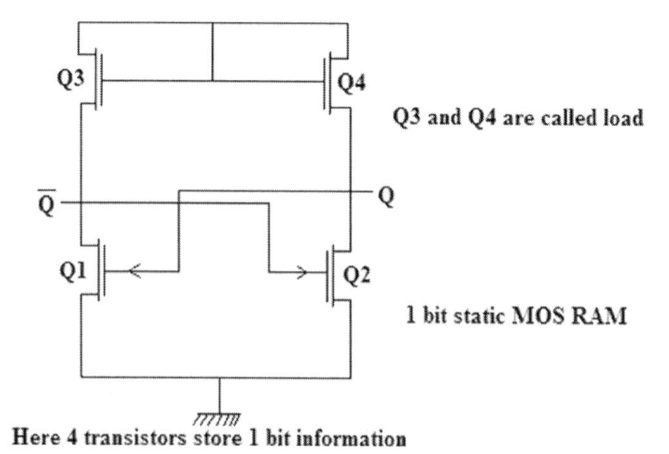

Figure 3.4 Static MOS RAM.

3.4.1 Advantages and Disadvantages of Bipolar Static R/W Memory

In dynamic RAM, the information is stored at the capacitor according to charging and discharging.

Advantage: The main advantage of bipolar static R/W memory is that it is of high speed.

Disadvantages

1. It has very low packing density (Few number of bipolar RAM can be accommodating in a small area), that is, the space required to fabricate a transistor is quite large.
2. It consumes more power (as it requires high supply voltage and its input impedance is not large) and it requires large voltage compared to MOS RAM.
3. It is quite expensive.

3.4.2 Advantages of Static MOS RAM

1. It has very high packing density, that is, space required to fabricate a MOS transistor is very small (as it requires high input impedance with high insulation at input).
2. It consumes less power.
3. It is quite inexpensive.
4. It can be operated with smaller voltage.
5. With the advent of improved IC technology, the speed of static MOS RAM is even greater than static bipolar RAM.
6. MOS RAM chip has on chip decoding circuits for which no external decoding circuits are needed. If the decoding circuits are external then the speed of static MOS RAM can be further increased.

In MOS technology we can have NMOS and PMOS technology. NMOS RAM has higher speed than PMOS RAM as the mobility of electrons is higher than that of holes. Mobility is defined as the drift velocity per unit applied electric field (E). As E of electron is less than hole so the mobility μ is more.

3.4.3 Dynamic MOS RAM

It is widely used as the primary or main memory of all recent microcomputer systems. Dynamic MOS RAM is shown in the Fig. 3.5.

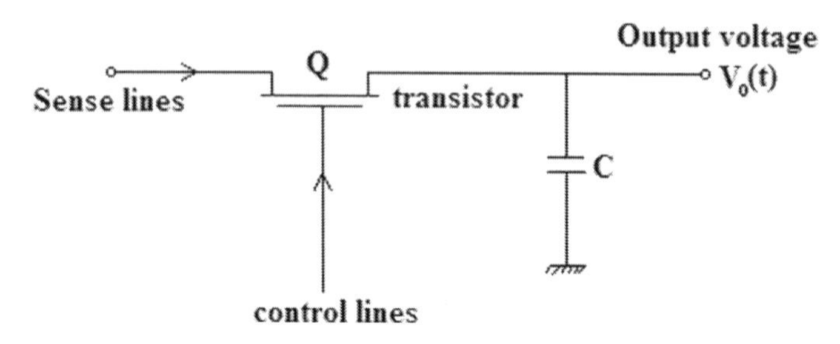

Figure 3.5 Dynamic MOS RAM.

The V_{CC} is the supply voltage for this transistor. When the capacitor charges to a voltage ≥2/3 V_{CC}, it implies that information 1 is stored as binary 1. When the capacitor voltage is <1/3 V_{CC}, it implies that information stored is binary 0. Information 0 is stored as binary 0.

Sense line	Control line	*Q*	*C*
1	1	on	charge toward $V_0 > 2/3\ V_{CC}$

So, information stored is 1.

DRAM requires periodic charging or discharging of capacitor for storing 1 and 0 information. This is called memory refreshing of the DRAM.

For DRAM, a memory refresh controlling circuit is used to refresh the capacitor periodically (The sense circuit checks whether the capacitor voltage is near 1 or 0).

3.4.3.1 Advantages of DRAM

1. Its packing density is very large even larger than static MOS RAM. To store 1 bit, it requires only one transistor, whereas in static MOS RAM it requires four transistors. 10 bits MOS static RAM requires 40 transistors whereas 10 bit dynamic MOS RAM needs 10 transistors and 10 capacitors.
2. It requires few standby power.
3. It is quite cheaper.

3.4.3.2 Disadvantages of DRAM

1. It requires external memory refresh controller circuit for periodic refreshing of each DRAM cell.
2. It is slower than static MOS RAM. (due to the capacitor discharging or charging time)

In a microcomputer system if the memory requirement is larger then DRAM is always used and for memory requirement is lower we can use static MOS RAM.

3.5 Memory Organization

By memory organization, we mean the details of various pins of a memory device. The number of pins for a memory device generally varies from 14 to 40 depending upon the capacity and organization of the memory device. The general block diagram of any memory device is as shown in Fig. 3.6.

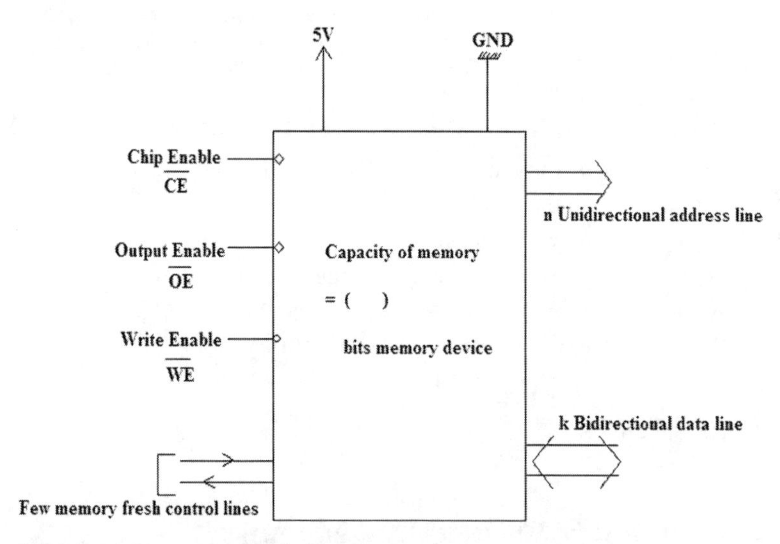

Figure 3.6 Memory organization.

1. n-unidirectional address lines will give us 2^n memory location and
2. K gives us the number of data stored in each memory location.

3. $\overline{CE} = 0 \Rightarrow$ chip enable signal, active low when $\overline{CE} = 0$, the memory device will be selected for either reading or writing.

If $\overline{CE} = 1 \Rightarrow$ the memory device is not selected.

4. $\overline{OE}$ = Output enable signal, when active low implies the reading from a particular location determined sign when $\overline{CE} = 1 \Rightarrow$ disabled reading/reading possible.
5. $\overline{WE} \Rightarrow$ Write enable signal.

When active low facilitates writing from a particular location determined by n address lines.
When $\overline{WE} = 1 \Rightarrow$ implies disable writing
Table 3.1 shows various modes of operations under different situations. For any particular memory device

Table 3.1 Operational table

$\overline{CE}$	$\overline{OE}$	$\overline{WE}$	Mode of operation
0	0	1	Read from a particular location determined by n
0	1	0	Write onto a particular location determined by n
1	X	X	High impedance state, that is, neither reading nor writing is possible
0	0	0	Not used or inhibit memory
0	1	1	Not used or inhibit memory

1. for which $K = 8$, whatever may be n, we call the device as byte organized memory
2. for any memory device for which $K = 1$, the memory device is called bit organized memory. Hence, realize 8 KB of memory by bit organized memory and byte organized memory. Find the number of n and K for both types of memory.

8 Kbytes = 8×1024 bytes
1 byte = 8 bits
so 8 Kbytes = $(8 \times 8 \times 1024)$ bits.

3.6 Bit and Byte Organized Memory

To realize the capacity of 8 KB, we must have $(2^n \times 1)$ bit = $(8 \times 8 \times 1024)$ bits as $K = 1024$ for bit organized memory.

$$2^n = (2^3 \times 2^3 \times 2^{10})$$
$$2^n = 2^{16}$$

or $n = 16$. So we require 16 address lines and 1 bidirectional line. So $n + K = 16 + 1 = 17$ pins are required for bit organized memory.

For byte organized memory $K = 8$, so realize 8 KB of memory by using a byte organized memory we have

$$2^n \times 8 = 8 \times 8 \times 1024$$
$$n = 13$$
$$(n + K) = (13 + 8) = 21$$

Pins are required. So, the cost of realization by using bit organized memory is less. So, it is used. So, for greater memory capacity, it is better to use bit organized memory. The DRAM is a bit organized memory. Generally, each memory location of DRAM chip is organized as 8 bit data + 1 bit parity (= 9 bits)

3.7 Different Memory Chips

Static MOS R/W memory RAM structure is shown in the Fig. 3.7 and corresponding Functional table is shown in the Table 3.2.

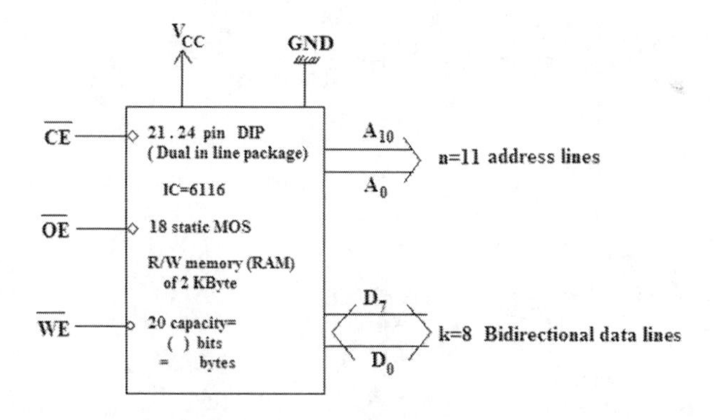

Figure 3.7 Static MOS R/W memory RAM.

$$= 2 \times 2^{10} \text{ bytes}$$
$$= (2 \times 1024) \text{ bytes}$$
$$= 2 \text{ KB as } (1024) \text{ bytes} = 1 \text{ KB}$$

Both 6116 and 2716 are standard memory devices (when full number of pins of EPROM is filled with data then it behaves as a ROM).

Table 3.2 Functional table

$\overline{CE}$	$\overline{OE}$	$\overline{WE}$	Operating mode
0	0	1	Read from any location 6116 can be possibly determined by *n*.
0	1	0	Write onto any location of 6116 can be possibly determined by *n*.
1	X	X	High impedance state
0	0	0	Not accepted
0	1	1	

EPROM = Erasable programmable read only memory 2716 and 6116 are pin compatible (all pins are same). The block diagram of 2 KB EPROM is shown in the Fig. 3.8.

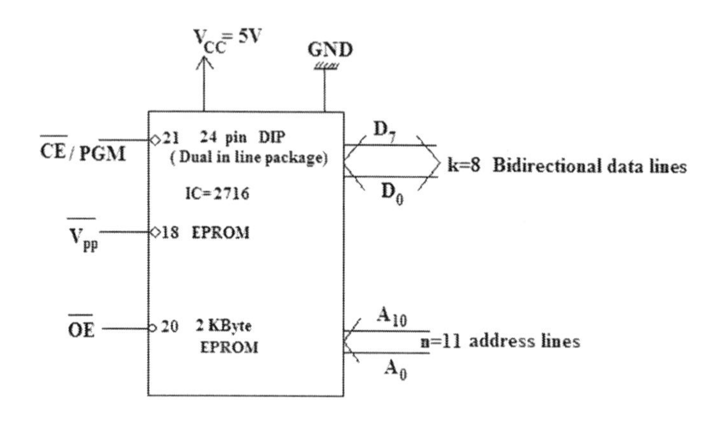

Figure 3.8 Block diagram of 2 KB EPROM.

3.7.1 Optical Windows

If it is exposed to ultra violet (UV) light the EPROM is erased and when EPROM is erased all its locations will contain FFH. The diagram of Optical Window has been shown in the Fig. 3.9 and its corresponding functional table is shown in the Table 3.3.

Programming an EPROM means some 1s are converted into 0s in each location—programming an EPROM is also called EPROM burning.

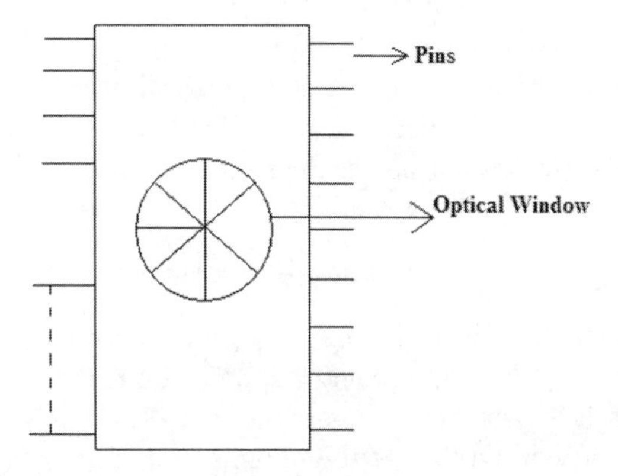

Figure 3.9 Diagram of optical window.

Table 3.3 Functional table

CE/PGM	OE	VPP	Mode of operation
0	0	5 V	Read from EPROM
1	1	25 V	Program EPROM, that is, writing date to EPROM (when EPROM is blank)
0	0	25 V	Program verify mode
1	1	5 V	Standby mode (neither read nor write onto EPROM)
1	0	5 V	Inhibit mode

3.8 Different Types of ROM

The classification of different types of ROM is pictorially represented by the Fig. 3.10.

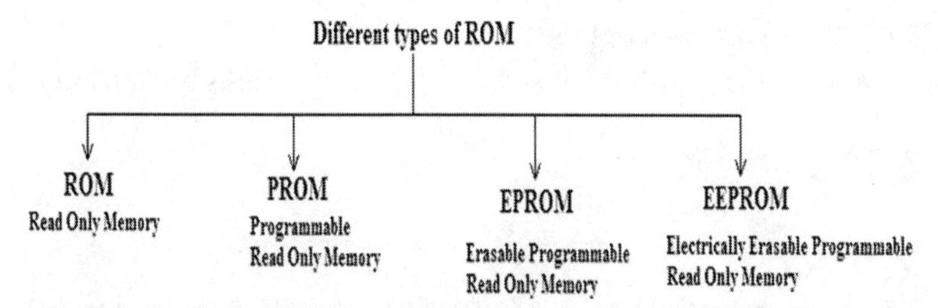

Figure 3.10 Different types of ROM.

3.8.1 Read Only Memory

The read only memory (ROM) is a nonvolatile memory which is prepared by the manufacturer according to the users specifications.

A special mask is used so that its contents can be permanently stored. It is quite expensive due to specially prepared mask and also relatively faster. It cannot be programmed, that is, its content cannot be changed. They have small access times (35–1200) ns.

3.8.2 Programmable Read Only Memory

The programmable read only memory (PROM) is also a nonvolatile memory. It can be programmed only once by using a PROM programmer. It is less expensive compared to ROM and also less slower than ROM. Once programmed its contents cannot be changed.

Example for one such device PROM is 74288 ($2^5 \times 8$ bits PROM).

3.8.3 Erasable Programmable Read Only Memory

The erasable programmable read only memory (EPROM) can be erased as many times required by exposing its optical window to UV light for 20 to 30 min. It can be programmed as many times as we want after erasing it.

3.8.3.1 Disadvantages of EPROM

1. Selective on board erasing memory location is not possible.
2. Selective on board programming cannot be done, that is, programming should be sequential. Programming pins for EPROM is large comparing to RAM memory.

3.8.3.2 Advantages of EPROM

1. Available from variety of sources and with various organization speed and capacity.
2. Erasing time is quite large = 20 to 30 min.
3. It is not so expensive.

3.8.4 Electrically Alterable PROM (EEPROM/EAPROM)

EEPROM stands for Electrically Erasable Programmable Read-only memory and it is also called as EAPROM. This type of memory is non-volatile in nature, which store small amounts of data.

3.8.4.1 Advantages of EAPROM

In this case we can selectively erase a particular location on board. Similarly, we can perform selectively on board programming and selective on board erasing requires a threshold voltage of 5 to 8 V.

Erasing time is 5 to 10 ms and programming time is 250 μs to 1 ms for a particular location.

3.8.4.2 Disadvantages of EAPROM

1. It is quite expensive.
2. It is not available from a variety of source.
3. It is not popular as this technology is still not matured.

3.8.5 Applications of ROM

1. ROM can be used to realize in particular combination or sequential logic design.
2. ROM can be used to store the microinstructions of a control unit in a microprogrammed control system.
3. ROM is used for code conversion.
4. ROM is used for storing the lookup tables (as in calculator).
5. It can be used to store the readymade subroutines or subroutine monitors programs.
6. ROM can be used in the character generator to store information of each character to be generated on the CRT screen. This is called character generator.
7. It can be used for effective emulation of other machine by microprocessors using the firmware approach with ROM. Dot pattern for each character is shown in Fig. 3.11.
8. The ROM can be used to store data. The OS is to protect it from computer virus.

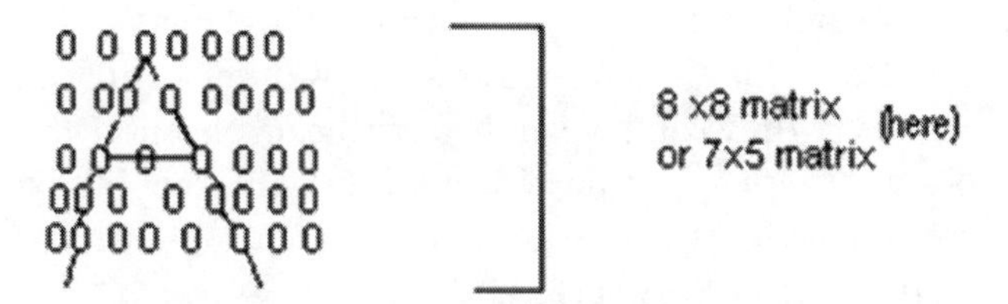

Figure 3.11 Dot pattern generation using ROM.

3.9 Ferrite Core Memories

This consists of ring shaped ferrite material and magnetizing property of ferrite is used for storing the data. Two types of wiring are used—one for writing and one for reading or sensing current in the direction produce by clockwise magnetization, implies binary 0 (direction of lines of force is clock wise) and sensing current in other direction, anti clockwise magnetization implies 1.

For reading the information, the material is also passed through read head and this detects whether the magnetization stored is clockwise or anticlockwise. Earlier it was used as primary memory but with the advent of semiconductor memory it is absolute.

3.9.1 Disadvantages of Ferrite Core Memories

1. It is not compatible with CPU.
2. Special type of wiring requirement makes it rather expensive.

3.10 Compact Disc-Read Only Memory (CDROM)

The CDROM surface at which the information is stored is obtained from a resin named polycarbonate, which is coated with aluminum to make the surface highly reflective. The information on a CDROM is stored as a series of microscopic bits and these bits are formed by injecting highly intense laser beam to the CDROM surface to form the bits. Therefore, CDROM is called optical disk memory. The information from a CDROM is restricted by inserting it on a CDROM drive unit which produces low power laser beams. The laser beam is aimed at the CDROM surface which is rotated with constant angular velocity (CAV) and when the laser beam encounter a bit on the CDROM surface, the reflected intensity from this bit changes optical sensors detect the changes in the reflected intensity as digital signal from the CDROM surface.

3.10.1 Main Advantages of CDROM

1. Its capacity is huge or very large (that is, of the order of 700 MB even more now a days) can be considered as a portable hard disc on which we can only read.
2. Its mass production is easier and faster.
3. It is flexible and portable and hence can be used as archival purpose, that is, data backup.
4. It is very much secure, that is, data on it cannot be corrupted by scratch or dirt.

Initially, the master disc is formed by striking of intense laser beam on the CDROM surface which creates microscopic bits. By using this master CD disc a die is produced from which several copies of this are produced (so its mass production is easier and faster).

The reason for (4) is that the top of CDROM surface is coated with clear lacquer (shiny surface) to protect it from any dirt and scratch.

3.10.2 Main Disadvantages of CDROM

1. Its contents cannot be updated or changed as it is ROM.
2. Its speed is quite slow compared to magnetic disc.

Other types of CDROM disk are write ones read many times CD (WORM CD) and erasable CD, we can erase the CD, that is, it can change and update the contents. But it is quite expensive and its drive is also expensive.

3.10.2.1 Constant linear velocity

If we use this CD with constant linear velocity (CLV) then the capacity of the CD will be large as the information is stored in several sectors cylindrical tracks.

3.11 ROM by Using Decoder and Gates

The circuit diagram of ROM using decoder and gates is represented by Fig. 3.12 and its working is shown in the Table 3.4.

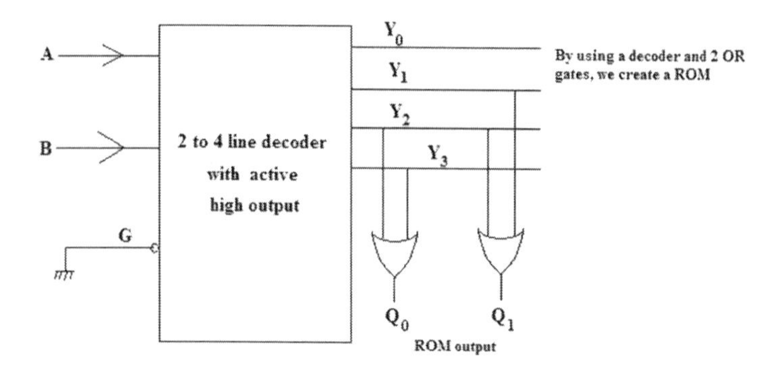

Figure 3.12 ROM realization by decoder and gates.

Table 3.4 Functional table

O_0	O_1	G	A	B	Y_0	Y_1	Y_2	Y_3
0	0	0	0	0	1	0	0	0
0	1	0	0	1	0	1	0	0
1	1	0	1	0	0	0	1	0
1	0	0	1	1	0	0	0	1

The circuit diagram of ROM using decoder and diodes is represented by Fig. 3.13 and its working is shown in the Table 3.5.

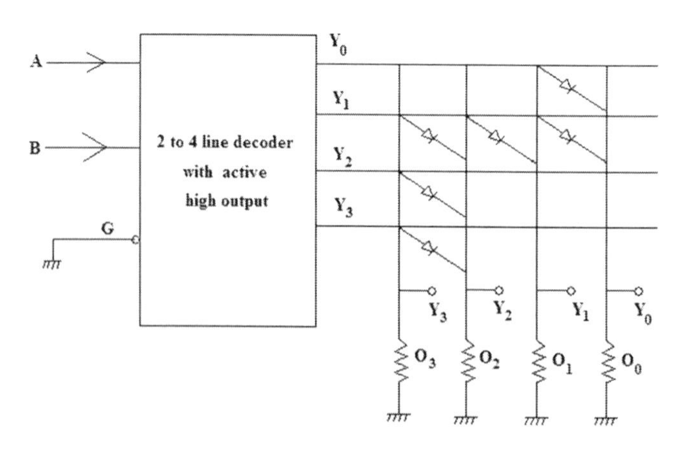

Figure 3.13 Diode ROM using decoder and diode.

Table 3.5 Functional table

G	A	B	Y_0	Y_1	Y_2	Y_3	O_3	O_2	O_1	O_0	$Data_0$
0	0	0	1	0	0	0	0	0	0	1	0001
0	0	1	0	1	0	0	0	1	1	0	0110
0	1	0	0	0	1	0	0	1	0	1	0101
0	1	1	0	0	0	1	0	1	0	0	0100

3.12 Function of 74189 RAM and 74288 PROM Chip

The structure of 74189 RAM is shown in the Fig. 3.14 and its functional table is shown in the Table 3.6.

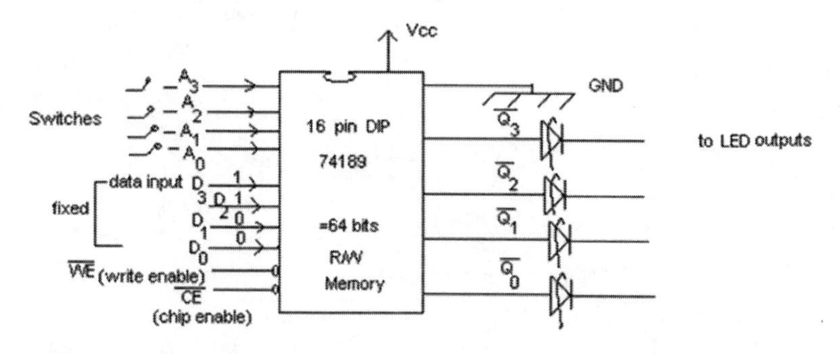

Figure 3.14 74189 RAM structure.

Table 3.6 Functional table of 74189 R/W memory

$\overline{CE}$	$\overline{WE}$	A_3	A_2	A_1	A_0	D_3	D_2	D_1	D_0	Q_3	Q_2	Q_1	Q_0
0	$1 \rightarrow 0$	0	0	0	0	1	1	0	0	0	0	1	1
0	$0 \rightarrow 1$	0	0	0	0	1	1	0	0	0	0	1	1
1	X	X	X	X	X	X	X	X	X	X	X	X	X

$74189 \rightarrow 16 \times 4$ bit R/W memory chip to be used in the lab.

organization = 16×4 bits

$=(2^4 \times 4)$ $n = 4$ and $K = 4$

We name as such as we can read as well write on to the 64 locations. It is a volatile memory and this can be verified by writing some data and then switch off and find that data is lost.

3.12.1 Operation

3.12.1.1 Write operation

The structure of 74288 PROM chip is shown in the Fig. 3.15 and its addressing schemes is shown in the Table 3.7. Output LED will not glow.

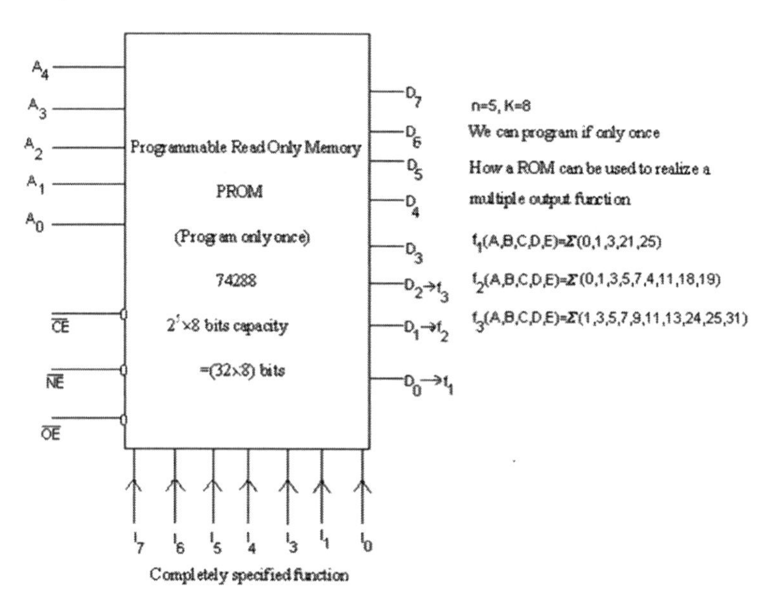

Figure 3.15 74288 PROM chip structure.

Table 3.7 Addressing schemes

$\overline{CE}$	$\overline{WE}$	$\overline{OE}$	Dec	A_4	A_3	A_2	A_1	A_0	D_7	D_6	D_5	D_4	D_3	D_2	D_1	D_0
0	0	1	0	0	0	0	0	0	0	0	0	0	0	0	1	1
0	0	1	1	0	0	0	0	1	0	0	0	0	0	1	1	1
0	0	1	2	0	0	0	1	0	0	0	0	0	0	0	0	0
0	0	1	3	0	0	0	1	1	0	0	0	0	0	1	1	1
.																
.																
.																
0	0	1	11	0	1	0	1	1	0	0	0	0	0	1	1	0
.																
.																
.																
0	0	1	31	1	1	1	1	1	0	0	0	0	0	1	0	0

3.12.1.2 Read operation

In this case output LED shows the status of the output is tristate. The structure of 74288 PROM chip is shown in the Fig. 3.16 and its addressing schemes is shown in the Table 3.7. We use output D_0 for f_1, D_1 for f_2, and D_2 for f_3.

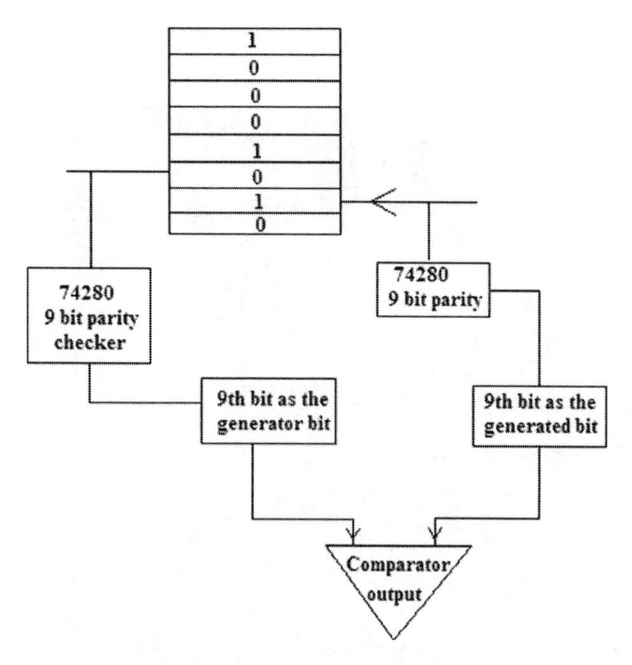

Figure 3.16 Method of RAM testing on a PC.

Realization of multiple output function by using PROM, that is, 74288 (32×8 bit memory) to verify programming use read operation from any memory location:

$\overline{CE}$	$\overline{WE}$	$\overline{OE}$
0	1	0

3.13 Method of RAM Testing on a PC

The PC contains primary memory.

In any 8 bit memory of a PC, the CPU will write 8 bit arbitrary data on 8 bits per RAM each location.

Flow diagram shown in Fig.3.17. Thus, checks the parity between whatever is written to whatever is read. If there is any error it is called RAM testing error.

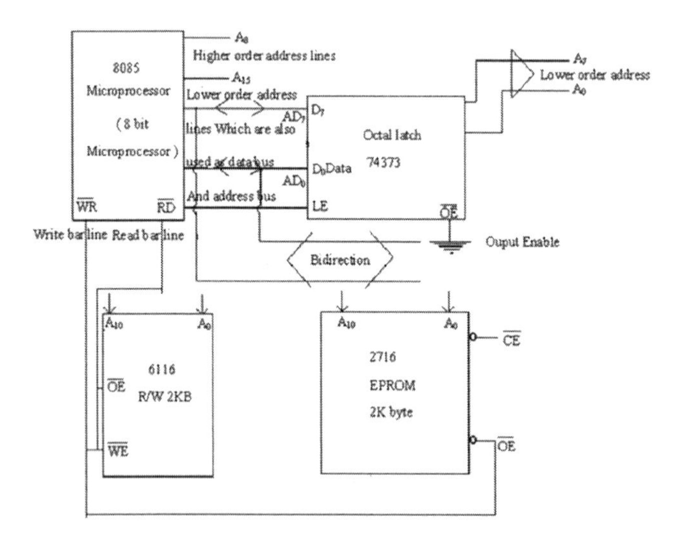

Figure 3.17 Memory interfacing by fully decoded addressing method.

$Y_1 = Y_2 \Rightarrow$ Then comparator input matches

$\Rightarrow$ Then no parity error

If $Y_1 \neq Y_2 \Rightarrow$ Comparator input do not match

match $\Rightarrow$ memory parity error

The computer will not boot until the memory error is eliminated. The RAM testing can be eliminated by pressing the escape key. While writing either use even or odd parity.

3.13.1 Memory Interfacing

Memory interfacing means the inter connection of more than one memory devices with a given microprocessor. However, microprocessor can access only one memory devices, when microprocessor try to access more than one memory devices then bus contention may occur which may damage the devices connected to the bus.

3.14 Memory Interfacing by Fully Decoded Addressing

Let 8085 microprocessor, we want to connect, one R/W memory devices 6116 (static MOS RAM, 2 KB capacity) and one EPROM device (IC 2716) of capacity 2 KB.
A_0–A_{15} → Out of 10 address lines 11 address lines are directly connected to memory devices and remaining five A_{11} to A_{15} are used to drive the chip, select signal for the two memory devices.

ALE = Address Latch Enable

ALE	LE	74373	Output of 74373
1	1	E	A_0-A_7
0	0	Disabled	Previous address is latched.

The Fig. 3.18 shows Memory Interfacing by fully decoded addressing method.

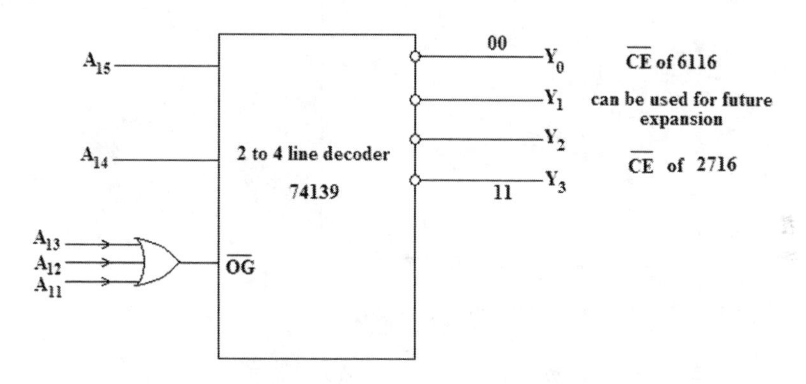

Figure 3.18 2 to 4 line decoder of 74139.

The Fig. 3.19 shows 2 to 4 line decoder of 74139 where all the address lines are used.

A_{15}	A_{14}	A_{13}	A_{12}	A_{11}	A_{10}				.... A_0	
0	0	0	0	0	0	0	0	0	0	→ Start address = SA=0000 H
0	0	0	0	0	1	1	1	1	1	= EA=07FF H

Since all the 16 address buses are used, so it is named as fully decoded addressing technique.

During time interval for which ALE is high say T_1 s, whatever input will appear at AD_0–AD_7 become lower order address lines (when $ALE = 1$) from A_0 to A_7 and

when $ALE = 0$ AD_0–AD_7 line is used as data lines, that is, D_0 to D_7. The EPROM memory structure is shown in the Fig. 3.19.

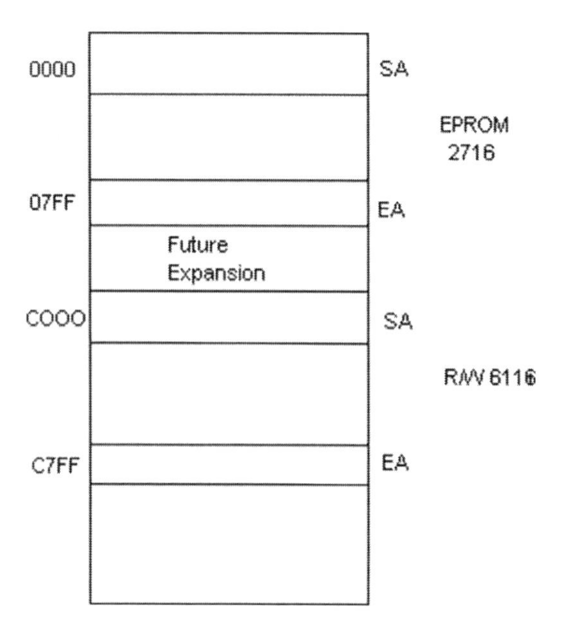

Figure 3.19 The EPROM memory structure.

For R/W memory
SA = C000 H same as EPROM
EA = C7FF H

3.14.1 Advantages

1. Its full address space is utilized.
2. The memory devices can be placed continuously.
3. Bus contention will never occur.
4. The two unused outputs of decoder can be used for future expansion.

Problems

1. What are the types of memories?
2. What is the basic difference between EPROM and EEROM?
3. Why are semiconductor memories used as main memory and magnetic/optical medium-based memories as secondary memory?
4. Define secondary memory with example.
5. What is the purpose of using cache memory?
6. What are the benefits of virtual memory?
7. What are the two types of primary memory? What is the basic difference between them?
8. To store 10 bit of information how many transistors are required in case of Static MOS RAM and Dynamic MOS RAM?
9. What control signals are used in the memory device?
10. What are the main applications of ROM?
11. How many pins are required to design 8-KB bit organized memory?
12. How many pins are required to design 8-KB byte organized memory?
13. What are the advantages of CDROM?
14. What are the disadvantages of CDROM?
15. Name the resin from which the information is obtained from the CDROM surface.
16. Which coating is used to make the CDROM surface highly reflective?
17. In which form Information are stored on a CDROM?
18. What is CLV?
19. How encountering bits are detected as digital signal?
20. What is memory interfacing?

Chapter 4

Timing Circuit

4.1 Introduction

Individual sequential logic circuits are used to design more complex circuits such as counters, shift registers, latches, or memories, etc., but to operate these types of circuits in a "Sequential" way, one requires the addition of a clock pulse or timing signal to cause them change their state. The clock pulses are generally square shaped waves that are produced by a single pulse generator circuit which oscillates between a "HIGH" and a "LOW" state and generally has an even 50% duty cycle, that is it has a 50% "ON" time and a 50% "OFF" time. The sequential logic circuits that use the clock signal for synchronization may also change their state on either the rising or falling edge, or both of the actual clock signals.

4.2 Multivibrators

These are devices for generating timing waveform or clock waveform. There are three types of multivibrators.

Foundation of Digital Electronics and Logic Design
Subir Kumar Sarkar, Asish Kumar De, and Souvik Sarkar

ISBN 978-981-4364-58-4 (Hardcover), 978-981-4364-59-1 (eBook)
www.panstanford.com

Those are

1. astable multivibrator.
2. monostable multivibrator.
3. bistable multivibrator or flip-flop is a prestable multivibrator.

4.2.1 Astable Multivibrator

Astable multivibrators are circuits with two quasi stable states and it switches between its two quasi stable states without requiring any trigger input. They are also called free running multivibrators or oscillators. There is a positive feedback and this does not require any input and the multivibrator switches from one quasi stable state to the other.

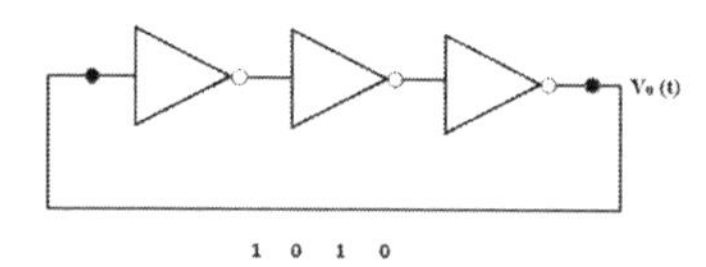

Figure 4.1 Astable multivibrator.

A NOT gate based scheme is shown in circuit shown in Fig. 4.1 where $V_0(t)$ represents the output.

So, the output oscillates between 0 and 1 even when the GND is removed. The output will change after $3t_{pd}$ where t_{pd} is propagation delay for each NOT gate as shown in Fig. 4.2.

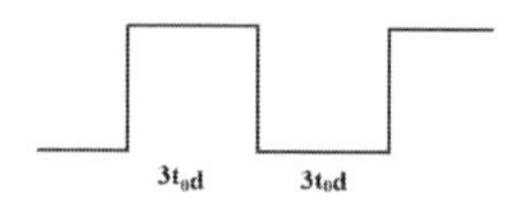

Figure 4.2 Clock pulse.

So, by observing the waveform $V_0\,(t)$ on a CRO (Cathode Ray Oscilloscope), the frequency f of the waveforms can be measured.

$$t_{pd} = 1/6f$$

If f is known we can find t_{pd} and if we use NAND gate instead of NOT gate, we can find its t_{pd}. To generate astable multivibrator waveform, by NOT or NAND gate we require odd number of NOT or NAND gates. If t_{pd} is very small then f is very large and the waveform and f count be measured accurately by the CRO.

4.2.2 Monostable Multivibrator

Monostable multivibrator is one, which has got a single stable state 0 or 1 that is low or high (not simultaneously 0 or 1) and it switches from stable state to quasi stable state when suitable trigger input is applied as shown in Fig. 4.3. How long it will remain in the quasi stable state depends upon the time constant (T_1) of the monostable multivibrator circuit.

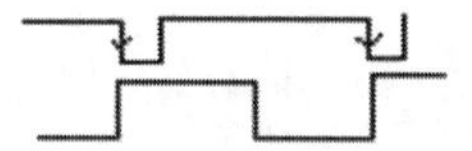

Figure 4.3 Clock pulse.

T_1 = RC = time constant for the circuit. So, at the time when trigger is applied, the stable state changes to quasi stable state.

4.2.3 Application

1. If we want a print out from a printer for 30 min, this can be achieved by using a monostable circuit whose duration of the quasi stable state is 30 min and the driving signal will make the print out for 30 min.
2. To erase an erasable programmable read only memory (EPROM) chip, its optic window can be exposed to ultraviolet for say 20 min to erase EPROM chip. The top has got an optical window which is used as a memory device.

Multivibrators can be designed by using logic gates, OP-AMPs, discrete components like transistors, 555 timer ICs and by using some other ICs.

4.3 555 Timer

The 555 Timer IC, as shown in Fig. 4.4, is a sequential logic device. It has 8 pin DIP. Its input and outputs are TTL and CMOS compatible. Timer output to input of CMOS and viceversa & also TTL output to input of the timer and viceversa.

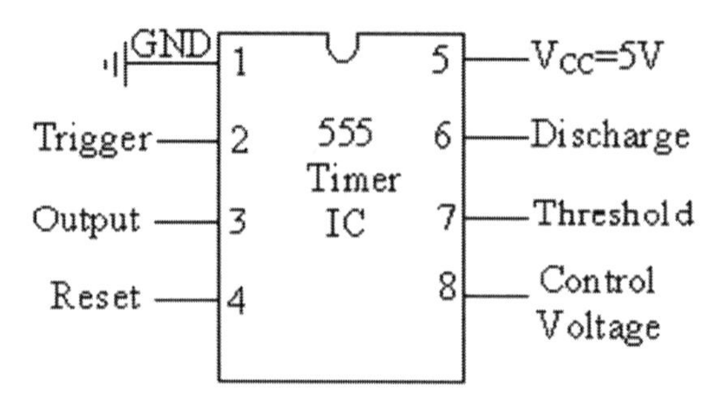

Figure 4.4 Block diagram of 555 timer IC.

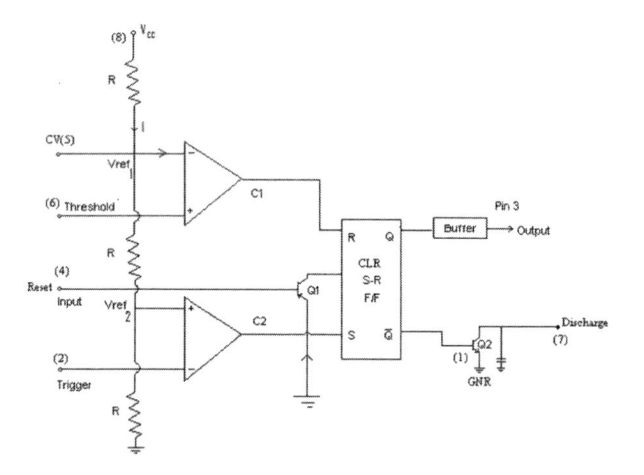

Figure 4.5 Internal circuit diagram of 555 timer.

In Fig. 4.5,

$$I = (V_{CC}/3R)$$
$$V_{ref1} = I\,(2R) = (V_{CC}/3R) * (2R) = \mu\, V_{CC}$$
$$V_{ref2} = I\,(R) = \mu\, V_{CC}$$

Two comparators C_1 and C_2 are connected in the circuit. The output of each comparator depends on both inputs. That is the output of C_1 is 1 if threshold voltage > V_{ref1}, otherwise, the output is 0. Similarly, the output of C_2 goes high with V_{ref2} > trigger voltage. In the S–R flip-flop, R and S cannot be simultaneously 1. So, C_1 and C_2 cannot be simultaneously 1.

555 timer can be operated in two modes:

(i) astable mode
(ii) monostable mode

4.3.1 Astable Mode Operation

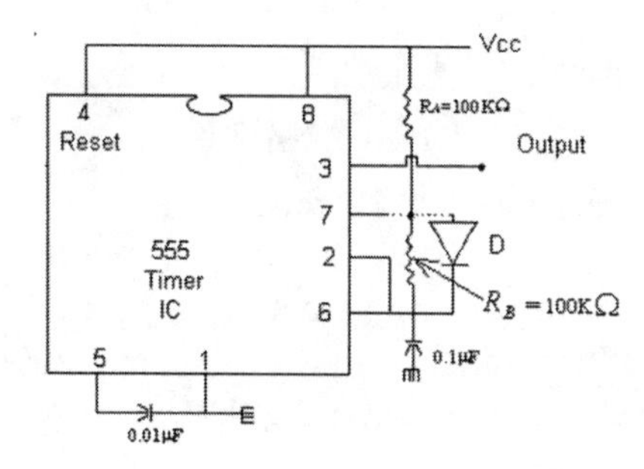

Figure 4.6 555 Timer as astable multivibrator.

The arrangement in Fig. 4.6 shows the reference voltage to the comparator is used to vary the timing characteristics of the signal generated electronically at output pin number 3 by applying a control voltage at pin number 5 that is at the same V_{ref1} input of comparator 1. When the external control voltage is not used, this pin is to be connected via a 0.01 μF capacitor to ground to reduce the power supply ripples and noise or power supply noise voltage. The capacitor used is the bypass capacitor.

Applying an input <0.4, the reset input which is independent of other conditions can clear the output of S–R flip-flop. So, $Clr = 0$, when reset input ≤0.4 V and $Q = 0$. If we do not want to clear the output of S–R flip-flop by using small reset input then the reset input is connected to $V_{CC} = 5$ V. (B–E voltage is removed. Biased, collector current is high, reset = Clr input is high, F/F is now enabled with S–R inputs).

We use this timer IC in laboratory to generate a square or rectangular waveform. The load is connected between 3 and V_{CC} or between 3 and GND. The waveform at output 3 is asymmetrical square wave. The R_A, R_B, C and C_1 are to be connected externally.

Let us assume initially, that $Q = 1$ and $\bar{Q} = 0$. Hence, $\bar{Q}_2$ will be off under this condition. The electron cannot go in the direction of pin number 7 and under this condition the capacitor C will charge exponentially towards V_{CC} with R_A and R_B and with a time constant proportional to $(R_A + R_B)$ C. As capacitor charges, V_C (say) increases. When V_C becomes greater than 2/3 V_{CC}, the output of the comparator C_1 (shown in Fig. 4.5) is 1, where R = 1 and $S = 0$. This will switch the flip-flop to $Q = 0$ and $\bar{Q} = 1$. As $\bar{Q} = 1$, this saturates the transistor Q_2. Now, the voltage of the capacitor will not discharge along pin number 7. The capacitor will stop charging but it now discharges via R_B and Q_2 with discharging time constant = $R_B C$ because when transistor is ON, its resistance is ideally 0.

$$\text{So, } T_{\text{d.t.c.}} = R_B C \alpha R_B \tag{4.1}$$

As V_C decreases and $V_C < 1/3\, V_{CC}$, this makes $C_2 = 1$ and $C_1 = 0$ and $S = 1$, $R = 0$. So, $Q = 1$ and $\bar{Q} = 0$ and again the earlier conditions are repeated without requiring any trigger pulse. This goes on for infinite period of time and as a result at the pin number 3 we get a continuous waveform as shown in Fig. 4.7.

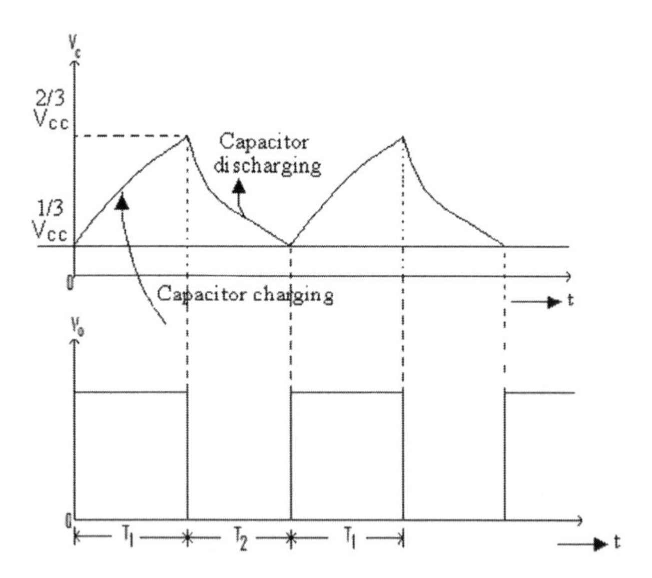

Figure 4.7 Waveform of astable multivibrator.

Here T_1 = Charging time constant and
T_2 = Discharging time constant
The charging and discharging time intervals are given by

$$T_1 = 0.7\ (R_A + R_B)\ C$$
$$\text{and } T_2 = 0.7\ R_B C$$

So, the output of the timer is an asymmetrical time waveform.
For charging a capacitor, charging equation of R–C circuit is

$$V = V_{CC}\ (1 - e^{-t_1/C(R_A + R_B)})$$
$$\text{At } t = t_1,\ V = \mu\ V_{CC} = V_{CC}\ (1 - e^{-t_1/C(R_A + R_B)})\ \ldots \qquad (4.2)$$
$$\text{or, } 1 - e^{-t_1/C(R_A + R_B)} = \tfrac{1}{3}$$
$$\text{or } e^{-t_1/C(R_A + R_B)} = \tfrac{2}{3}$$
$$\text{At } t = t_2,\ V = \mu\ V_{CC}\ \ldots \qquad (4.3)$$
$$(\frac{2}{3}) = 1 - e^{-t_2/C(R_A + R_B)}$$
$$e^{-t_2/C(R_A + R_B)} = \frac{1}{3}$$
$$e^{(-t_1 + t_2)/C(R_A + R_B)} = 2$$
$$\frac{e^{(t_2 - t_1)}}{C(R_A + R_B)} = 2$$
$$t_2 - t_1 = T$$

So $\dfrac{t_2 - t_1}{c(R_A + R_B)} = \text{In } 2$

T_1 is the time interval during which output is high

$$T_1 = 0.693C\ (R_A + R_B)$$
$$T_1 \approx 0.7C\ (R_A + R_B) \qquad (4.4)$$

For discharging the capacitor C from a voltage V_{CC},

$$V = V_{cc} e^{-\frac{t}{R_B C}}$$

At $t = t_2$

$$V = \frac{2}{3} V_{cc}$$

$$\frac{2}{3} V_{CC} = V_{CC} e^{\frac{-t_2}{R_B C}}$$

$$\frac{2}{3} = e^{\frac{-t_2}{R_B C}} \qquad (4.4a)$$

At $t = t_3$ $$V = \frac{1}{3}V_{cc}$$

$$\frac{1}{3}V_{cc} = V_{cc}e^{\frac{-t_3}{R_B C}} \tag{4.4b}$$

Dividing (4.4a) by (4.4b) we have,

$$2 = e^{(+t_3 - t_2)/CR_B}$$

$t_3 - t_2 = T_2 =$ off period of the timer output
$t_2 - t_1 =$ on period of the timer output
So $t_3 - t_2 = \ln 2CR_B$

$$T_2 = 0.7\ CR_B \tag{4.5}$$

As $T_1 \neq T_2$, so the generated square wave at the timer output is asymmetrical.

$$\text{Duty Cycle} = \frac{T_{on}}{\text{Total time period}} = \frac{T_1}{T_1 + T_2} = \frac{0.7(R_A + R_B)C}{0.7(R_A + 2R_B)C}$$

$$\text{So the duty cycle } = \left(\frac{R_A + R_B}{R_A + 2R_B}\right)$$

$$\%\text{duty cycle} = \left(\frac{R_A + R_B}{R_A + 2R_B}\right) \times 100 \tag{4.6}$$

To make the generated waveform periodic, connect a diode D parallel to R_B.

Under this condition, $T_1 = 0.7C\ (R_A + R_B)$

$$T_1 = 0.7R_A C \rightarrow \text{charging time}$$
$$T_2 = 0.7R_B C \rightarrow \text{discharging time}$$

So, as the diode is reverse biased, if we make $R_A = R_B$ and $T_1 = T_2$ then we will get a symmetrical square wave frequency of unsymmetrical square wave as

$$\frac{1}{T_1 + T_2} = \frac{1}{0.7(R_A + 2R_B)}\ (\text{as } R_A = R_B = R)$$

Frequency of unsymmetrical square wave is

$$\frac{1}{0.7(2RC)} \tag{4.7}$$

4.3.1.1 Generation of symmetrical waveform using J–K flip-flop (without using diode)

The circuit connection and symmetrical waveform representing how symmetrical waveform is generated using the diode is depicted in Figs. 4.8 and 4.9 respectively.

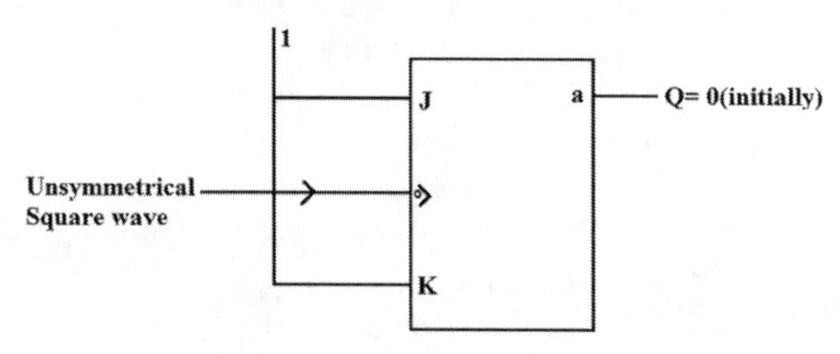

Figure 4.8 Negative edge triggered J–K flip-flop with the asymmetrical wave form as the pulse input.

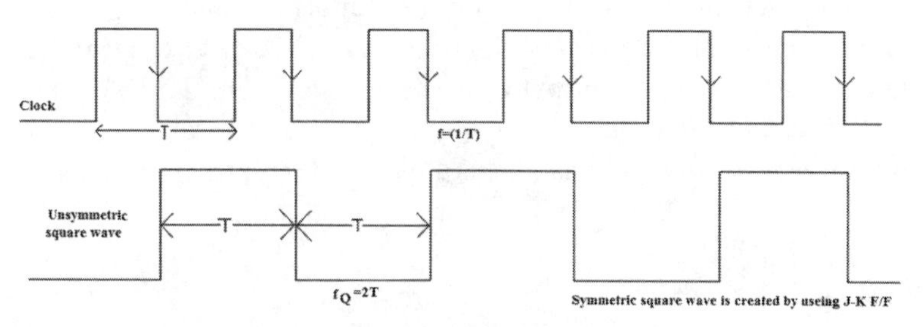

Figure 4.9 Output of the J–K flip-flop.

4.3.2 Monostable Mode Operation

The circuit connection and output of symmetrical wave form created by J–K flip flop are depicted in Fig. 4.8 and Fig. 4.9.

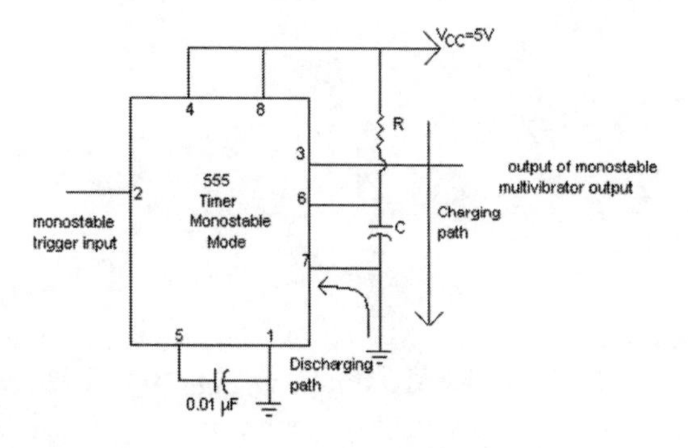

Figure 4.10 A monostable multivibrator with 555 timer.

Let initially $Q = 0$ and $\bar{Q} = 1$. Under this condition the transistor Q_2 is on (refer to Fig. 4.5).

Now capacitor is under full discharge condition. If trigger input is high, as $C_1 = C_2 = 0$. So $R = S = 0$. This makes $Q = 0$ and $\bar{Q} = 1$. That is the output of timer is at stable state 0.

If a negative going trigger is applied at pin number 2 then $C_2 = 1$ and $C_1 = 0$ and that makes $S = 1$ and $R = 0$ for which $Q = 1$ and $\bar{Q} = 0$, that is the timer output switches from stable state 0 to quasi stable state 1.

As $\bar{Q} = 0$, transistor Q_2 is off and the capacitor gradually charges toward V_{CC} via R and C with time constant RC.

When V_C is > 2/3 V_{CC} then $C_1 = 1$ and $C_2 = 0$. This makes $R = 1$ and $S = 0$ for which $Q = 0$ and $Q^{-} = 1$ and thus, the timer output returns to the stable state 0 from quasi stable state 1 and thereby the capacitor C discharges quickly through Q_2 transistor. The charging will take more time than discharging time which is small. When negative trigger is applied, the output of timer changes from 0 stable state to 1 state and depending on the RC time constant it states at 1. The output waveform of monostable multivibrator is shown in Fig. 4.11.

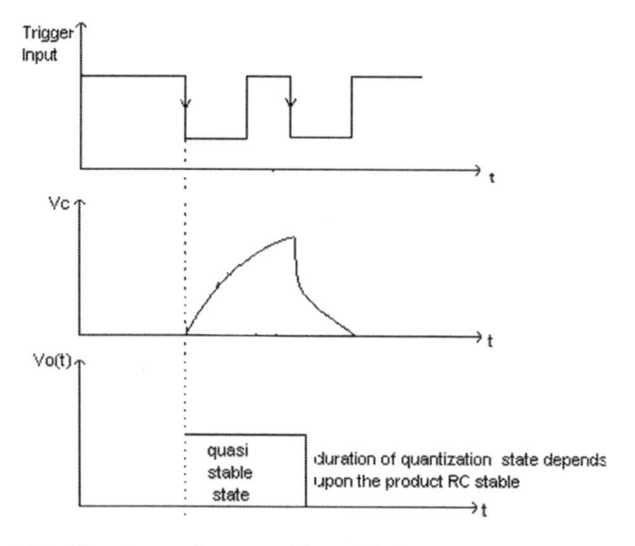

Figure 4.11 Waveform of monostable multivibrator.

We have not used the control voltage until now. But if the control voltage signal $x(t)$ is used and a time varying capacitor voltage (V_{CV}) is applied with respect to a fixed V_{ref} then variation of V_{CV} changes V_{ref}

and pulse width also changes and at the output, we get a frequency varying pulse width (on time and off time pulse width will vary) and duration of quasi stable state (T_1) is 0.7 *RC*.

4.4 Monostable Multivibrator Using Logic Gate

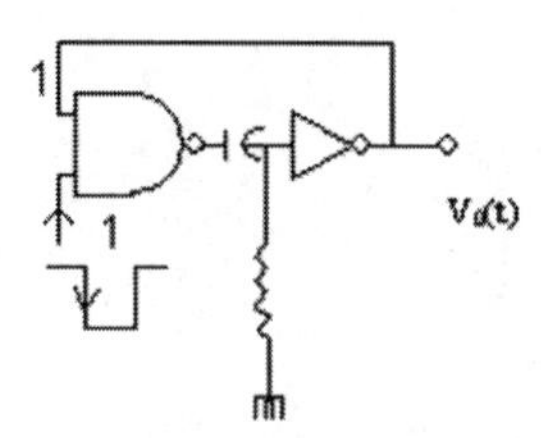

Figure 4.12 Monostable multivibrator using NAND.

In the circuit shown in Fig. 4.12, the capacitor discharges from 5 V to 0 V if the trigger input is high and thus, the stable state output $V_0(t)$ is high or logical 1. As negative edge trigger appears at input of NAND gate, the output of NAND gate is high = 5 V. But capacitor will take time to charge. The equivalent circuit is represented by Fig. 4.13.

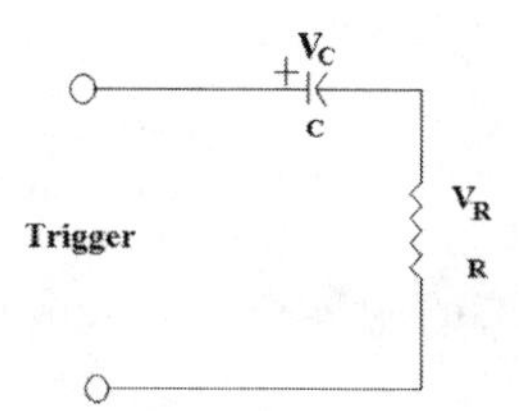

Figure 4.13 Equivalent circuit of monostable multivibrator using NAND gate.

$$5\text{ V} = V_C + V_R$$

or at $t = 0+$, $5\text{ V} = 0 + V_R$ (NOT output is 0 and NAND output is 1)

Now V_C rises and reaches $V_{C,MAX}$ then V_R will fall to 0 and so output of NOT gate is 1 and NAND gate is 0.

Here the charging time constant = 0.7 *RC*

The equivalent output waveform is shown in Fig. 4.14.

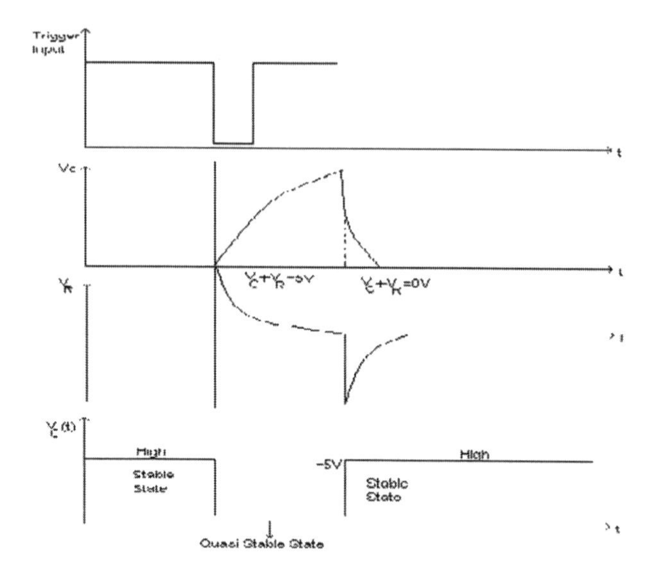

Figure 4.14 Output waveforms.

4.5 Generation of Timing Waveform Using OP-AMP

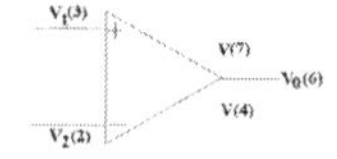

Figure 4.15 OP-AMP.

For an OP-AMP (shown in Fig. 4.15), the output is represented as

$$V_0 = A\ (V_2 - V_1) \tag{4.8}$$

where

A = Gain of the OP-AMP
V_1 = Noninverting voltage and
V_2 = Inverting voltage

Ideal OP-AMP characteristics are as follows:

1. infinite gain
2. infinite bandwidth

3. infinite common mode rejection ratio

$$CMRR = P = A_d/A_c$$
$$= \text{(Differential gain/common mode gain)}$$

4. infinite input impedance
5. zero output impedance
6. Very good slew rate

Generally, A has negative high value:

$$V_0 = +\text{ve or } +V_{Sat}$$

$$V_0 = -\text{ve or } -V_{Sat}$$

OP-AMP requires dual supply voltage ($V+$ and $V-$)

If the supply voltage of the OP-AMP is ± 15 V, V_{Sat} is generally ±2 volt lesser than the supply voltage

Hence, for supply voltage = ±15 V

$$+V_{Sat} = 15 - 2 = 13 \text{ V}$$
$$-V_{Sat} = -15 + 2 = -13 \text{ V}$$

Just by using the output, we can compare the magnitudes of V_1 and V_2.

4.5.1 Generation of Symmetric Waveform

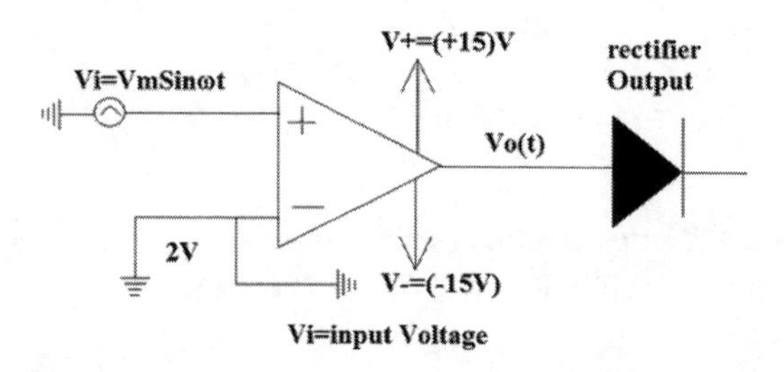

Figure 4.16 Generation of timing waveform.

In the circuit shown in Fig. 4.16, a sinusoidal wave is applied to the inverting input of the OP-AMP and the non-inverting terminal is grounded. The output signal swings from $+V_{sat}$ to $-V_{sat}$ as shown in Fig. 4.17. The output of the OP-AMP V_0 is then connected to a diode, which acts as a rectifier which can be clearly seen in Fig. 4.16. The output of the rectifier (Shown in Fig. 4.18) is represented as V_1(t).

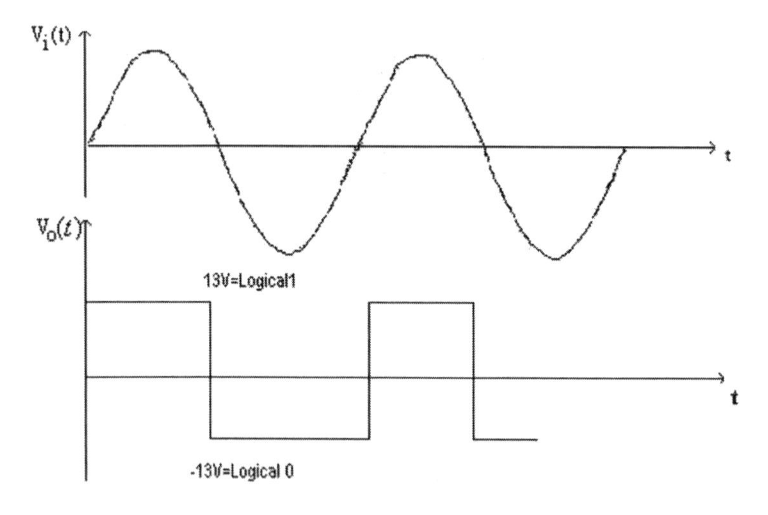

Figure 4.17 Output waveform prior to the rectifier.

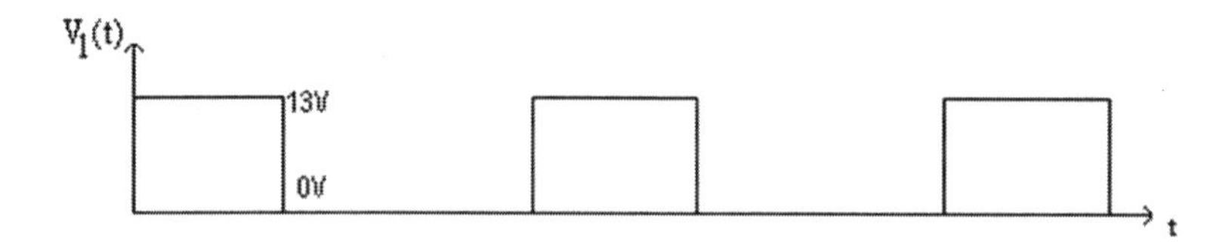

Figure 4.18 Output of the rectifier.

The reverse diode circuit connection and its corresponding waveform are shown in Figs. 4.19 and 4.20 respectively.

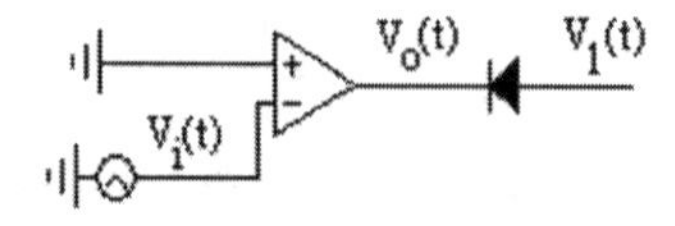

Figure 4.19 Generation of the timing waveform with reverse diode connection.

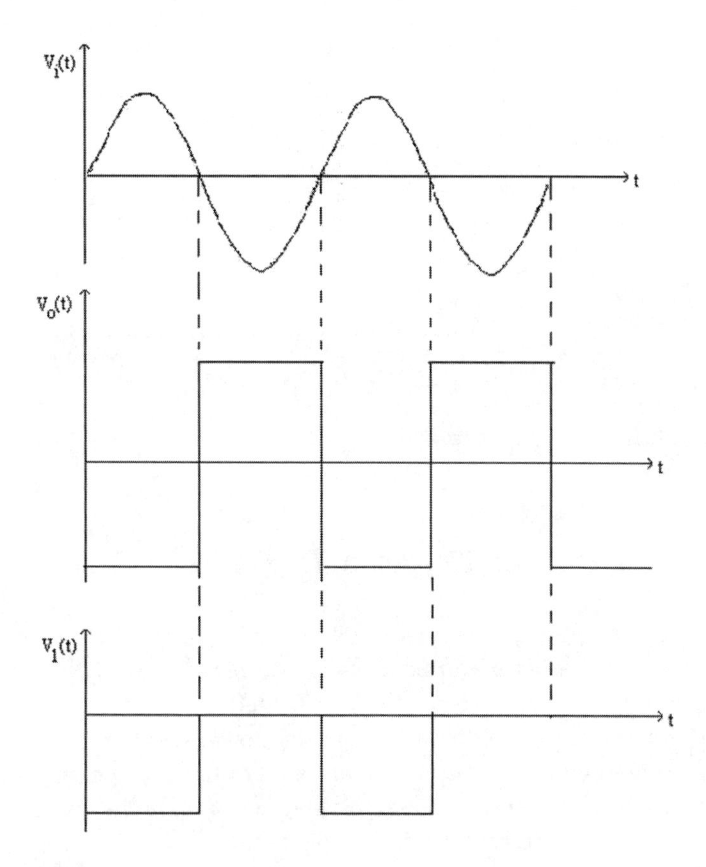

Figure 4.20 Output waveforms.

4.5.2 Generation of Asymmetric Waveform

Here sinusoidal signal is applied to the noninverting node and a constant voltage of 2 V is applied to inverting terminal. This circuit is a comparator, which compares both the inputs and produces the output signal V_o as shown in Fig. 4.22.

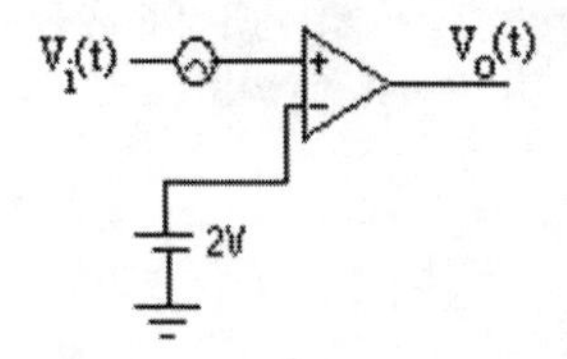

Figure 4.21 Generation of waveform without the diode.

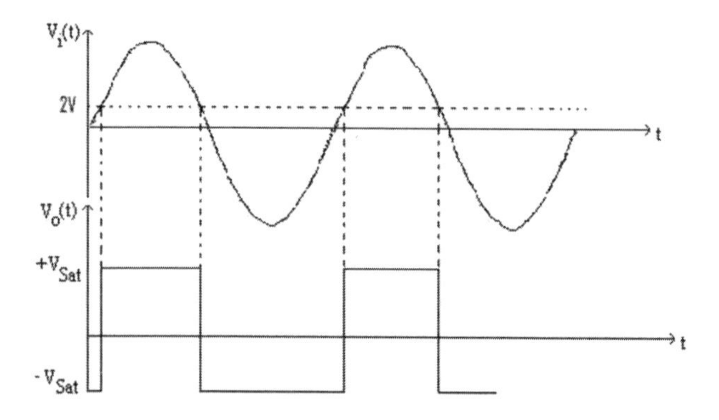

Figure 4.22 Asymmetric waveforms.

4.6 Registers and Types of Registers

Register is a sequential device, which can be used for storage of binary information or through which the binary information can be shifted either toward right or toward left. As we know a single flip-flop can store 1 bit of binary information, a suitable cascade of *n* such flip-flops can be used to store n bit of information. Register can also be considered as a memory device. Counter always passes through some specific states (not Haphazard States) with the arrival of the clock pulse. But unlike counter, the register does not pass through some specified sequence of states except in some special purpose registers.

There are different types of register or shift registers (data in register can be shifted either to the right or left, not both simultaneously).

1. serial in serial out shift register (SISO).
2. serial in parallel out shift register (SIPO).
3. parallel in parallel out shift register (PIPO).
4. parallel in serial out shift register (PISO).
5. bidirectional shift register.
6. rotate right shift register.
7. rotate left shift register.

All seven operations can be verified by using a single IC chip with mode select inputs.

4.6.1 Serial In Serial Out Scheme

The number of bits shift depends on the number of flip-flops used in the particular shift register as shown in Fig. 4.23 (Example for four flip-flops for 4 bits).

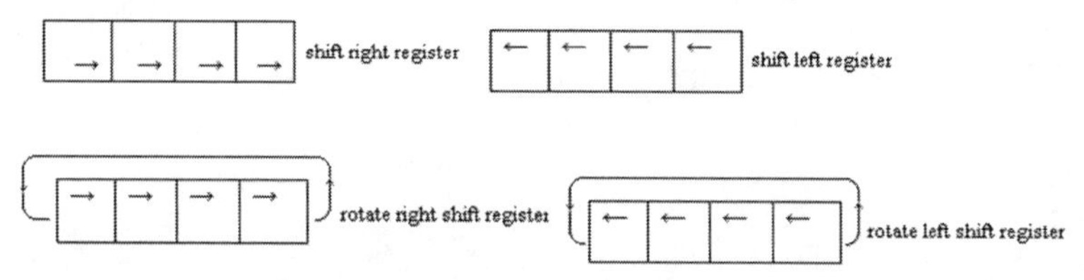

Figure 4.23 Serial in serial out scheme.

4.6.2 4 Bit Serial In Serial Out Shift Register

The shift register shown in Fig. 4.24, is designed with four D flip-flops connected in series. The external input is connected to D_0 (input of the first flip-flop) and the final output is taken at Q_3. Application of the pulse at the clock input shifts the data bits from left to right.

Initially, all the outputs of flip-flops are set to 0 by applying $Clr_i = 0$ and $Pr_i = 1$ where $i = 0, 1, 2, 3$.

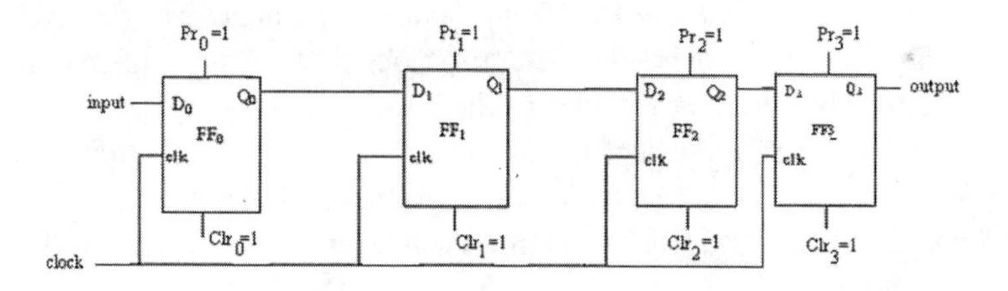

Figure 4.24 4 Bit SISO.

Table 4.1 Shifting of the input bits through the application of pulses

Clk	*SI* data at D_0	Q_0	Q_1	Q_2	Q_3	
0	1	0	0	0	0	Initial condition
1	1	1	0	0	0	0 is output
2	0	1	1	0	0	0 is output
3	1	0	1	1	0	0 is output
4	0	1	0	1	1	0 is output
5	0	0	1	0	1	1 as first serial output data from Q_3.
6	0	0	0	1	0	
7	0	0	0	0	1	
8		0	0	0	0	

To complete 4 bit serial in serial out operation we require $2 \times 4 = 8$ clock pulses. The serial in serial out is a very slow operation. Last bit is delayed by 8 clock pulse, next-to-last bit by 7 clock pulses, and so on. The shifting operation has been shown in tabular form in Table 4.1.

4.6.3 Serial In Parallel Output

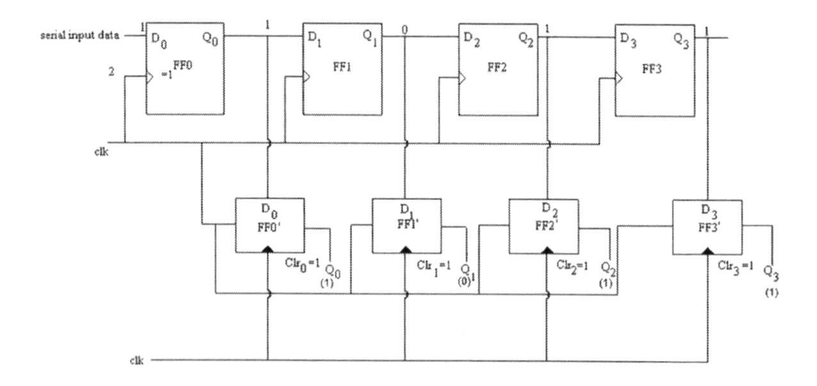

Figure 4.25 4 Bit SIPO.

In SIPO register shown in Fig. 4.25, the input data bits are shifted serially through the upper four D-flip-flops and parallel through lower four flip-flops after each pulse. After 4th clock pulse, 4 bit serial in data appears at the output of 4 flip-flops. The 4 bit serial in parallel operation is completed in (4 + 1) clock pulses. This is a much faster operation than SISO. So n bit serial in to parallel out operation is completed in $(n + 1)$ clock pulses. After the 4th clock pulse the output is available in parallel.

4.6.4 Parallel In Parallel Out Operation

This is the fastest data transmission technique as it requires only one clock pulse and the time for parallel data transfer is independent of the of the number of bits transferred in parallel as shown in Fig. 4.26. So, for 4 bit PIPO operation to be completed we require 1 clock pulse. For n bit PIPO we also require 1 clock pulse. However, communication between two computers to transfer the data (8 bits), we require eight such wires and for longer distance between them, longer wires are required. So, this scheme is not economical. All long distance data transfer is taking place through SISO (though it is time consuming, hardware cost is less).

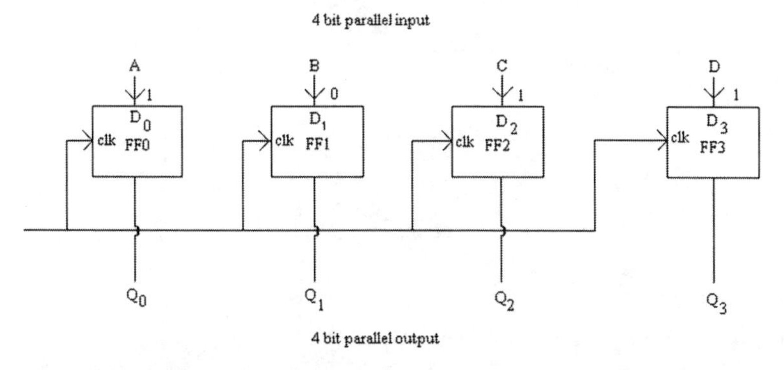

Figure 4.26 4 Bit PIPO.

The circuit shows a 4 bit PIPO shift register, where the inputs A to D are connected parallely to all the four D flip-flops. After triggering the first pulse at the pulse end we can get all the 4 bits simultaneously.

The ABCD input data have to be held constant to get the same output (by attaching switches).

The PIPO operation is shown in the table below (Table 4.2)

Table 4.2 Truth table

Clk	*A*	*B*	*C*	*D*	Q_0	Q_1	Q_2	Q_3	
0	1	0	1	1	0	0	0	0	Initially by $Pr_S = 1$ and $Clr = 0$
1	1	0	1	1	1	0	1	1	

So there is only a delay of *T* sec of the time period of 1 clock pulse (which is very fast can be considered negligible.).

4.6.5 Parallel In Serial Out

The above design, as shown in Fig. 4.27 is a 4 bit PISO shift register. The whole operation of the circuit is control by the shift/load pin. That is at a time the circuit can perform either loading data bits or shifting the data bits. The shifting operation is done with the help of the AND gates 1 to 3 and loading is through the AND gates 4 to 6. In this circuit the input bits A to D appears at output after 1 clock pulse.

LOAD/SHIFT = 0 = Parallel data in to parallel data out operator.
= 1 = Gives serial shifting of 4 bit parallel data to the right through the successive flip-flop.

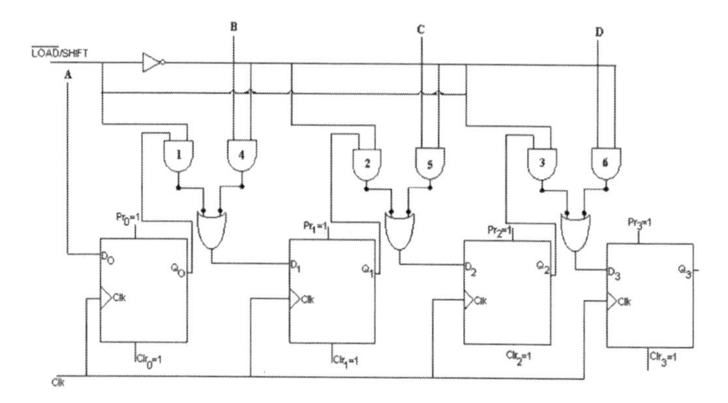

Figure 4.27 4 Bit PISO.

When clocks are applied the data are successively shifted from the output of one flip-flop through the output of the next flip-flop. This operation is described below in Table 4.3.

Table 4.3 Shifting of the input bits through the application of pulses

Sr. no.	Load/Shift	A_1	A_2	A_3	A_4	A_5	A_6	A	B	C	D	Q_0	Q_1	Q_2	Q_3
0	0	D	E	D	E	D	E	1	0	1	1	0	0	0	0
1	1	E	D	E	D	E	D	1	0	1	1	1	0	1	1
2	1	E	D	E	D	E	D	1	0	1	1	1	0	1	1
3	1	E	D	E	D	E	D	1	0	1	1	1	1	1	0
4	1	E	D	E	D	E	D	1	0	1	1	1	1	1	1
5	1	E	D	E	D	E	D	1	0	1	1	1	1	1	1

For 4 bit PISO to computer we require 1 + 4 = 1 + 4 = 5 clock pulse. For an n bit PISO, we require $(n + 1)$ clock pulses. This is a right shift operation. To convert it to left shift we have to reverse the connection toward left and serial output data is obtained from Q_0.

4.6.6 Bidirectional Shift Register

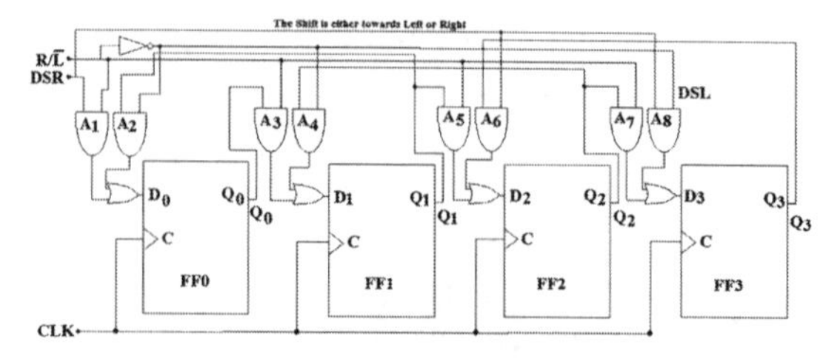

Figure 4.28 4 Bit bidirectional shift register.

The above Fig. 4.28 shows a bidirectional shift register.
Here, left/right is the control input.
DSR = Right shifted serial input data.
DSL = Left shifted serial input data.

Right shift operation occurs when left/right control input is 1. We get right shift operation from Q_0 toward Q_3 and the final right shifted serial out data will be obtained from the output of Q_3.

When left/right = 0, we get left shift operation from Q_3 toward Q_0 and the final left shifted serial data will be collected from the output of Q_0.

When left/right = 1, we get A_1, A_3, A_5, A_7 are enabled and A_2, A_4, A_6, A_8 are disabled. Under this condition, the bits applied at DSR will successively arrive (1101) at the input of each flip-flop through AND gate A_i and OR gate where $i = 1, 3, 5$, and 7 and with each clock pulse the successive serial input data at DSR input will pass toward right from the output of one flip-flop to the input of succeeding flip-flop. This gives serial right shifting of DSR.

In the left shift operation, left/right = 0, and A_2, A_4, A_6 and A_8 will be enabled. These gates help to pass the bits left from output of one flip-flop to the input of succeeding flip-flop. Since, the serial data bit appears at the output at last flip-flop after 4 Tsec, so, for an n bit bidirectional shift register the first serial data appears at the output of last flip-flop after n Tsec, where n = number of flip-flops and T = clock period.

So, this scheme can be used as a delaying mechanism. *Pr* and *Clr* should be connected to 1 to enable the flip-flops. A change in the output of flip-flop is in synchronism with clock pulse.

4.6.7 74194 IC (4 Bit Universal Shift Register)

This IC, shown in Fig. 4.29 is called 4 bit universal shift register IC because using this we can perform all types of shift register operation such as

1. 4 bit SISO operation
2. 4 bit SIPO operation
3. 4 bit PIPO operation
4. 4 bit PISO operation
5. bidirectional shift register
6. rotate right
7. rotate left

It has four synchronously operated D-type flip-flops with active high clock.

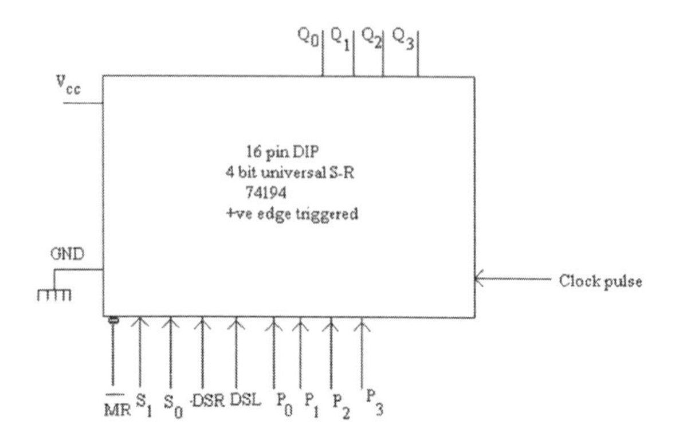

Figure 4.29 4 Bit universal shift register.

Here S_1 and S_0 are called mode controlling inputs, *MR* is the master reset input, *DSR* is the serial input data for right shift *DSL* is the serial input data for left shift, P_0, P_1, P_2, P_3 are 4 bit parallel inputs and Q_0 Q_1 Q_2 Q_3 is the 4 bit output and CP is clock pulse.

The mode select operation of universal shift register is described in the Table 4.4.

Table 4.4 Mode select table for 74194

$\overline{MR}$	S_1	S_0	*DSR*	*DSL*	P_0	P_1	P_2	P_3	Q_0	Q_1	Q_2	Q_3	
0	X	X	X	X	X	X	X	X	0	0	0	0	Output Cleared
1	0	1	1	X	X	X	X	X	*1*	P_0	P_1	P_2	Right shift operation P_3 out
1	0	1	0	X	X	X	X	X	0	P_0	P_1	P_2	Right shift operation P_3 out
1	1	0	X	1	X	X	X	X	P_1	P_2	P_3	1	Left shift Operation P_0 out
1	1	0	X	0	X	X	X	X	P_1	P_2	P_3	0	Left shift Operation P_0 out
1	1	1	X	X	P_0	P_1	P_2	P_3	P_0	P_1	P_2	P_3	Parallel Load During +ve edge
1	0	0	X	X	X	X	X	X	Q_0	Q_1	Q_2	Q_3	change With clock pulse → Output Holds.

Rotate right connection of IC 74194 and its corresponding truth table is shown in Fig. 4.30 and Table 4.5 respectively.

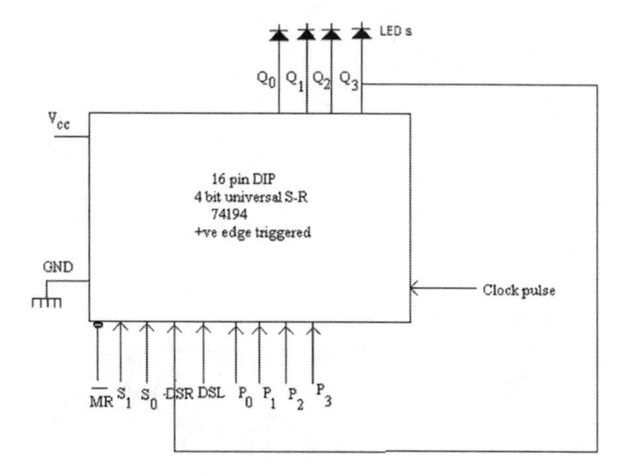

Figure 4.30 Rotate right connection of 74194.

Table 4.5 Truth table

Clk	*DSR*	Q_0	Q_1	Q_2	Q_3	S_1	S_0	R
0	1	0	1	1	1	0	1	1
1	1	1	0	1	1	0	1	1
2	1	1	1	0	1	0	1	1
3	0	1	1	1	0	0	1	1
4	1	0	1	1	1	0	1	1

4.7 Ring Counter and Johnson Counter

Shift registers can also be used as shift register counters, such as

1. ring counter.
2. Johnson counter or switch tail ring counter.

4.7.1 4 Bit Ring Counter

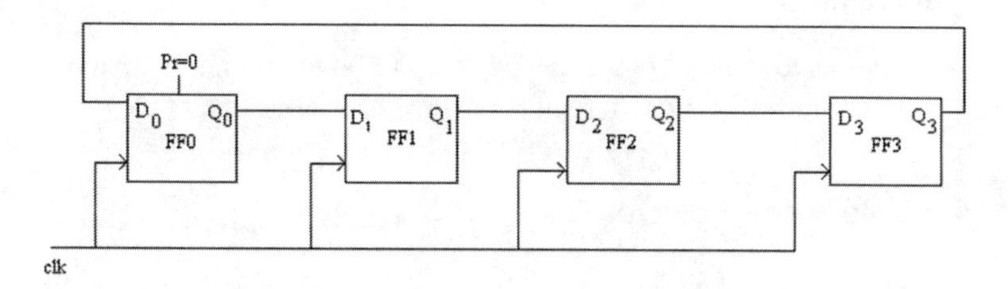

Figure 4.31 4 Bit ring counter.

In the circuit shown in Fig. 4.31, all the flip-flops connected serially and the last flip-flop's output is connected as the input of the first one. So it does not require any external inputs and the content of the flip-flops shift like a ring with the application of clock pulse. All the clock outputs cannot be set to 0 initially. So, for ring counter, initially we set only one output of flip-flop to logical one. Clock is applied simultaneously.

The corresponding truth table is presented in Table 4.6.

Table 4.6 Truth table of ring counter

Clk	D_0	Q_0	Q_1	Q_2	Q_3
0	0	1	0	0	0
1	0	0	1	0	0
2	0	0	0	1	0
3	1	0	0	0	1
4	1	1	0	0	0

To realize Mod 4 ring counter, we require four flip-flops which are uneconomical, as generally for Mod 4 counter we require two flip-flops.

Disadvantages of Ring Counter

1. Uneconomical use of flip-flops to realize a counter of definite modulus.
2. We cannot start with initial state with all zeroes (Output change cannot be detected).

Advantages

1. No decoder is required to read the count as for the ring counter only one output is 1 at a time.
2. It is quite faster in operation.

The first disadvantage of a ring counter can be overcome by using a Johnson counter or a switch tail ring counter.

4.7.2 Johnson Counter

Johnson counter can be designed by connecting the output $\bar{Q}_3$ to the input D_0 as shown in Fig. 4.32. As the feedback is taken from the tail of the cascaded connection (Q_3), it is also called switch tail counter.

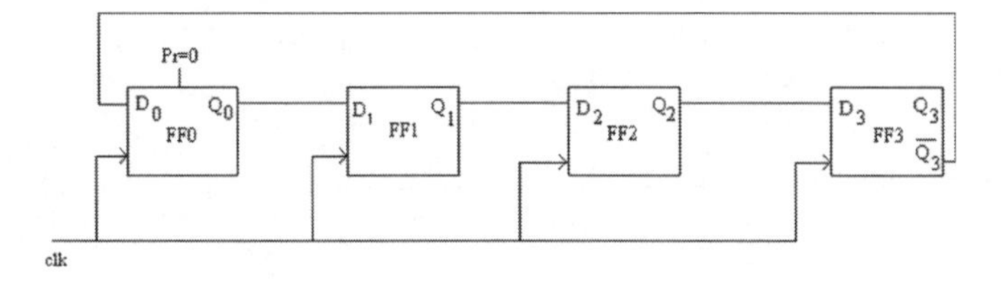

Figure 4.32 4 Bit Johnson counter.

The operation can be summarized by the truth table shown below in Table 4.7.

Table 4.7 Truth table of Johnson counter

Clk	D_0	Q_0	Q_1	Q_2	Q_3	$\overline{Q_3}$
0	1	0	0	0	0	1
1	1	1	0	0	0	1
2	1	1	1	0	0	1
3	1	1	1	1	0	1
4	0	1	1	1	1	0
5	0	0	1	1	1	0
6	0	0	0	1	1	0
7	0	0	0	0	0	0
8	1	0	0	0	0	1

4.7.2.1 Advantages

1. Decoding is easier.
2. Faster in operation.

4.8 Multiplexer as a Parallel In Serial Out Register

Multiplexer can be used as PISO converter and demultiplexer can be used as SIPO converter as shown in the following Figs. 4.33 and 4.34 respectively. The operations can be understood from the corresponding truth tables (Tables 4.8 and 4.9 respectively).

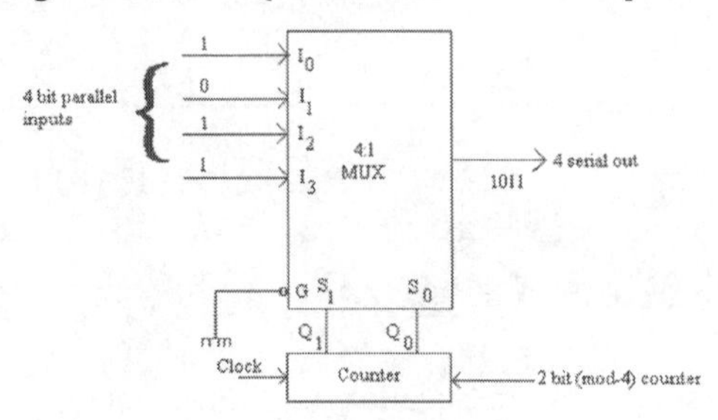

Figure 4.33 4 Bit parallel in to serial out converter using 4 : 1 multiplexer.

Table 4.8 4 Bit parallel in to serial out conversion using 4 : 1 multiplexer

Clk	Q_1	Q_0	
0	0	0	$I_0 = 1$
1	0	1	$I_1 = 0$
2	1	0	$I_2 = 1$
3	1	1	$I_3 = 1$

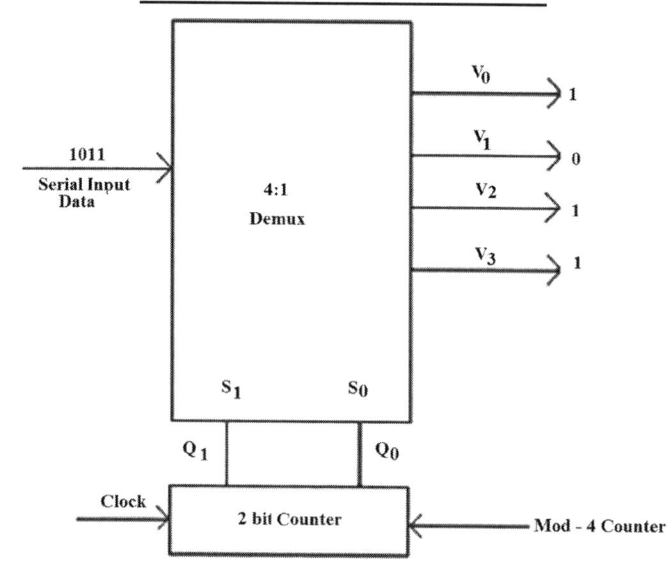

Figure 4.34 4 Bit serial in parallel out converter using 1 : 4 demultiplexer.

Table 4.9 4 Bit serial in to parallel out conversion using 1 : 4 demultiplexer

Clk	S_1 Q_1	S_0 Q_2	Serial data input	Y_0	Y_1	Y_2	Y_3
0	0	0	1	1	X	X	X
1	0	1	0	X	0	X	X
2	1	0	1	X	X	1	X
3	1	1	1	X	X	X	1

After every four clock pulse, we get the output as 1011.

4.9 Application of Register

All computers accept data parallely and output data parallely. One way to communicate between them is to draw 8 bit parallel lines.

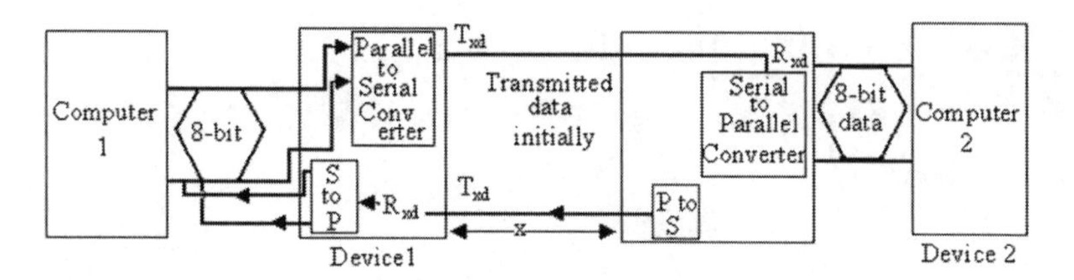

Figure 4.35 Connection between two computers.

But the parallel lines will be much costly as distance is very large, but there is single line transmission in the region "*x*". This is the essence of serial transmission. The interfacing devices are called universal asynchronous receiver transmitters (UART).

4.9.1 Advantage

Hardware complexity is less.

4.9.2 Disadvantage

These devices are also called universal synchronism asynchronous receiver transmitters (USART) and by using USART, the data transmission has much higher speed and the cost is also less.
Delayed serial transmission.

The device 1 and 2 are Intel 8251 USART.

4.9.3 Applications of Registers

1. Registers are used as a temporary data storage device in a microprocessor based system.
2. Registers can also be used as a status indicator in a microprocessor based system. The status indicator in microprocessor can perform only logical or arithmetic operation. The carry generation, parity will be reflected by the status indicator.
3. It can be used to generate time delays.

Set of registers can be used as a stack memory which are organized in a last in first out (LIFO) [(LIFO) that is the data, which is stored or inputted last can be retrieved first] manner as explained in Fig. 4.36.

4. The right shift and left shift registers are used to perform binary multiplication and division by repeated addition and shift and by repeated subtraction and shift, respectively.

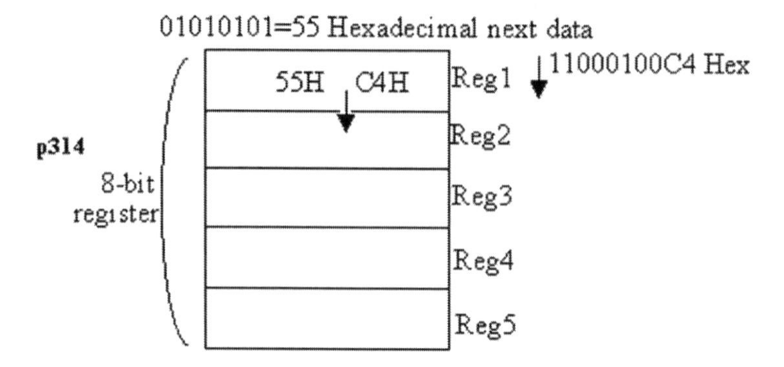

Figure 4.36 The LIFO scheme.

5. The SIPO shift registers and PISO shift registers are used for a long distance serial transmission between two remote computers.
6. Shift registers can be used to generate several shift register counters such as ring counter and Johnson counter or switch tail ring counter.
7. Shift registers can be used to realize a moving display.
8. Shift registers can be used to scan the keyboard in a keyboard encoder (in the keyboard encoder matrix).

4.10 Design of a Sequence Generator

It is a system or device which generates a prescribed sequence of bits in synchronism with clock pulses as can be seen in Fig. 4.37.

Applications of sequence generator:

1. It can be used as counter.
2. It can be used as frequency divider in a timing circuit.
3. It can be used for code generation.

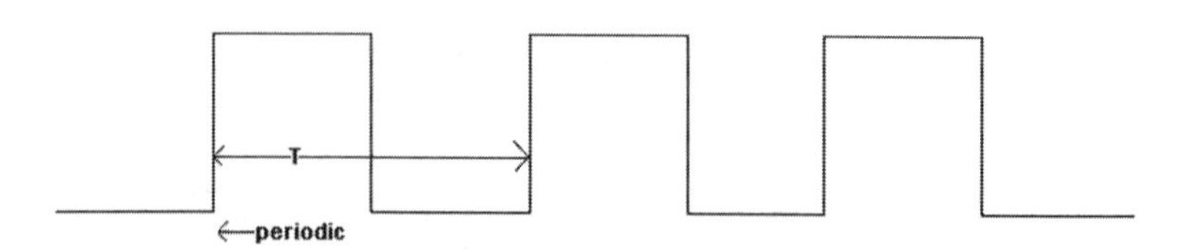

Figure 4.37 Prescribed sequence of bits like a periodic waveform.

Let us consider, a prescribed sequence of 4 bits—1101, 1011, 0111, 1110, and we get each sequence at the output of different flip-flops (Fig. 4.38).

Let us represent 1101 by several pulses.

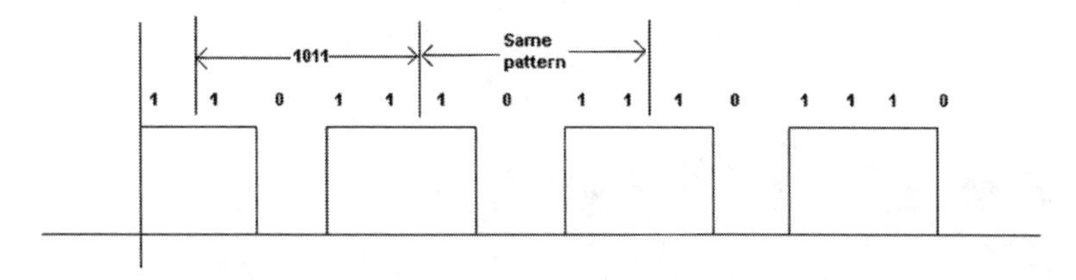

Figure 4.38 Pulse representation of 1101 sequence.

So, at the output of one flip-flop this pattern is repeated after regular interval of time that is after 4 *T*sec. The 1101 and 1011 are the same sequence, but one is delayed with respect to the other by a time *T*sec. Again 0111 is another pattern and again 1110 is the last pattern. So, for same sequence, we can have four patterns which are the outputs of a sequence generator. The basic structure of a sequence generator using D-flip-flop is shown in Fig. 4.39.

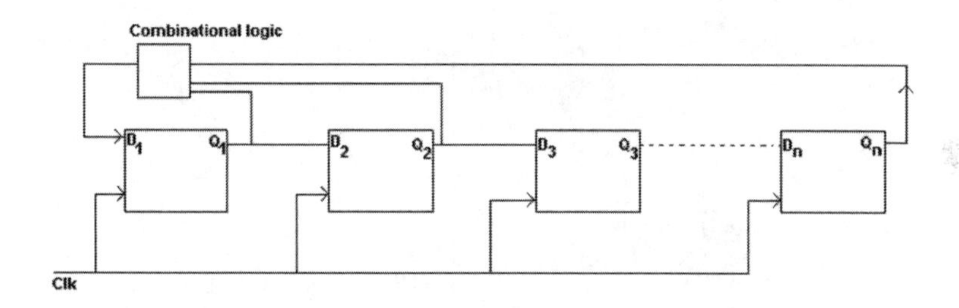

Figure 4.39 Sequence generator using D–flip-flops.

Here, $D_1 = f(Q_1, Q_2, ..., Q_n)$

If a prescribed sequence is of length S then such sequence can be generated from minimum number of N flip-flops using the relationship.

$S \le 2^N - 1$, where N is the minimum number of flip-flop required

Let, for a sequence 1101 that is $S = 4$.

$4 \le 2^N - 1$

If we take $N = 2$ then $4 \le 3$ (not satisfied)

If $N = 3$ then $\le 2^3 - 1 \rightarrow 4 \le 7$.

So, minimum number of flip-flops required is 3.

Table 4.10 Sequence generation using D–flip-flops

Q_1	Q_2	Q_3	D_1	Decimal
1	1	0	1	6
1	1	1	0	7
0	1	1	1	3
1	0	1	1	5
1	1	0	1	6
1	1	1	0	7
0	1	1	1	3
1	0	1	1	5
1	1	0	1	6
1	1	1	0	7
0	1	1	1	3
1	0	1	1	5

After every four clock pulse, we get the same sequence.

$$D_1 = f(Q_1, Q_2)$$

From the state transition table (Table 4.10) for a sequence generator $D_1 = 0$ for $Q_1\,Q_2\,Q_3 = 111$ and

$$D_1 = 1 \quad \text{when input is 3, 5, 6.}$$
$$D_1 = f(Q_1, Q_2, Q_3)$$
$$= \prod M\ (7) + d(0, 1, 2, 3, 4)$$

Thus the K-map representation of D_1 is given in Fig. 4.40.

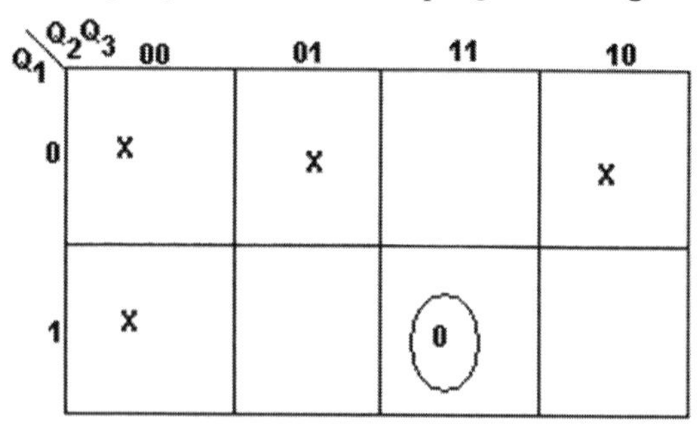

Figure 4.40 Karnaugh map representation of D_1.

$$D_1 = \overline{Q_1} + \overline{Q_2} + \overline{Q_3}$$
$$= \overline{Q_1\,Q_2\,Q_3}$$

4.11 Error Detection and Correction Codes

Methods— of Error Detection and Correction:

1. parity check error detection scheme.
2. block parity error detection and correction code (BPED and CC).

3. Hamming code.
4. checksum scheme.
5. cyclic redundancy code.

4.11.1 Parity Check Error Detection Scheme

Through this scheme one extra bit is added to the four bits of information. This is one bit error detections scheme using horizontal parity.

The scheme is described pictorially in Fig. 4.41.

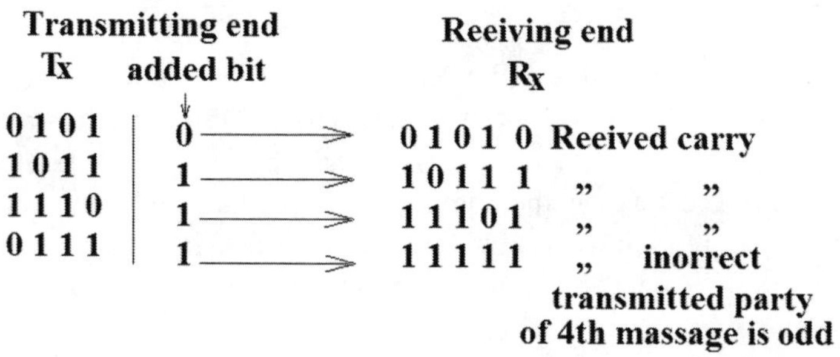

Transmitting end | Reeiving end
Tx added bit | Rx

0101 | 0 → 01010 Reeived carry
1011 | 1 → 10111 " "
1110 | 1 → 11101 " "
0111 | 1 → 11111 " inorrect
transmitted party
of 4th massage is odd

Figure 4.41 Parity check error detection scheme.

If T_x and R_x parity are not identical then R_x informs T_x to transmit to same message again. However, if the received message has two bit error then the parity error cannot be detected. So, multiple errors cannot be detected. The immunity of digital system is greater than analog system, hence, the probability of error is small that is of 10^{-6}. This particular method is effectively used and thus is used in the memory testing in computer.

4.11.2 Block Parity

This uses horizontal and vertical parity and can detect error as well as select bit error. In BPED (Block Parity Error Detection) and CC (Correction Code) these messages are transmitted as a matrix of rows and columns, and to each row and column one extra bit is added so, that all the rows and columns must have either even parity or odd parity. This scheme is briefly represented in the following Fig. 4.42.

Transmitting end (T_X)		Receiving end (R_X)	
0 1 0 1 1	0	0 1 0 1 1 0	← row parity ok
0 1 1 0 1	0	0 1 1 0 1 0	row parity ok
0 1 1 0 0	1	0 1 1 0 0 1	row parity ok
1 1 1 1 0	1	1 1 1 0 0 1	row parity not ok
0 1 1 1 0	1	0 0 1 0 1 1	Horizontal parity ok
0 0 1 0 1	1		

Figure 4.42 Block parity scheme.

4.11.3 Hamming Code

This is a distance 3 code and it can detect and correct one bit error. The minimum distance between two successive Hamming codes differ at least by 3 bit and hence, it is called distance 3 code. Here $d_{min} = 2t + 1$, where t is the number of errors that can be corrected and d_{min} is the minimum number of bits by which two successive numbers (when encoded into Hamming code) differ.

Example for

if $d_{min} = 3$

$$3 = 2t + 1$$

$$2t = 2$$

$$t = 1$$

Like that if $d_{min} = 7$ then $t = 3$

if $d_{min} = 5$, then $t = 2$, and so on

In Hamming code, K number of information bit is encoded into an n bit code (when $n > k$) and in each n bit Hamming code there are $(n-k) = m$ number of parity bits.

If $M_n \ldots M_2M_1 = n$ bit Hamming code

The K parity check bits in an n bit Hamming code will occupy the positions 1, 2, 4, 8, 16 ... upto 2^{K-1}

$M_8\ M_7\ M_6\ M_5\ M_4\ M_3\ M_2\ M_1$ Check sum

The parity bits depend:

$C_4\ I_4\ I_3\ I_2\ C_3\ I_1\ C_2\ C_1$ — parity check bit C_1 upon the number of information bit.

↓

We can encode K information bits into n bit Hamming code using either even parity or odd parity. If we use even parity to determine all C_1, C_2, C_3, C_4, ... and so on. parity should be unique.

If we use even parity in Hamming code then to determine C_1 we must have

$$M_1 \oplus M_3 \oplus M_5 \oplus M_7 \oplus M_9 \oplus M_{11} \oplus \ldots = 0$$

and $M_1 \oplus M_3 \oplus M_5 \oplus M_7 \oplus M_9 \oplus M_{11} \oplus \ldots = 1$ for odd parity.

M_1, M_3, M_5 are inserted in the information bits and

$$C_1 \oplus I_1 \oplus I_2 \oplus I_4 \oplus I_5 = 0$$

So, if I_1, I_2, I_4, I_5 are known then we can find out C_1. Like that C_2, C_3 and C_4 can be calculated as

$$C_2 = (M_2 \oplus M_3) \oplus (M_6 \oplus M_7) \oplus (M_{10} \oplus M_{11}) \oplus \ldots = 0$$
$$C_3 = (M_4 \oplus M_5 \oplus M_6 \oplus M_7) \oplus (M_{12} \oplus M_{13} \oplus M_{14} \oplus M_{15}) \oplus \ldots = 0$$
$$C_4 = (M_8 \oplus M_9 \oplus M_{10} \oplus M_{11} \oplus M_{12} \oplus M_{13} \oplus M_{14} \oplus M_{15}) \oplus (M_{24} \oplus \ldots) = 0$$

Example: Encode a string 10101 into Hamming code having odd parity.

$$\begin{array}{ccccccccc} M_9 & M_8 & M_7 & M_6 & M_5 & M_4 & M_3 & M_2 & M_1 \\ \downarrow & \downarrow & \downarrow & \downarrow & \downarrow & \downarrow & \downarrow & \downarrow & \downarrow \\ I_5 & C_4 & I_4 & I_3 & I_2 & C_3 & I_1 & C_2 & C_1 \\ 1 & C_4 & 0 & 1 & 0 & C_3 & 1 & C_2 & C_1 \end{array} \qquad \begin{array}{ccccc} I_5 & I_4 & I_3 & I_2 & I_1 \\ 1 & 0 & 1 & 0 & 1 \end{array}$$

To find C_1

$$M_1 \oplus M_3 \oplus M_5 \oplus M_7 \oplus M_9 = 1 \text{ (condition)}$$
$$C_1 \oplus I_1 \oplus I_2 \oplus I_4 \oplus I_5 = 1$$
$$C_1 \oplus 1 \oplus 0 \oplus 0 \oplus 1 = 1$$
$$C_1 = 1$$

To find C_2

$$M_2 \oplus M_3 \oplus M_6 \oplus M_7 = 1$$
$$C_2 \oplus I_1 \oplus I_3 \oplus I_4 = 1$$
$$C_2 \oplus 1 \oplus 1 \oplus 0 = 1$$
$$C_2 = 1$$

To find C_3

$$(M_4 \oplus M_5 \oplus M_6 \oplus M_7) = 1$$
$$C_3 \oplus I_2 \oplus I_3 \oplus I_4 = 1$$
$$C_3 \oplus 0 \oplus 1 \oplus 0 = 0$$
$$C_3 = 0$$

To determine C_4

$$M_8 \oplus M_9 = 1$$
$$C_4 \oplus I_5 = 1$$
$$C_4 \oplus 1 = 1$$
$$C_4 = 0$$

Now inserting all the bits in place of M_1, M_2, M_4, M_8 we have

$M_9\ M_8\ M_7\ M_6\ M_5\ M_4\ M_3\ M_2\ M_1$
1 0 0 1 0 0 1 1 1 = 9 bit Hamming code

So, 5 bit message is encoded into 9 bit Hamming code (9–5) bits = 4 parity check bits checksum.

T_X 9 bit Hamming code $\qquad\qquad$ R_X

$\qquad\qquad\qquad\qquad\qquad\qquad$ $M_9\ M_8\ M_7\ M_6\ M_5\ M_4\ M_3\ M_2\ M_1$

100100111 $\rightarrow$ 0 0 0 1 0 0 1 1 1

$\uparrow$

bit in error. We take here that only 1 bit error occurs at the receiving end (MSB)

So, to detect the error in the receiver also, we have to test the parity error check for C_1, C_2, C_3, C_4 as is done in T_X to find C_1, C_2, C_3, C_4.

To check for C_1 at receiver we must have:

$$\bar{M}_1 \oplus \bar{M}_3 \oplus \bar{M}_5 \oplus \bar{M}_7 \oplus M_9 = 1$$
$$\text{LHS} = 1 \oplus 1 \oplus 0 \oplus 0 \oplus 0 = 0 \neq 1$$

Since there is an error, take $C_1 = 1$ as $\bar{M}_1$ is the position for C_1.

Rule: If there is an error take that value of C as 1; otherwise take it as 0.

To check for C_2

$$M_2 \oplus M_3 \oplus M_6 \oplus M_7 = 1$$
$$\text{L.H.S} = 1 \oplus 1 \oplus 1 \oplus 0 = 1$$

as there is no error for C_2 so denote $C_2 = 0$

To check for C_3

$$\bar{M}_4 \oplus \bar{M}_5 \oplus \bar{M}_6 \oplus \bar{M}_7 = 1$$

$$0 \oplus 0 \oplus 1 \oplus 0 = 1$$

As there is no error for C_3 so, $C_3 = 0$

To check for C_4

$$\bar{M}_8 \oplus \bar{M}_9 = 1$$

$$\text{LHS} = 0 \oplus 0 \neq 1$$

So, error in C_4 and $C_4 = 1$

Now $C_4 = 1$, $C_3 = 0$, $C_2 = 0$ and $C_1 = 1$

$C_4\ C_3\ C_2\ C_1$

1 0 0 1 = 9 decimal equivalent $\Rightarrow$ that 9th bit in the receiver is incorrectly received. So to correct error change 9th bit from 0 to 1 or from 1 to 0 whatever it may be.

4.11.4 Checksum Method

Let we want to transmit the four messages as

Message	T_X
M_1	01010
M_2	10101
M_3	01110
M_4	10001

Find checksum of messages $= M_1 \oplus M_2 \oplus M_3 \oplus M_4 \oplus \ldots = 0\,0\,0\,0\,0$

Along with each transmitted message, the checksum of all the messages are also transmitted.

R_X

$\rightarrow$ 01010

$\rightarrow$ 11101 At the receiver end also the receiver checksum is calculated.

$\rightarrow$ 01010

$\rightarrow$ 00001

11100 $\leftarrow$ receiver checksum

$\uparrow$

The receiver checksum is now compared with the transmitted checksum. If received checksum = transmitted checksum then there is no error. So, checksum method can be used for multiple error detection but it does not have the capability of correcting the error. This is used where the probability of error is very small. [It is used to detect the condition of EPROM (whether blank or filled)].

4.11.5 Cyclic Redundancy Code

In cyclic redundancy code (CRC), the transmitted bit sequence is:

T_X	R_X
Series data + CRC data	data + CRC data

The transmitted CRC is compared with the R_X CRC and if they match then there are no errors. If they do not match then error is there.

4.12 Display Devices

Now we discuss about display devices which are adjacent parts of digital electronics.

There are two types of display devices.

1. light emitting diode (LED)/laser injection diode (LID)
2. liquid crystal display (LCD)

4.12.1 Light Emitting Diode (LED)

If we forward bias the diode, light is emitted. In gallium arsenide some energy is required to create electron hole pairs.

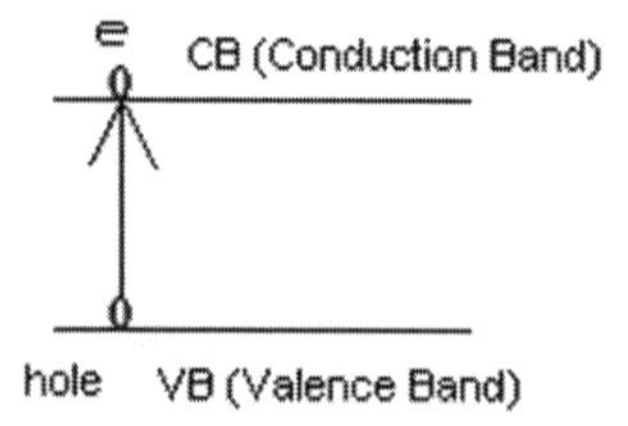

Figure 4.43 Energy band.

At very low temperatures semiconductor behaves as insulator and all the electrons occupy the valence band. But if we supply any external energy, the "*e*" is raised to conduction band as shown in Fig. 4.43. This vacancy of "*e*" is called a hole. So, to create electron hole pair some external energy has to be supplied. If electron and hole recombine with each other, energy will be released or liberated. Any energy liberated will not give light unless it falls in the visible range. In Ge or Si electron hole pair recombines through traps and hence, the liberated energy goes into the crystal as that and no light is emitted. But in gallium arsenide and some other semiconductor devices, the electron and hole can directly recombine and the liberated energy in a *p*–*n* junction diode is in the infra red range which is invisible (higher wavelength, lower frequency side). But by using suitable coating or coloring of the junction (*p*–*n*) the infrared is converted into visible light. When LED is forward biased as can be seen in the circuit representation of LED in Fig. 4.44, some electrons move from VB to CB and after that they recombine to emit light.

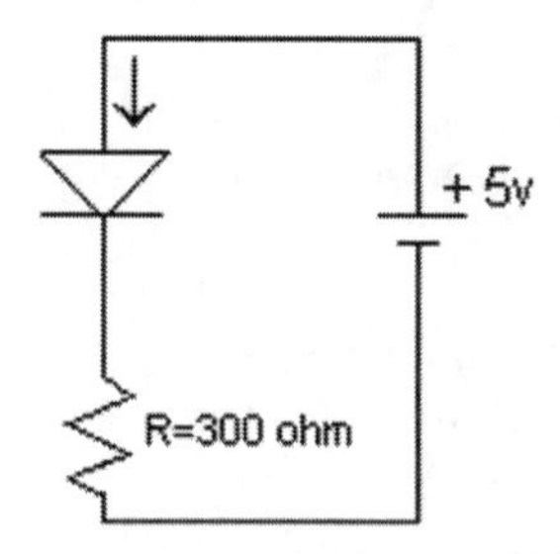

Figure 4.44 Forward biased diode.

4.12.1.1 Advantages of LED

1. Can be operated with low voltage of the order of 1 to 2 V.
2. Fast on–off switch.
3. Wider operating temperature.
4. TTL and CMOS compatible.
5. It is visible in dark.
6. It is available in wide variety.

4.12.1.2 Disadvantages of LED

1. consumes more power
2. draws large current of the order of mA
3. It cannot be given variety of shape.
4. Difficult to fabricate

This emission efficiency of LED depends upon the injection current and lowering of temperature. If the emitted light is coherent (light wave has the same state of polarization and phase) in source *p–n* junction diode then it will be called injection laser diode.

4.12.2 Laser

It will be really an incomplete discussion if we do not explain laser here. Hence, a brief account of laser is given here to satisfy the readers. But for a complete look it is advised to go through a specific to laser (Light wave amplification by stimulated emission of radiation). The resistance R is required to protect the diode from excessive current. So, R is called the limiting resistance which saves diode from over current and overheating.

4.12.3 Liquid Crystal Display (LCD)

The LCD is made up of liquid crystal filler (10 to 12 microns or 0.005 inch) and is sandwiched between two glass plates. The two glass plates are grooved and the grooves in the two glass plates are perpendicular to each other. When liquid crystal filler is pressed it takes a rod shape. Spiral staircase configuration and this special nature of liquid crystal rotate light or cause polarization effect.

The LC technology is possible due to the following facts:

1. The ability of the crystal to polarize light (as crystal is a piezoelectric substance, it obstructs vibration along certain direction and allow vibration in certain direction).
2. The orientation of the crystal can be changed by applying electrostatic field.

The LCDs are of two types:

1. dynamic scattering type.
2. field effect type.

4.12.3.1 Dynamic scattering type

In this type the LC molecule scrambles whenever electric field is applied (the molecules between two plates come closer). This produces etched glass looking light characters against dark background.

4.12.3.2 Field effect type

It was polarization to absorb light and it produces dark character against silver gray background.

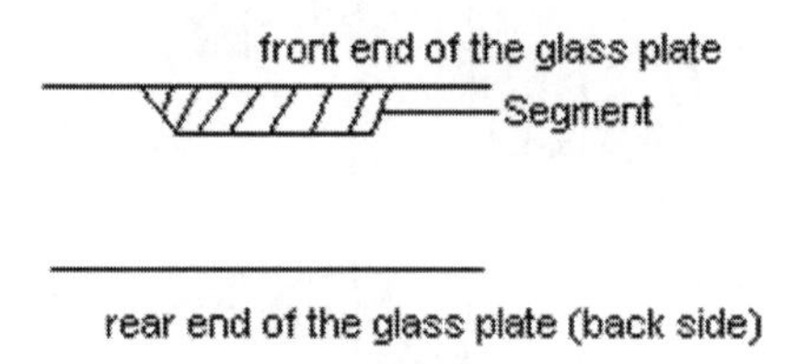

Figure 4.45 Structure of field effect type LCD devices.

The rear end of the glass plate is coated with transparent conducting material. This rear end of the glass plate is called back plane. The front end of the glass plate is called segment after giving the particular shape of the character is also coated with same transparent conducting film. When a voltage is applied between a segment and the back plane, an electric field is created in the region under the segment. This electric field changes the transmission of light through the region under the segment film. To activate a LCD display, it should be driven by segment driven signal which is 2–3 volt square wave in the frequency range 30 to 150 Hz. When a segment is not activated that is off it reflects light and hence invisible against background (the segment is activated or not activated depends on the decoder driver code). When a segment is activated it does not reflect light (but absorbs light) and appears dark and visible against background. The TTL is recommended for LED but not for LCD because when TTL output is low it may damage the liquid display device. The reason for this is that LCDs rapidly and irreversibly deteriorate if a steady DC voltage of about more than 50 mV is applied between the segment and back plane. So, for LCD we have to drive it via CMOS logic as low state of CMOS is less than 50 mV.

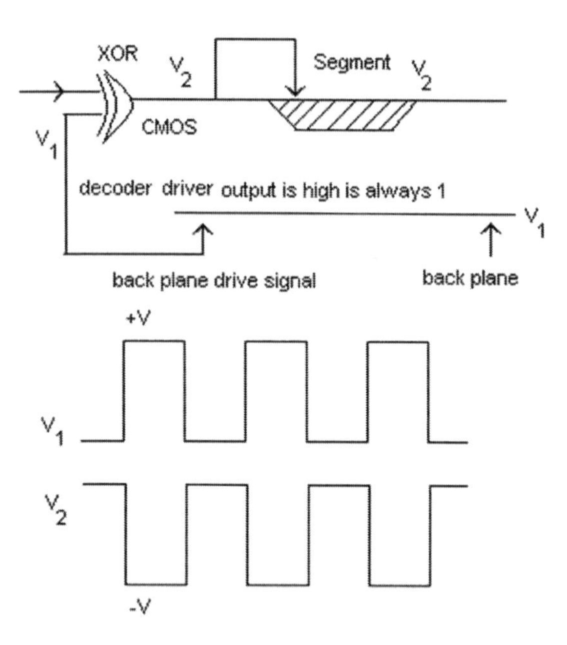

Figure 4.46a Operation of field effect type LCD.

In Fig. 4.46a since there is a voltage difference between V_1 and V_2 electric field is not equal to 0 and then segment will be activated. To prevent DC buildup on the segments, the segment drive signals for LCD must be a square with a frequency of 30–150 Hz.

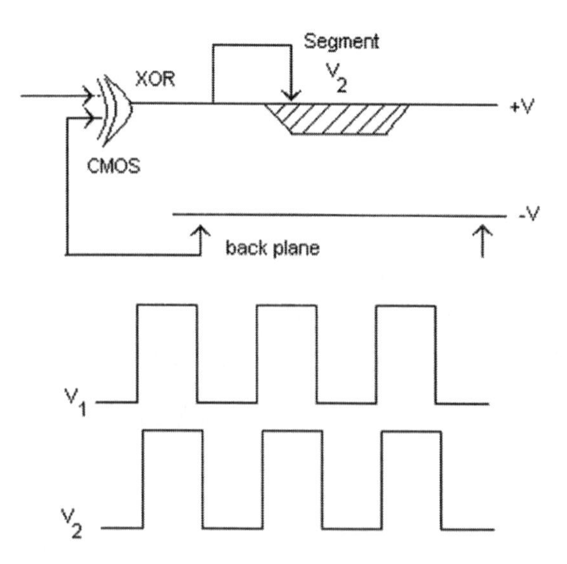

Figure 4.46b Operation of field effect type LCD.

Here, in the Fig. 4.46b, V_1 and V_2 are same and in phase so no electric field and thus the segment will be off not activated.

4.12.3.3 Advantages of LCD Over LED

1. It consumes very less power (CMOS). The MOS can be operated with very small voltage, less power input.
2. LCDs are thinner and lighter.
3. It is cheaper.
4. It can be given variety of shapes as the segment region is coated with the desirable character shape (even alphabets).
5. It is easy to fabricate.
6. It draws a very small current of the order of nano ampere.
7. It does not suffer from radiation hazards.

4.12.3.4 Disadvantages of LCD

1. It is not visible in dark. But LED is visible in dark since it emits light.
2. TTL is not recommended for LCD.
3. Smaller operating temperature.
4. Resolution of LCD display is not very good.

Problems

1. Which kind of circuit can be used to change data from special code to temporal code?
2. How many states does a ring counter consisting of five flip-flops have?
3. In which code does the successive numbers differ from their preceding number by single bit?
4. Shifting a register content to left by one bit position is equivalent to which mathematical operation?
5. The group of bits 11001 is serially shifted (right-most bit first) into a 5-bit parallel output shift register with an initial state 01110. After three clock pulses, the register contains ______.
6. Assume that a 4-bit serial in/serial out shift register is initially clear. We wish to store the nibble 1100. What will be the 4-bit pattern after the second clock pulse? (Right-most bit first.)
7. Which error detection method uses one's complement arithmetic?

8. Which error detection method consists of just one redundant bit per data unit?
9. In cyclic redundancy checking, what is CRC?
10. In cyclic redundancy checking, the divisor is _____ the CRC.
11. To guarantee the detection of up to 5 errors in all cases, the minimum Hamming distance in a block code must be _____.
12. In a linear block code, the _____ of any two valid codewords creates another valid codeword.
13. What is the role of the variable resistor and capacitor?
14. The 555 timer has a DIL layout. What does DIL mean?
15. How many types of multivibrator circuits are there?
16. Define Monostable Multivibrator.
17. What is the practical use of astable multivibrator?
18. Define Register. How many types are there?
19. List the applications of 555 timer in monostable mode of operation.
20. List the applications of 555 timer in Astable mode of operation.
21. Mention some applications of 555 timer.

Chapter 5

Logic Family

5.1 Introduction

In less than four decades the advances in microelectronics shows the rapid growth of the digital integrated circuit (IC) technology from small scale integration (SSI) through medium scale integration (MSI), large scale integration (LSI), very large scale integration (VLSI) to ultra large scale integration (ULSI). Now the technology is tending toward giant scale integration (GSI), where millions of gate equivalent circuits can be integrated on a single chip. The reliability of the use of ICs has improved because of reduction of power consumption, cost, and overall size of the digital systems. The only drawback of the ICs is that they cannot handle very large voltage and electrical devices like transformers, inductors, and so on that cannot be implemented on chips. The ICs are fabricated using different technologies like RTL, TTL, ECL, CMOS, and so on.

Foundation of Digital Electronics and Logic Design
Subir Kumar Sarkar, Asish Kumar De, and Souvik Sarkar

ISBN 978-981-4364-58-4 (Hardcover), 978-981-4364-59-1 (eBook)
www.panstanford.com

5.2 Logic Parameters

The important parameters by which one can choose a digital IC (combinational and sequential) are as follows:

1. propagation delay
2. power dissipation
3. speed power product
4. noise margin
5. fan-in and fan-out

5.2.1 Propagation Delay

The difference in interval of obtaining an output in response to an input of a logic gate is known as propagation delay (Fig. 5.1). Speed of a gate is determined by the propagation delay t_{pd}. Smaller is the propagation delay (t_{pd}) greater is the speed. High speed and low speed gate can be chosen on the basis of t_{pd}.
There are two types of propagation delay:

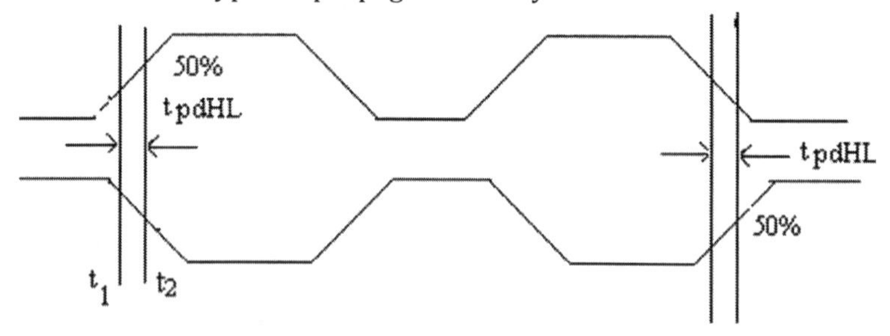

Figure 5.1 The diagram of propagation delay.

t_{pdHL} → This is defined as the time interval between the reference point of input and output signal (of logic gate) when the output changes from high to low state reference point is taken at time in both input and output signal at which input or output is 50% of its maximum value.

Similarly when output changes from low to high state again there is a propagation delay.

t_{pdHL} → This is defined as the time interval between the reference point of input and output signal (of logic gate) when the output changes from low to high state.

In general $t_{pdHL} \neq t_{pdLH}$.

If $t_{pdHL} = t_{pdLH}$ then propagation delay can either of the two t_{pdLH} or t_{pdHL}.

If $t_{pdHL} \neq t_{pdLH,}$ then propagation delay $(t_{pd}) = t_{pdHL}$ if $t_{pdHL} > t_{pdLH} = t_{pdLH}$ if $t_{pdLH} > t_{pdHL}$.

5.2.2 Power Dissipation (Pd)

Power dissipation (Pd) of a logic gate is defined as the power required by the gate to operate with 50% duty cycle at a specified frequency and the Pd can be determined as the product of supply voltage V_{CC} and average supply current I_{CC}.

So, power dissipation = $V_{CC}\, I_{CC}$.

5.2.3 Average Supply Current

Here I_{CCL} is the supply current of a gate when its output is low and I_{CCH} is the supply current of a gate when its output is high.

$\therefore I_{CCL} > I_{CCH}$ (Fig. 5.2) (as the differences in potential matters)

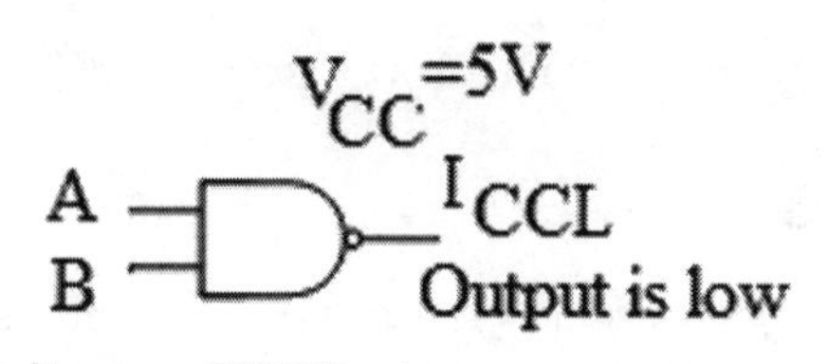

Figure 5.2 The diagram of NAND gate.

So, the average supply current (I_{CC})

$$= \frac{I_{CCL} + I_{CCH}}{2}$$

For example, for digital circuits using 50 gates and each having $I_{CC} = 2$ mA and supply voltage = 5 V.

Then the power dissipation (Pd) = $(V_{CC} \times 50 \times I_{CC})$

$$= (5 \times 50 \times 2 \times 10^{-3})$$
$$= 0.5 \text{ watt}$$

Total I_{CC} for 50 gates $= 50 \times 2$ mA $= 100$ mA.

If the digital circuits need current for 2 hr then average ampere hour of the battery $(100 \times 10^{-3} \times 2) = 0.2$ Ah. Unless, the battery has got this much of ampere hour that battery cannot supply to a digital circuit (unit is Joules/sec or watt).

5.2.4 Speed Power Product (SPP)

Speed power product of a logic gate is specified by the manufacturer as the product of propagation delay and power dissipation at a particular frequency (f_r). It is specified from a particular frequency, if f_r is changed.

$$\text{So SPP} = (t_{pd} \times \text{Pd})f_r = f_c$$

When we want to compare two gates

$$(\text{SPP})_{\text{Gate 1}} = K_1 \text{ at } f = f_c$$
$$(\text{SPP})_{\text{Gate 2}} = K_2 \text{ at } f = f_c \; K_1 > K_2$$

So, 2nd gate is desired as it has less t_{pd} and less Pd.

5.2.5 Fan-In and Fan-Out

Fan-in is defined as the number of independent inputs that the gate is designed to handle without hampering its normal operation.

For example: A three input NAND gate has got a fan-in of 3.

A 13 input NAND gate has got a fan-in of 13.

Fan-out also known as the loading factor of a logic gate is defined as the number of loads that the output of a gate can drive.

5.2.6 Noise Margin

In digital circuits, the circuit's ability to tolerate noise voltages at its inputs is called noise immunity. A quantitative measure of noise immunity is called noise margin.

Noise margins are of two types:

i. DC noise margin
ii. AC noise margin

The DC noise margin of a logic gate is a measure of its noise immunity that is its ability to withstand voltage fluctuations at the input of a logic gate without deviating from the specified output. Common sources of noise are variations of the DC supply voltage, ground noise, magnetically coupled voltages from adjacent lines and radiated signals.

Let for a logic gate

V_{IL} = low level input voltage
V_{IH} = high level input voltage
V_{OL} = low level output voltage
V_{OH} = high level output voltage

For example, if $V_{IL\,(max)} = 0.5$ V then input voltage ≤ 0.5 V represents logic 0 input and if $V_{IH\,(min)} = 4$ V then input voltage ≥ 4 V represents logic 1 input. So, it operates with discrete set of voltages.

For TTL these parameters are same, but these may vary with different logic family like CMOS.

Noise margin can be divided into two different categories

(i) Low state noise margin
(ii) High state noise margin

Low state noise margin = $V_{IL\,(max)} - V_{OL\,(max)}$
High state noise margin = $V_{OH\,(min)} - V_{IH\,(min)}$

Low state noise margin is the difference between the largest possible low output and the maximum input voltage. High state noise margin is the difference between the lowest possible high output and the minimum input voltage required.

5.3 Resistor Transistor Logic (RTL)

The NOR gate is the basic circuit of RTL logic (Fig. 5.3). The above circuit is a two input NOR gate. Here, each input is applied through one resistor (400Ω) and one transistor. For operation of this circuit two logical levels are selected as 5 V (logic 1) and 0.2 V (logic 0). If any one input (either X or Y) or both the inputs are high then the output goes low. But if both the inputs are low then both the transistors are cut off. Hence, the output approaches to V_{CC}, which represents logic 1. The propagation delay of the RTL is 25 ns and the power dissipation is about 12 mW.

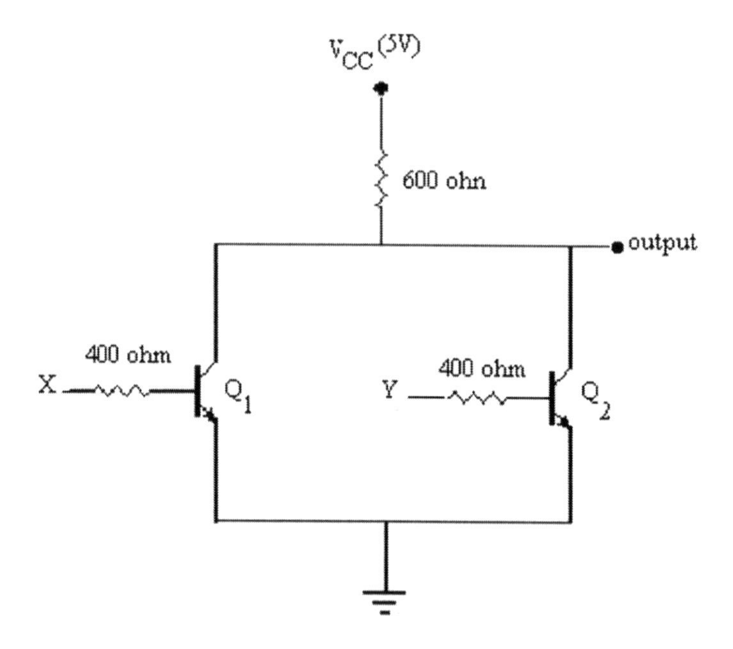

Figure 5.3 Two input NOR.

5.4 Diode Transistor Logic (DTL)

This above circuit functions as three input NAND gate (Fig. 5.4).

Let, anyone input is low ($V_A \approx 0.2$ V) then the diode D_A will be forward biased and voltage at P will be

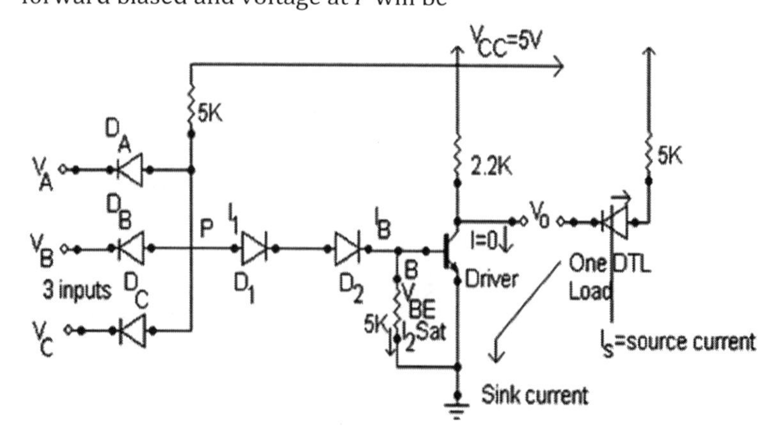

Figure 5.4 Three input DTL.

$$V_P = V_{DA} + V_A = 0.7 + 0.2 = 0.9 \text{ V}$$

To saturate transistor Q we require a min voltage at

$$\begin{aligned} P &= V_{D1} + V_{D2} + V_{BE\,Sat} \\ &= 0.7 + 0.7 + 0.8 \\ &= 2.2 \text{ V} \end{aligned}$$

But as $V_P = 0.9$ V (which is less than 2.2 V). So, transistor Q will be at cut off and the output voltage $V_0 \approx 5$ V (logical 1).

When all the three inputs are high that is logical 1 then all the input diodes D_A, D_B, D_C are reversed biased. Then voltage at P will rise toward $V_P = V_{CC} = 5$ V but the voltage at P will clamp at 2.2 V and transistor Q will saturate. Thus, when all inputs are high, output voltage rises toward $V_{CE\,Sat}$, that is $V_0 \approx 0.2$ V (logical 0). Hence, it acts like NAND gate.

Since, the circuit contains diode and transistor it is called DTL logic.

5.4.1 Calculation of the Minimum Forward Current Ratio ($h_{fe\,min}$) for Transistor Q

When all the inputs are high, voltage at P (V_P) = 2.2 V. So currents

$$I_1 = \frac{(V_{CC} - V_P)}{5\text{K}} = \left(\frac{5 - 2.2}{5}\right) = \frac{2.8}{5} \times 10^{-3} \text{ A} = 0.56 \text{ mA}$$

$$I_2 = \left(\frac{V_{BESat}}{5}\right) = \left(\frac{0.8}{5 \times 10^3}\right) = 0.16 \text{ mA}$$

Applying Kirchhoff's current law at point B

$$\begin{aligned} I_1 - I_B - I_2 &= 0 \\ I_B &= I_1 - I_2 \\ &= 0.56 \text{ mA} - 0.16 \text{ mA} \\ &= 0.40 \text{ mA (base current for the transistor)} \end{aligned}$$

When Q is saturated during all inputs high then the saturated collector current

$$I_{\text{CQSat}} = \left[\frac{V_{\text{CC}} - V_{\text{CESat}}}{2.2\text{K}}\right] = \left(\frac{5-0.2}{2.2}\right) = \frac{4.8}{2.2} = 2.18 \text{ mA}$$

$$h_{\text{fe}}(I_{\text{B}}) \geq I_{\text{CQ Sat}}$$

$$h_{\text{fe}_{\text{min}}} = \left(\frac{2.18}{0.4}\right) = 5.45$$

Use of the two diodes D_1 and D_2 gives us greater noise margin. When at least one input is low, $V_P = 0.9$ V. In that case, if only one diode either D_1 or D_2 is used then to saturate Q with single diode at point P it requires minimum voltage (0.7 + 0.8 = 1.5 V).

So, $V_{NL} = (1.5 - 0.9 \text{ V}) = 0.6$ V

But with the introduction of second diode, the noise margin and the voltage requirement at P to saturate Q becomes 0.7 + 0.7 + 0.8 = 2.2 V

So, with both the diodes $V_{NL} = (2.2–0.9)$ V = 1.3 V.

Hence, the noise margin increases.

5.4.2 Noise Voltage with Negative Polarity

To find the fan-out of DTL gate let the output DTL drives N similar load gates. When the output of Driver DTL is high, the input diode of load DTL gate is–ve biased and thus, the driver source a small reverse saturation current to each DTL load. When the output of the driver DTL gate is low then the driver sinks a current I from each DTL load gate, because as when V_0 is low, the input diode for the load gate is forward biased. This current is sink current defined by the driver and this sink current is greater than the source current as sink current arising due to the forward bias and source current due to reverse bias. So, DTL is called the current sink logic (reverse case of RTL).

The sink current will add to driver saturation current. Thus, the total collector current through the driver when driver output is low and drives n similar DTL load will be

$$I_{\text{C total}} = I_{\text{CQSat}} + NI \tag{1}$$

(for N loads)

Sink current from each DTL load is

$$I = \frac{V_{CC} - V_D - V_{CESat}}{5\text{ K}} = \frac{5 - 0.7V - 0.2\ V}{5\text{ K}} = \frac{4.1}{5} = 0.82\text{ mA}$$

From Eq. (1) $I_{C\,total} = 2.18 + N \times 0.82$

As $h_{fe}\,(I_B) \geq I_{C\,total}$ and $I_B = 0.4$ mA, for saturated Q

$$h_{fe} \times 0.4 \geq 2.18 + 0.8 \times N$$

$$N = \left(\frac{12 - 2.18}{0.82}\right) = 11.97 \approx 12$$

For a high gain transistor $h_{fe} = 30$,

$$\therefore\ 30 \times 0.4 = 2.18 + 0.82\ N.$$

5.4.3 Propagation Delay

When the output of the driver is high (that is $V_0 \approx 5$ V) and the input diodes of the load provides a reverse bias capacitance C_L.

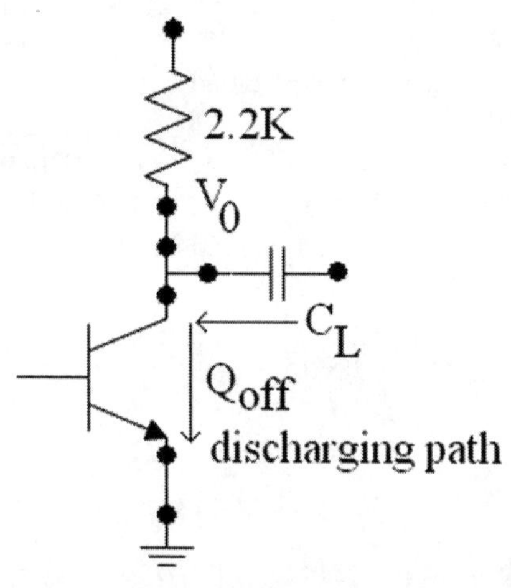

Figure 5.5 The circuit diagram of discharging path.

The capacitor will charge and the charging time constant is $(2.2 \times C_L \times N)$ if there are N such loads. When voltage V_0 goes from low to high state, capacitor will charge and this shows propagation delay is quite large. During high to low state of output C_L discharges (Fig. 5.5). The discharging time constant is $(N \times C_L \times R_{SQ})$, where R_{SQ} is the saturation resistance of the transistor Q. To increase N, we can increase h_{fe} by changing transistor, and also by increasing I_B of the same transistor.

5.4.4 Modified DTL

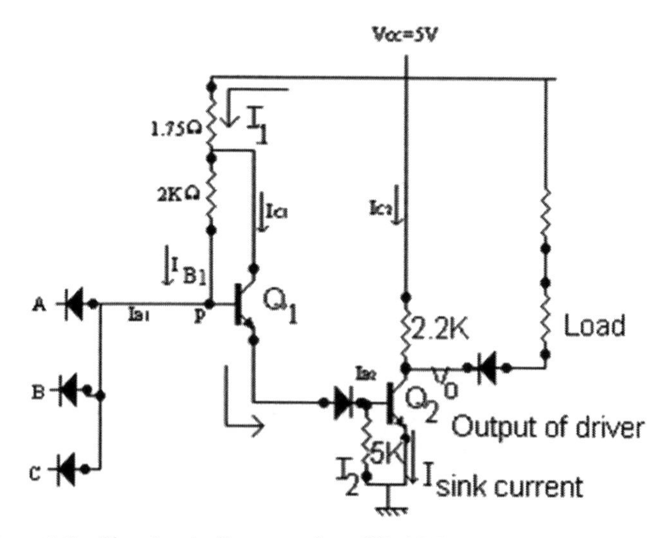

Figure 5.6 The circuit diagram of modified DTL.

The above modified DTL circuit (Fig. 5.6) is used to increase fan-out of DTL gate. The diode D_1 in earlier circuit is replaced by another transistor Q_1 due to which, the I_B will increase for Q_2.

Applying KCL at Q_1

$$I_1 - I_{C1} - I_{B1} = 0$$
$$I_1 = I_{C1} + I_{B1}$$
$$I_1 = h_{fe} \times I_{B1} + I_{B1}$$
$$I_1 = (1 + h_{fe})\, I_{B1}$$

When, all the inputs are high then voltage at $P = V_P = 2.2$ V

Applying KVL

$$V_{CC} - V_P = (I_1 \times 1.75 + I_{B1} \times 2\text{ K})$$
$$5 - 2.2 = (1 + h_{fe})\, I_{B1} \times 1.75 + 2 \times I_{B1}$$
$$\text{if } h_{fe} \approx 30$$
$$2.8 = [31 \times 1.75 + 2] \times I_{B1}$$

$$I_{B1} = \frac{2.8}{31 \times 1.75 + 2} = 0.05 \text{ mA}$$

Now $I_1 = (1 + h_{fe})I_{B1}$

$$I_1 = (31 \times 0.05)$$
$$= 1.543 \text{ mA}$$

When Q_2 is saturated

$$I_2 = \frac{V_{BESat}}{5\text{K}} = \frac{0.8}{5} = 0.16 \text{ mA}$$

Applying KCL at Base of the Q_2

$$I_1 - I_{B2} - I_2 = 0$$
$$I_{B2} = (I_1 - I_2)$$
$$I_{B2} = (1.54 - 0.16)$$
$$I_{B2} = 1.38 \text{ mA}$$

Hence, the presence of the transistor Q_1 gives a better base drive circuit.

Sink current

$$I = \frac{V_{CC} - V_0 - V_{CESat}}{3.75\text{K}} = \frac{5 - 0.7 - 0.2}{3.75} \text{ mA} = 1.09 \text{ mA} \approx 1.1 \text{ mA}$$

For N DTL load $= N \times 1.1 \text{ mA} = I_{1total}$

$$I_{CQ\,Sat} = \frac{V_{CC} - V_{CESat}}{2.2} = 2.18 \text{ mA}$$

Total collector circuit for Q_2

$$I_{C\,total} = I_{CQ2\,Sat} + NI = 2.18 + 1.1\,N$$

As $h_{fe}\ (I_{B2})\ I_{c\,total}$

$(30)\ (1.38 \text{ mA}) \geq (2.18 + 1.1\,N)$

$$N = \frac{30 \times 1.38 - 2.18}{1.1} \approx 35.65$$

So, the fan-out increases by more than three times (Fig. 5.7) and also base circuit increases more than three times.

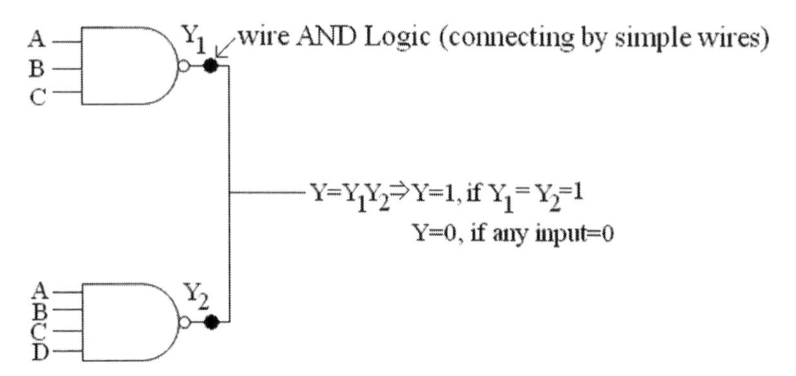

Figure 5.7 The circuit diagram of wire AND logic.

5.5 High Threshold Logic (HTL)

It (Fig. 5.8) is used in industry using motors, high voltage switching circuit to provide greater noise margin. The power supply is increased from 5 to 15 V and resistances are increased to make circuit in each branch equal.

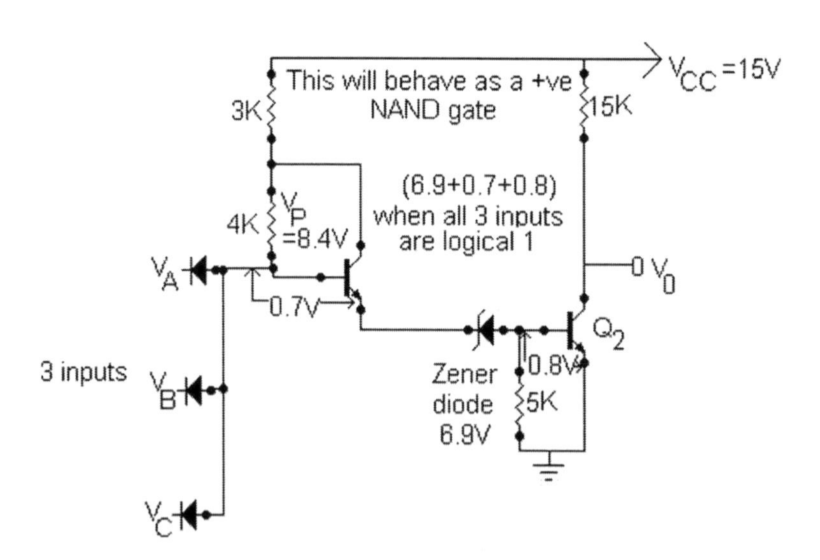

Figure 5.8 The circuit diagram of HTL.

5.6 Transistor Transistor Logic (TTL)

This is the fastest saturated logic gate family. During saturation, some charge is stored in the base region and when the transistor is again charged we have to remove the stored base charge, which is removed fastest in TTL. The TTL (Fig. 5.9) is a modified version of DTL, which gives greater fan-out, greater noise margin and smaller propagation delay than DTL.

The TTL is obtained from DTL logic family by

(i) replacing input diodes by multiple emitter base junction of the transistor Q_1

(ii) replacing the diode D_1 by the base collector junction of the same transistor Q_1

(iii) replacing the diode D_2 by base emitter junction of another transistor Q_2

(iv) representing the transistor Q by transistor Q_3

In integrated circuit, multiple emitter junctions are there on single base multiple emitter transistors. There are three diodes connected in the emitter base junction of the transistor.

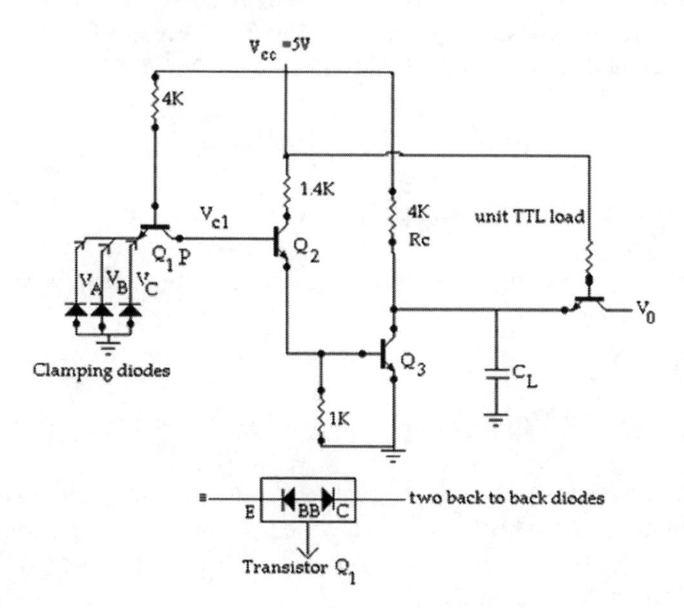

Figure 5.9 The circuit diagram of three inputs TTL NAND gate.

When all the inputs are high all E–B junctions are reverse biased. V_P rises toward 5 V and it is clamped to 2.1 or 2.2 V and this V_P is (0.8 + 0.7 + 0.7) sufficient to saturate Q_2 and Q_3. So, V_0 is low (logical 0). In this condition the collector voltage of Q_1 is

$$\begin{aligned} V_{C1} = V_{B2} &= (0.8 + 0.8) \\ &= (V_{BE2\,Sat} + V_{BE3\,Sat}) \\ &= 1.6\ \text{V} \end{aligned}$$

When any input become low then the E–B voltage of Q_1 is 0.2 V and under this condition E–B is forward biased and C–B junction is reverse biased. Thus, the transistor Q_1 operates in the active region and that makes Q_2 and Q_3 cut off.

Hence, the output voltage goes high (logic 1). Since, the collector current of Q_1 is very large which removes the stored base charge from Q_2 and Q_3 faster. Thus, Q_1 helps Q_2 and Q_3 to come out of saturation point.

For high output, each TTL offers a load capacitance C_L. When $V_0 = 1$ (Q_3 is off) C_L will charge by a time constant = $R_C\ C_L$ and when the output goes from high to low C_L discharges by a time constant = $C_L R_{SCQ3}$. To reduce propagation delay we have to reduce charging time constant to reduce R_C. But if R_C is reduced then large current will pass through R_C and can damage Q_3.

$$V_{RC} = \text{Voltage across } R_C = (V_{CC} - V_{CE\,Sat}) \tag{5.1}$$

So, reducing R_C will increase power dissipation and Q_3 may be damaged. So, to reduce the charging time constant, the output circuit of TTL should be modified and the modified output circuit is called Totem pole output configuration.

5.7 Totem Pole (TTL)

The circuit shown in Fig. 5.10 is a three input TTL NAND gate, where Q_3 and Q_4 are connected in Totem pole fashion. At any time one of them will be conducting that is when Q_2 and Q_3 are saturated Q_4 is off and when Q_2 and Q_3 are off Q_4 is on.

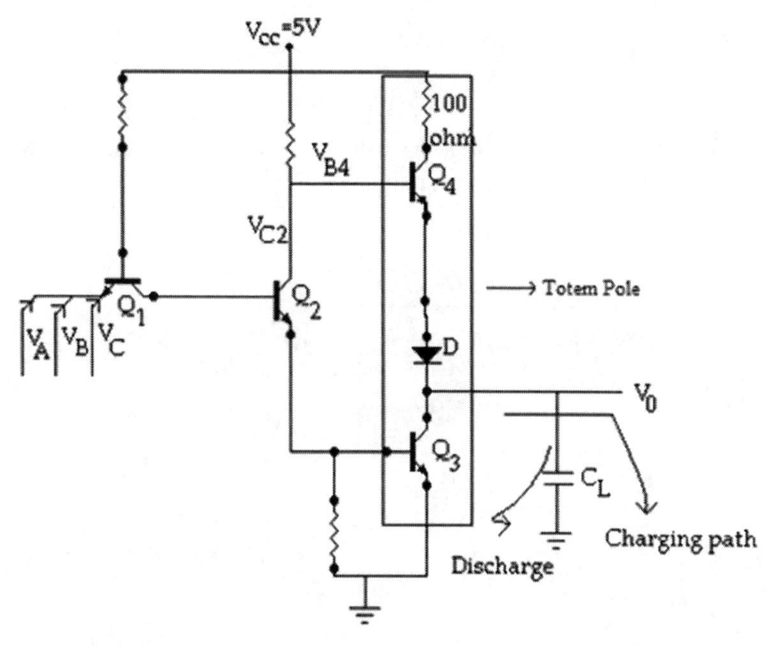

Figure 5.10 Totem pole TTL.

The charging time constant of the $C_L = (100\ \Omega + R_{SRQ_4} + R_{FBD})\ C_L$, where R_{FBD} is forward bias resistance of diode. Since, R saturation of Q_4 and R_{FBD} are very small, C_L is fixed. So, time constant reduces. When Q_2 and Q_3 are on:

$$\begin{aligned} V_{C2} = V_{B4} &= V_{CE2\ Sat} + V_{BE3\ Sat} \\ &= 0.2 + 0.8 \\ &= 1\ V \end{aligned}$$

If the diode is not present 1 V is sufficient to forward bias Q_4, but with presence of diode this is not possible as (0.8 + 0.7) > 1 V. Load capacitance charges when Q_2 and Q_3 are off. With this condition $V_{C2} = V_{B4} \approx 5$ V which clamped at (1.5) V = (0.7 + 0.8) V. This voltage is sufficient to saturate Q_4 and forward bias D and hence the capacitor charges. The clamping diodes protect the E–B junctions against excessive forward bias which may damage Q_1. Totem pole output gives active (as transistor is an active element) pull out and the previous are given passive pull out as RC pulls the output from low to high state.

5.8 Open Collector Output

This is one type of TTL (Fig. 5.11) where the transistor Q_4 and the diodes are not connected. In this TTL gate in order to get the proper high and low logic levels an external pull up resistor is connected to the V_{CC}

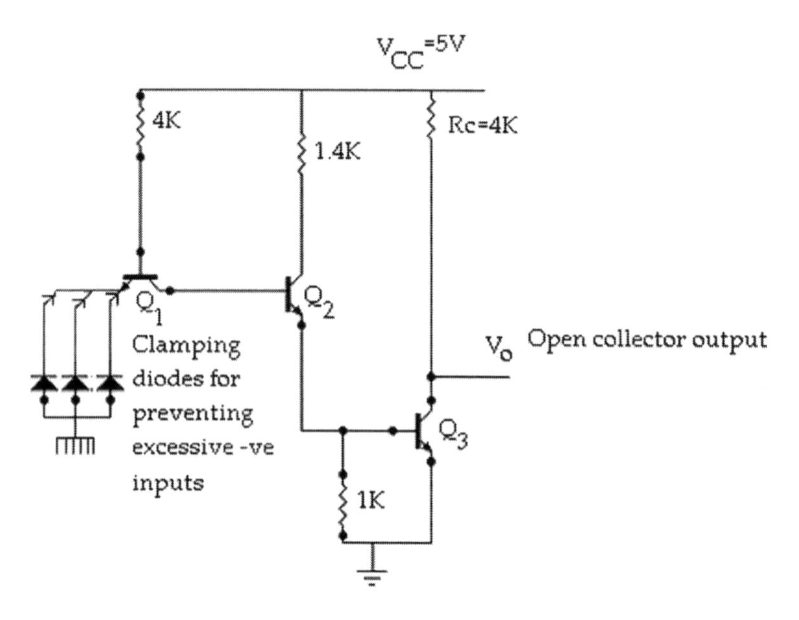

Figure 5.11 Open collector of TTL.

from the collector terminal of the Q_4. The open collector connection is much slower than the Totem pole connection. The speed of open collector circuit can be increased only a little bit by selecting smaller resistance.

5.8.1 TTL Families

74S00 → Schottky TTL
74LS00 → Low power Schottky TTL
74ALS00 → Advance low power Schottky TTL (it will never be driven into saturation)
74H00 → High power TTL
7400 → Normal TTL

5.8.2 Advantages of Totem Pole Output

1. fast switching time
2. low power dissipation

5.8.3 Disadvantages of Totem Pole Output

1. large current spike during switching from low to high state

5.9 Wire and Logic for Open Collector TTL

Two or more open collector outputs of TTL gate can be directly connected to get AND logic and this is called wire AND logic (Fig. 5.12).

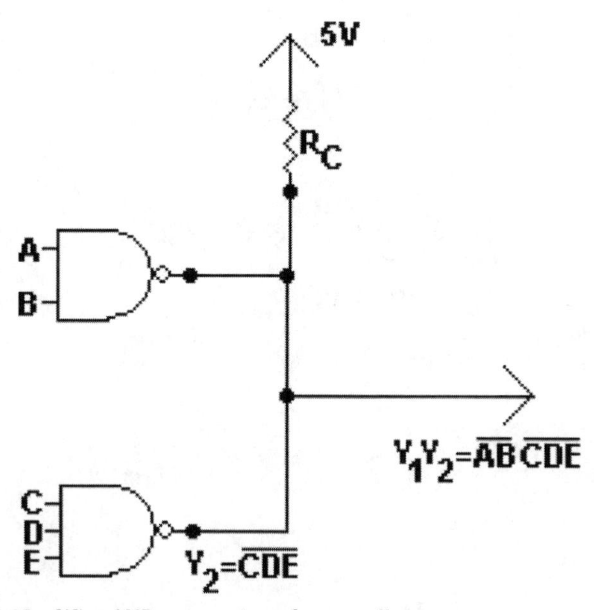

Figure 5.12 Wire AND connection of open collector gate.

A common R_C is connected to both the output of both NAND gates. This type of wire AND logic is impossible in Totem pole output and for this we have to slightly modify the circuit. The added transistor Q_4 is called active pull up output Q_2 and Q_3 are ON, Q_4, and D are off. This will reduce the charging time constant.

5.9.1 Modified Totem Pole Circuit

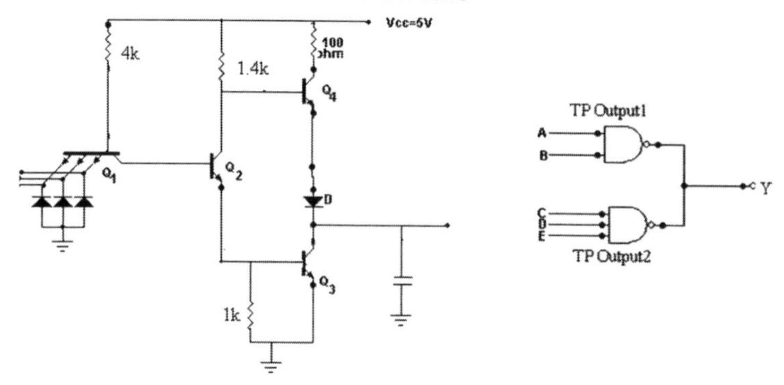

Figure 5.13 Totem pole output configuration of TTL.

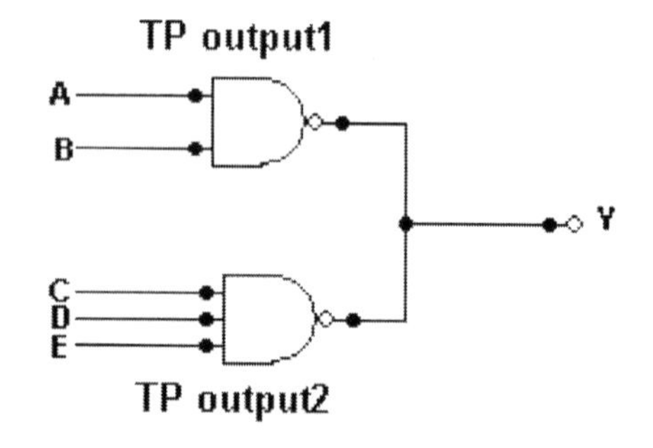

Figure 5.14 Wire AND connection of Totem pole gate.

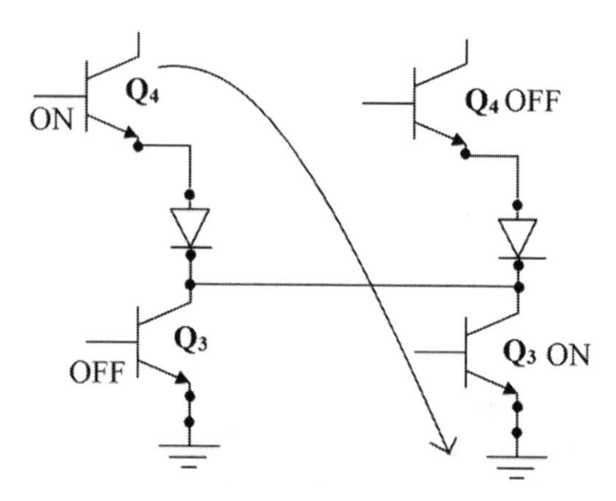

Figure 5.15 Current flow in Totem pole connection when wire ANDed.

In Totem pole circuits (Fig. 5.13) for some period of time, when output switches from OFF to ON heavy current may pass (30–50 mA) through so the transistors (Fig. 5.14) may be damaged. So, two Totem pole outputs for TTL cannot be inter connected to realize wire AND logic (Fig. 5.15).

5.10 Tristate TTL Inverter

It is called tristate TTL, because it allows three possible outputs (HIGH, LOW, and HIGH impedance-high Z). It utilizes both the Totem pole and open collector advantages i.e high speed operation of Totem pole and wire ANDding of open collector configuration. In the high Z state both the transistors of the Totem pole arrengement are turned off, so the output terminal is a highimpedance to V_{CC} or ground. The Tristate inverter is shown in the Fig. 5.16. Where two inputs are there, one is A which is the normal input and the other one is the enable input E that can produce high Z. Table 5.1 shows the truth table for tristate inverter.

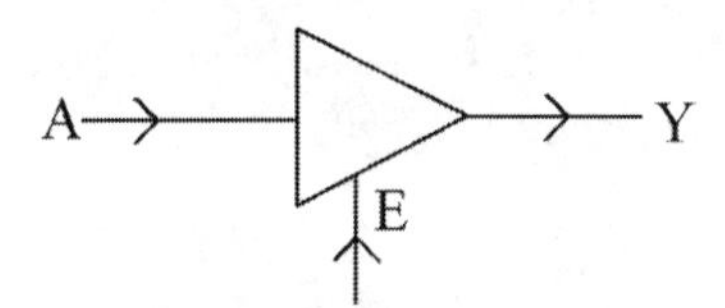

Figure 5.16 Traditional symbol of tristate inverter.

Table 5.1 Truth table for the tristate inverter

E	I	Y
1	0	1
1	1	0
0	X	High Z

When E = High, the circuit operates like an inverter as a normal inverter because high voltage at E does not put any effect on Q_1 and Q_2. Hence, in this condition this circuit behaves like an inverter. But when the E goes low it forward biases E–B junction of Q_1. So, Q_2 turns off, which in turn makes Q_4 off. This low E also forward biases the

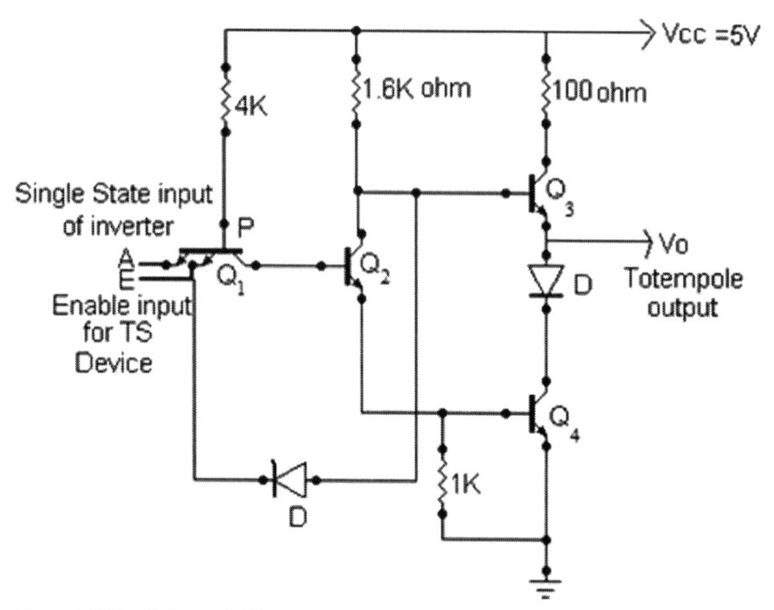

Figure 5.17 Tristate TTL inverter.

diode to shunt current away from the base of Q_3. So, Q_3 also goes off. Though both the Totem pole transistors are at off state, the output terminal is now open circuited (Fig. 5.17).

5.11 QUAD D-Type F/F with Tristated Output (74LS373)

LE = Latch enable input.
OE = Output enable input.

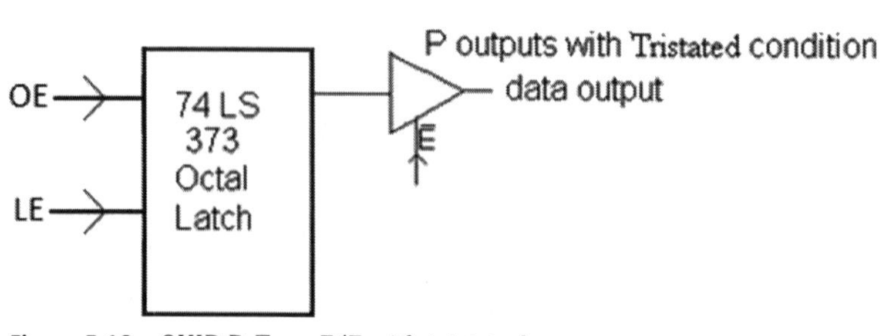

Figure 5.18 QUID D-Type F/F with tristated output.

20 Hex at input = 20 H at output in case 1
30 Hex at input = 20 H at output in case 3
This (Fig. 5.18) is used for time division multiplexer for lower and higher order. The truth table is shown in Table 5.2.

Table 5.2 Truth table for the QUID D-type F/F with tristated output

LE	OE	Data input	Data output
1	0	New data input	Output = new data input
1	1	X	High Z
0	0	New data	Previous data at the output

5.12 Integrated Injection Logic (IIL)

This is the new logic and is popular in LSI and VLSI circuits. It is also otherwise known as current injection logic. These logic gates are constructed using only BJTs. It is not suitable for discrete gate ICs. Like TTL and MOS no transients are produced in IIL, because in IIL the currents are constant.

5.12.1 Advantages

1. low power consumption.
2. easily fabricated and economical.

5.12.2 Disadvantage

It requires one more step in its manufacturing process than those used in MOS.

5.12.3 IIL Inverter

Figure 5.19 shows an IIL inverter where two BJTs are used (one PNP and one NPN). The Q_1 serves as a constant current source which injects current to node *P*. When the input is low, injected current flows into input diverting current from the base of Q_2 which makes Q_2 OFF. Hence, the output goes high (V_{CC}–logic 1). If the applied input is high then that makes Q_2 ON by diverting the current toward the base of Q_2 which makes the output low (logic 0).

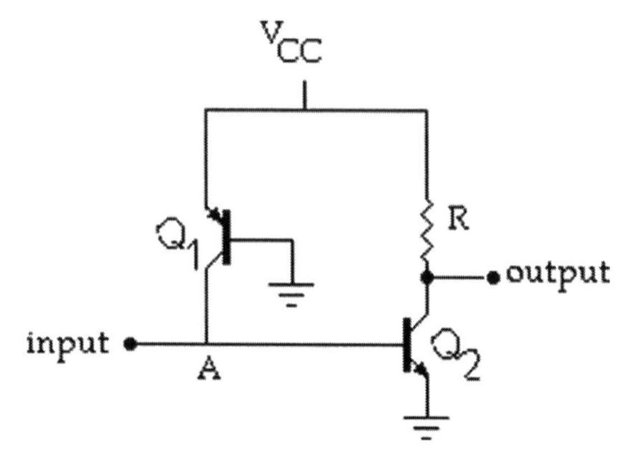

Figure 5.19 The circuit diagram of inverter.

5.12.4 IIL NAND and NOR Gate

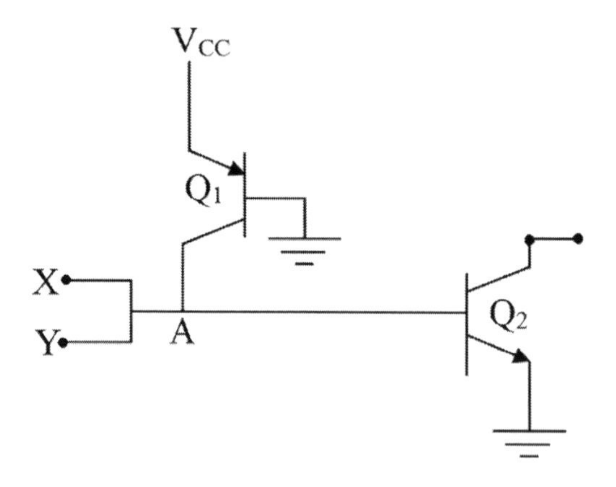

Figure 5.20 IIL NAND.

IIL as NAND (Fig. 5.20) and NOR (Fig. 5.21) gates are shown here. In NAND a simple inverter is presented with two inputs. When either X or Y or both are low then the injected current flows into the inputs resulting the transistor Q_2 OFF. So the output goes high (logic 1). If both the inputs go high then that makes Q_2 ON. Hence the output shows logic 0.

In the second circuit two inverters are used separately and their outputs tied together. If both or any one input is high then the corresponding transistor goes ON. So the output is logic 0. If both the

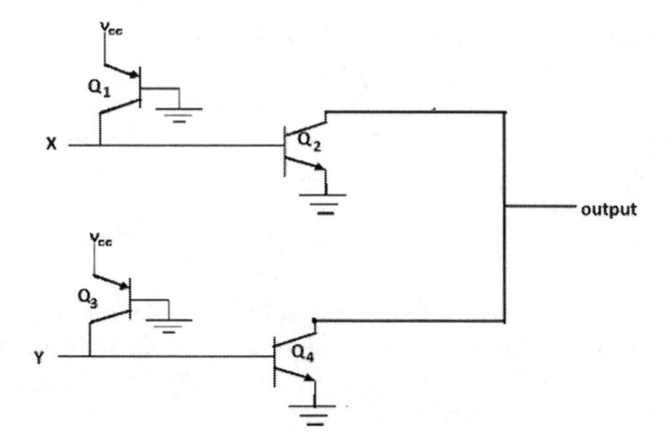

Figure 5.21 IIL NOR.

inputs are low then both Q_2 and Q_4 go OFF. Hence, that results high (logic 1) output.

5.13 Emitter Coupled Logic (ECL)

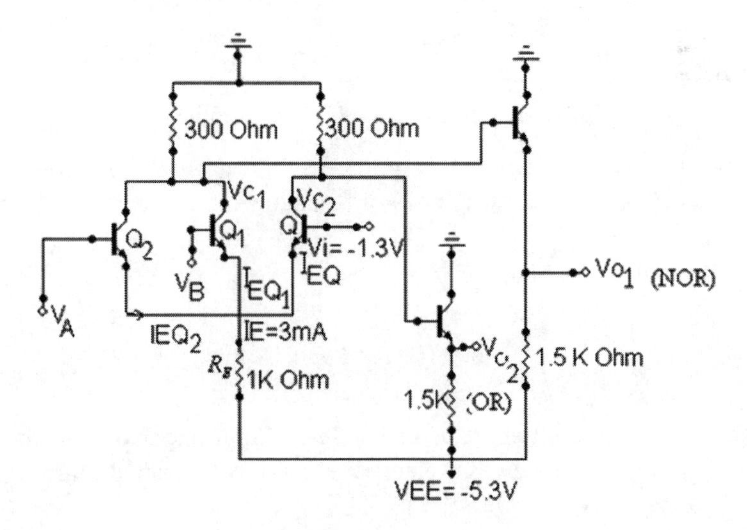

Figure 5.22 Emitter coupled logic circuit.

The configuration given here in Fig. 5.22 is called ECL, because the BJT's that are coupled at their emitters. The basic configuration of ECL is a differential amplifier configuration with Q_1 and Q. Here emitters of Q, Q_1, and Q_2 are coupled through R_E and the current flowing through R_E is the combinations of all three transistors. That is total current $= I_E = I_{EQ2} + I_{EQ1} + I_{EQ}$, which is held constant. So it is also known as current mode logic (CML). In ECL none of the transistors are driven to saturation and the collector side is grounded and emitter to V_{EE} to reduce noise voltage. All the transistors are operating either in the active region or in the cut off region. In ECL both the inputs V_A and V_B are given negative. Two logical levels of the input voltages selected as

(i) logic 1 (−0.8 V)
(ii) logic 0 (−1.7 V)

The truth table is shown in Table 5.3.

Table 5.3 Truth table for the ECL

$V_A(V)$	$V_B(V)$	Q_1	Q_2	Q	V_{01}	V_{02}	V_{C1}	V_{C2}
0	0	OFF	OFF	Active (as transistor will never saturate)	1	0	1	0
0	1	OFF	Active	OFF	0	1	0	1
1	0	Active	OFF	OFF	0	1	0	1
1	1	Active	Active	OFF	0	1	0	1
					NOR operation	OR operation		

The functions of emitter follower in ECL are as follows:

(a) It provides necessary voltage level shift to generate proper ECL logic level at the output of emitter follower

$$V_{01} + V_{BE\,Sat} = V_{C1}$$
$$V_{01} = (V_{C1} - V_{BE\,Sat}) = (V_{C1} - 0.8)$$
$$V_{02} + V_{BE\,Sat} = V_{C2}$$
$$V_{02} = V_{C2} - V_{BE\,Sat}$$
$$V_{02} = (V_{C2} - 0.5)$$

(b) Since the emitter follower has low output impedance of the order of 7 Ω it gives greater fan-out as well as small charging time constant (at least 25).

V_{01} satisfies NOR and V_{02} satisfies OR logic.

5.13.1 Characteristics of ECL Logic Family

1. It is the fastest unsaturated logic having minimum $t_{pd} = 1$ ns as none of the transistors is driven into saturation.
2. All its voltage levels are negative.

$$\text{Logic } 1 = -0.8 \text{ V}$$
$$\text{Logic } 0 = -1.7 \text{ V}$$

3. It provides two complementary output that is OR and NOR output.
4. Its power dissipation is nearly 25 mW, which is comparable with TTL logic.
5. Worst case noise margin is 250 mV ≈ 0.25 V which is small and hence, ECL logic is not suitable in industrial environment.
6. Fan-out of ECL ≈ 25 (which is fairly high as emitter follower output stage gives small impedance).
7. Since, the current through R_E is fairly constant independent of logic state so drainage current through supply voltages are not much.

5.13.2 Advantages of ECL

1. It is the fastest unsaturated logic.
2. It provides minimum propagation delay.
3. The stray capacitances can be quickly charged and discharged because of high currents and low output impedances.

5.13.3 Disadvantages of ECL

The ECL input and output are not electrically compatible for direct connection with any other logic family (as the inputs of other logic family are positive).

5.14 Wired-OR Connection of ECL Logic

Figure 5.23 shows an IIL inverter where two BJTs are used (one PNP and one NPN). Q_1 serves as a constant current source which injects current to node P.

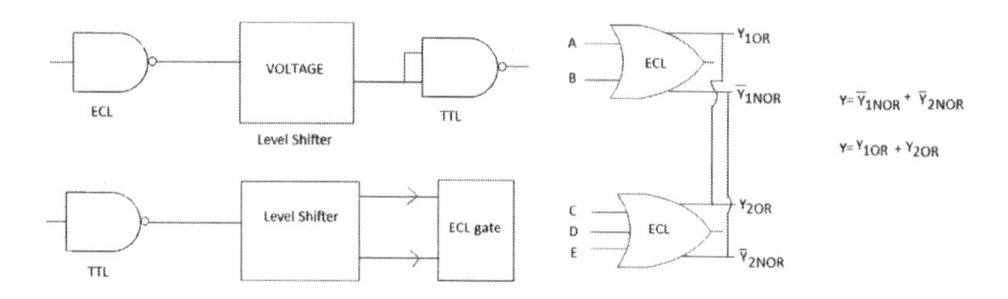

Figure 5.23 Wired-OR connection of ECL logic.

5.15 Wired-OR MOS Logic Family

The MOS logic family uses metal oxide semiconductor FET, which are unipolar that is either N-MOS or P-MOS field effect transistor.

The MOSFET are of two types:

1. enhancement type MOSFET.
2. depletion type MOSFET.

Enhancement type MOSFET represents a nonconducting channel between drain and source. For N-channel enhancement MOS if $V_{GS} \geq V_T$ (threshold voltage) then we get conducting channel between source (S) and drain (D).

For N-MOS

If $V_{GS} > V_T = 1.5$ V then $R_{ON} = 1000\ \Omega = 1\ K\Omega$ and

If $V_{GS} < V_T$, $R_{OFF} = 10^{10}\ \Omega$. Under off state, T_r will provide very high resistance.

For P-channel enhancement MOSFET it is just reverse.

If $V_{GS} <= -1.5$ V then $R_{ON} = 1000\ \Omega = 1\ K\Omega$ and

If $V_{GS} > -1.5$ V then $R_{OFF} = 10^{10}\ \Omega$.

5.15.1 N-MOS Inverter

Here two transistors are connected as per the circuit, where Q_1 is always ON as $V_{DD} = 5$ V and source is at lower voltage and V_{GS} is always positive. The Q_2 switches on ON to OFF in response to the input V_A. As per the input applied at V_A, the output gives the logical levels

(Fig. 5.24). If the input applied as 0 volt (logic 0) then it makes Q_2 OFF. Hence, its R_{OFF} is 10^{10} Ω
Then

$$V_0 = \frac{V_{DD} R_{OFF}(Q_2)}{R_{ON}(Q_1) + R_{OFF}(Q_2)} = 5\,V(\text{logic}1) \tag{5.2}$$

But if the input is 5 V (logic 1) then that makes Q_2 ON.
So, R_{ON} is 1 KΩ then

$$V_0 = \frac{V_{DD} R_{OFF}(Q_2)}{R_{ON}(Q_1) + R_{OFF}(Q_2)} = 0\,V(\text{logic}0) \tag{5.3}$$

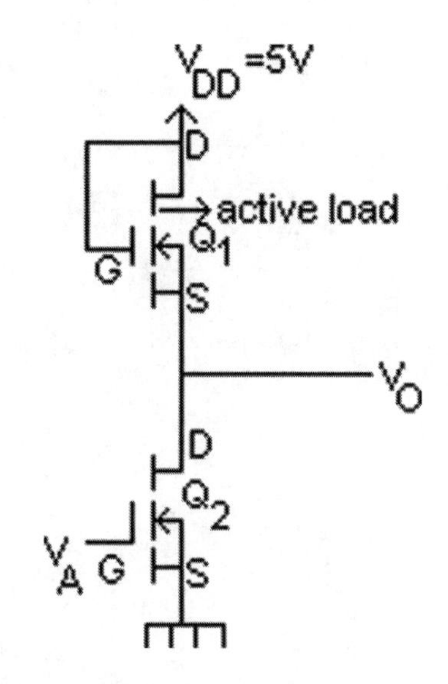

Figure 5.24 NMOS inverter.

Table 5.4 Truth table for the NMOS inverter

V_A	Q_1	Q_2	V_0
0 V	ON	OFF	V_{DD} = 5 V = Logical 1
5 V	ON	ON	0 V ≅ Logical 0

5.15.2 Two Input N-MOS NAND and NOR Gates

Two figures show the two input NMOS NAND (Fig. 5.25) and NOR (Fig. 5.26) gates. In both three transistors are used where Q_1 is always ON (as per the inverter configuration). As per the inputs supplied both the transistors Q_2 and Q_3 go ON and OFF reflecting two resistance levels (either $10^{10}\,\Omega$ or 1 KΩ). Hence, gives the output (Shown in the Table 5.5.).

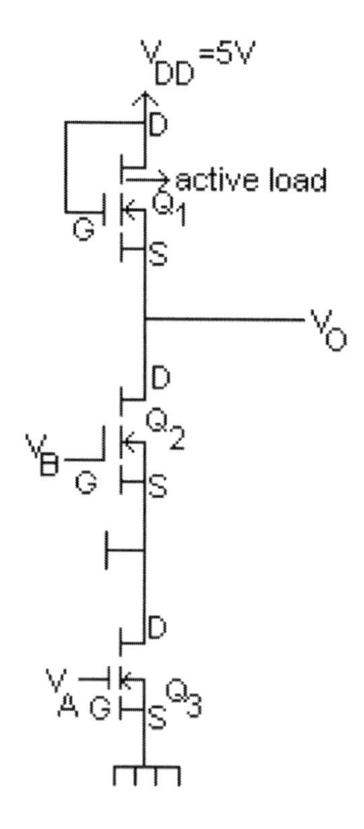

Figure 5.25 NMOS NAND.

Table 5.5 Truth table for NMOS NAND

V_A	V_B	Q_1	Q_2	Q_3	V_0
0 V	0 V	ON	OFF	OFF	5 V
0 V	5 V	ON	ON	OFF	5 V
5 V	0 V	ON	OFF	ON	5 V
5 V	5 V	ON	ON	ON	0 V

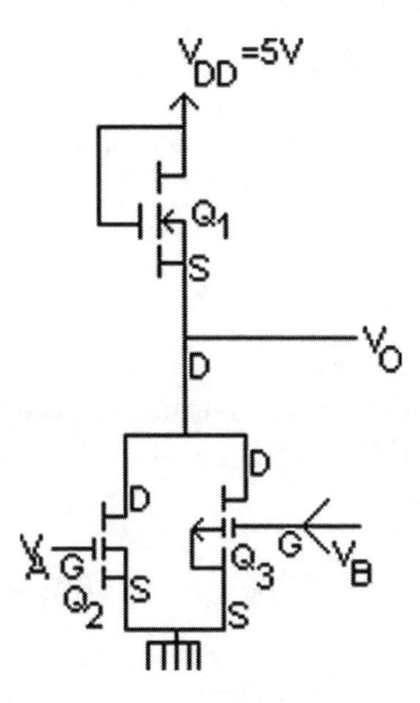

Figure 5.26 NMOS NOR.

Truth table of NMOS NOR is shown in Table 5.6.

Table 5.6 Truth table for NMOS NOR

V_A	V_B	Q_1	Q_2	Q_3	V_0
0 V	0 V	ON	OFF	OFF	5 V
0 V	5 V	ON	OFF	ON	0 V
5 V	0 V	ON	ON	OFF	0 V
5 V	5 V	ON	ON	ON	0 V

5.15.3 Characteristics of MOS Logic

1. It is easier to fabricate MOS circuit without any resistor.
2. It is having high packing density. Hence, MOS logic finds high application in calculator chip, microprocessor, and microcomputer chip being LSI or VLSI technology.
3. It has got very low power dissipation, as its input impedance is very large of the order of 10^{12} Ω (signal loss v^2/R is very less).
4. Propagation delay is large as output impedance is large.
5. Fan-out is fairly high (can be infinity but with increase of load, capacitance increases which increases charging time and so fan-out is limited to only 50).

5.16 Complementary MOS Logic (CMOS)

In CMOS both N-channel and P-channel MOS transistors are used. The output logical levels are calculated as per the conditions of both types of the transistors (either ON or OFF).

5.16.1 Advantages

1. Extremely low power dissipation is even smaller than MOS logic.
2. The propagation delay is quite small as output impedance is small.

5.16.2 Disadvantages

1. Difficult to fabricate on IC.
2. Low packing density (so they are not recommended for LSI or VLSI circuits).

5.16.3 CMOS as Inverter

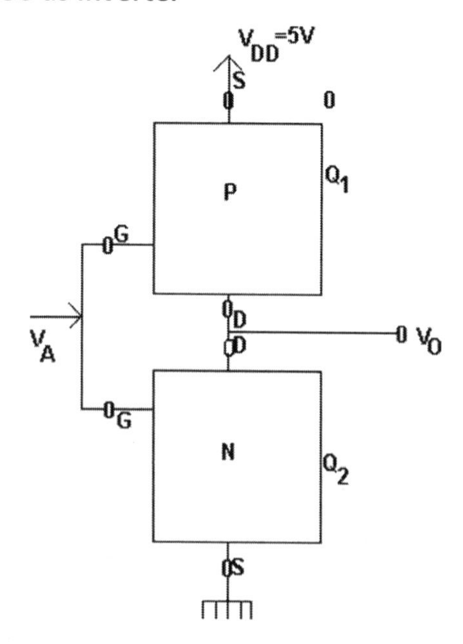

Figure 5.27 CMOS inverter.

Figures 5.27 and 5.28 show a CMOS inverter, where one NMOS and one PMOS are connected. As per input applied either a transistor goes ON or OFF. The output logical levels are calculated as per the conditions of both the transistors (shown in the truth table in Table 5.7).

Table 5.7 Truth table of CMOS inverter

V_A	Q_1	Q_2	V_0
0 V	ON	OFF	V_{DD} = 5 V = Logic 1
5 V	OFF	ON	V_{DD} = 0 V = Logic 0

5.16.4 Two Input NAND using CMOS

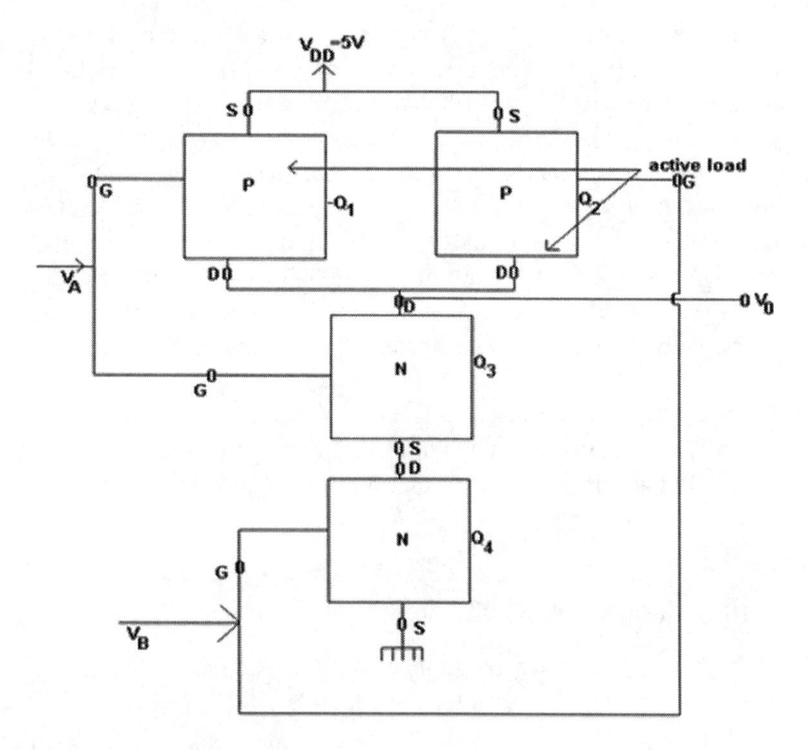

Figure 5.28 CMOS NAND.

Truth table is shown in Table 5.8.

Table 5.8 Truth table for CMOS NAND

S	V_B	Q_1	Q_2	Q_3	Q_4	V_0
0 V	0 V	ON	ON	OFF	OFF	5 V (V_{DD})
0 V	5 V	ON	OFF	OFF	ON	5 V (V_{DD})
5 V	0 V	OFF	ON	ON	OFF	5 V (V_{DD})
5 V	5 V	OFF	OFF	ON	ON	0 V

5.16.5 CMOS Series

When pin configuration of both ICs are same 74 series and 54 series. The 2716 and 6116 two different ICs provide the same function. Both gives quad (TTL) logic 7400 and 40C00 (CMOS logic) input NAND and also have same function electrically compatible IC can be directly connected to each other. The ECL and TTL cannot be connected directly depending upon these terms. These are 4000/14000 series of CMOS. This gives low power dissipation, large t_{pd}, and small at output capability. It is neither electrical compactable functionality nor it has pin to pin compatibility with some TTL ICs. The 740HC/40HC TXX series of CMOS is high speed CMOS ICs, provides pin compatibility, and functional equivalence with some TTL ICs. The IC 740HCT is also electrical compatible with some TTL ICs but 40HC is not electrically compatible. This series gives greater output current capability. 40ACXX/40ACTXX is similar to 40HCT. 740HCT is called advance CMOS ICS.

Both series have operating voltage range of 2 to 6 V. The 40ACXX are pin to pin compatible with some TTL ICs but not 40ACTXX but is electrical compatible.

5.16.6 Bi CMOS Series

This uses a combination of bipolar and CMOS circuits (Fig. 5.29). The advantages of this circuit are as follows:

1. t_{pd} of CMOS < MOS
2. Power dissipation is smaller but with increase in frequency, power dissipation increases.

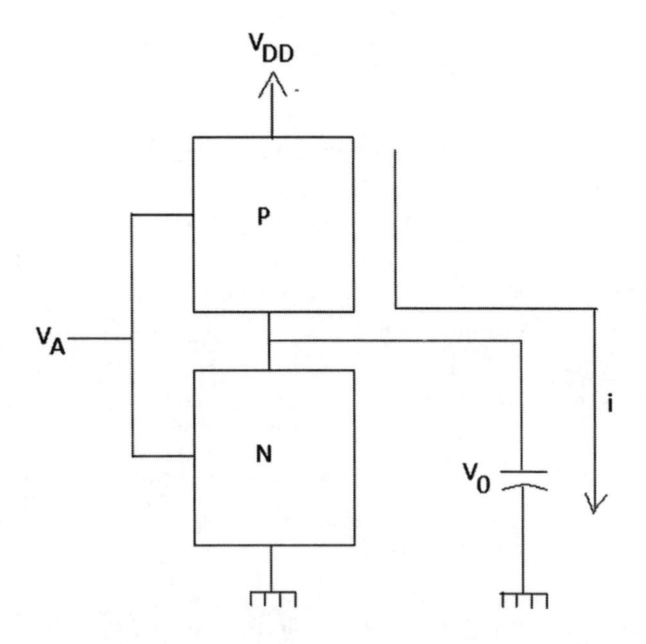

Figure 5.29 Bi CMOS circuit.

5.17 Comparison of Logic Families

The comparison of logic families are represented in Table 5.9.

Table 5.9 Truth table for comparison of logic families

Logic family	Propagation delay (ns)	Power dissipation (mW)	Noise margin (V)	Fan-in	Fan-out	Cost
TTL	9	10	0.4	8	10	Low
ECL	1	50	0.25	5	10	High
MOS	50	0.1	1.5	8	10	Low
CMOS	<50	0.01	5	10	50	Low
IIL	1	0.1	0.35	5	8	Very low

Random access memory—Read and write access time are same.

FFH = 11111111

When an EPROM is blank, all its locations will contain FFH EPROM can be erased by specially devised also program.

EPROM chips = 2716 – 2 KB
2732 – 4 KB
2764 – 8 KB

The program means 1s will be converted to 0s.

11111111 = FFH

00110000 = 30 H

Address is given through switches in ROM.

EEPROM = we can erase on board.

Any unconnected terminal should be connected to power supply CMOS, as CMOS is very disreputable to electrostatic charge from outside, so to protect CMOS ICs we connect it to suitable power supply.

Lock Out: If a counter, while pasing through its possible states acquires a nonpossible stste and does not return to its desired state, such sitiation is called lock out.

Problem 1:

A photo all triggers an assembly line bottle filter and the maximum filling rate in 10 bottles per min. How many counter stages are registered to count a 1 hr output?

Solution

In 1h the maximum number of bottles can be fixed,

10 bottles × 60 = 600 bottles

$\therefore 600 = 2^N - 1$

$\Rightarrow N \, log_{10} 2 = log \, 601 \; (\because 2^N = 601)$

$\Rightarrow N = 9.23$

Of course building 0.23 F/F is impossible. Since more than 9 F/Fs are needed, a 10 bit counter is required. The maximum count for a counter with N F/Fs is given by max <= $2^N - 1$.

Problem 2:

The typical 0.5 in LED-7 segment display draws 10 mA per segment from a 5 V power supply. The same size LCD draws 1 A per segment. Compare the maximum power dissipation.

Solution

The maximum power is used when all seven segments are activated. For the LED display:

$P_{\max} = 5\ \text{V} \times 7\ \text{segments} \times 10\ \text{mA/segment} = 350\ \text{mV}$

For the LCD display:

$P_{\max} = 5\ \text{V} \times 7\ \text{segments} \times 1\ \mu\text{A/segment} = 35\ \mu\text{W}$

Thus, the given example shows that for the same size display, an LED draws 10,000 times as much power as an LCD (Fig. 5.30).

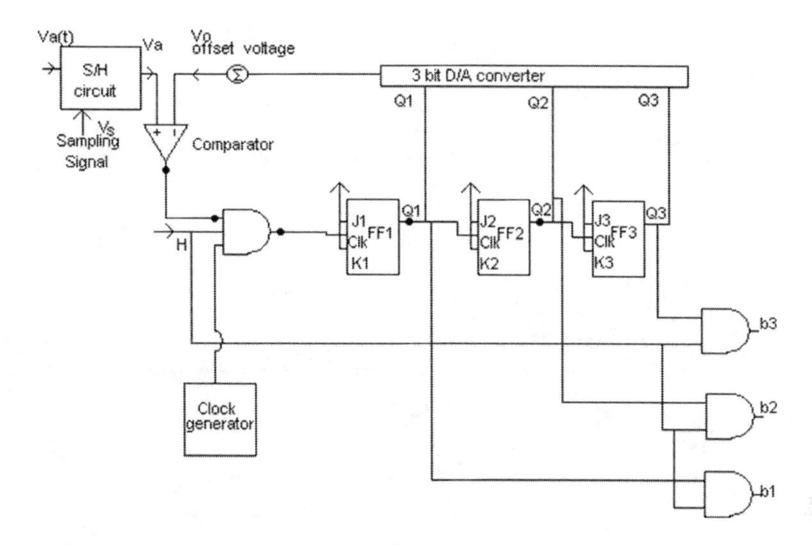

Figure 5.30 The circuit diagram of digitalized system.

5.17.1 High Level DC Noise Margin

The high level DC noise margin is considered when the load gate (Figs. 5.31–5.32) is driven by high output from the driver gate.

Driver gate $V_{OH(min)}$ $V_{IH(min)}$ Load gate

$$(V_{\text{OH min}})-(V_{\text{IH min}})$$
$$= V_{\text{NH}}$$
$$= 4.5 - 4$$
$$= 0.5 \text{ V}$$

If noise voltage is added with similar polarity, no problem as $(4.5 + \chi) > 4$ V but with opposite polarity then 4.5 – (+ve noise voltage) $< (V_{\text{IH}})_{\text{min.}}$

5.17.2 AC Noise Margin

By this we mean that the noise appears at the input of a logic gate for time duration, which is very much shorter than the response time

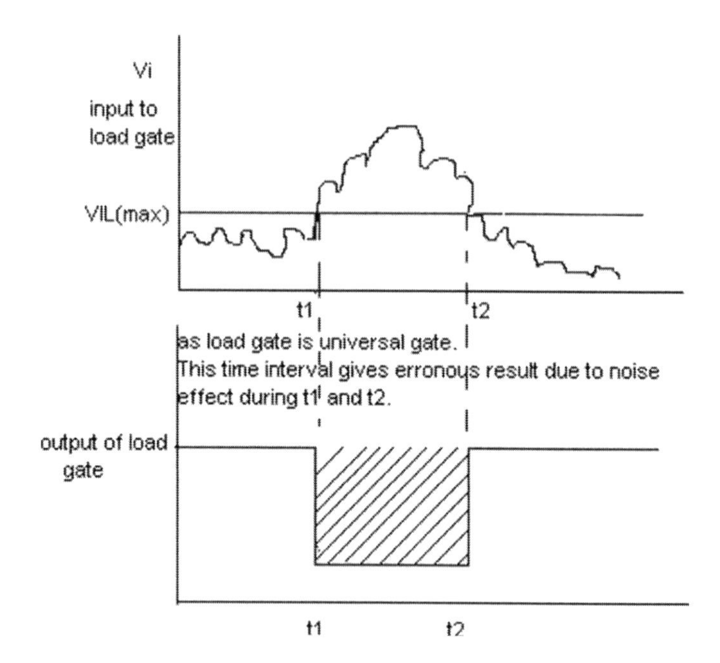

Figure 5.31 The input–output waveform of load gate.

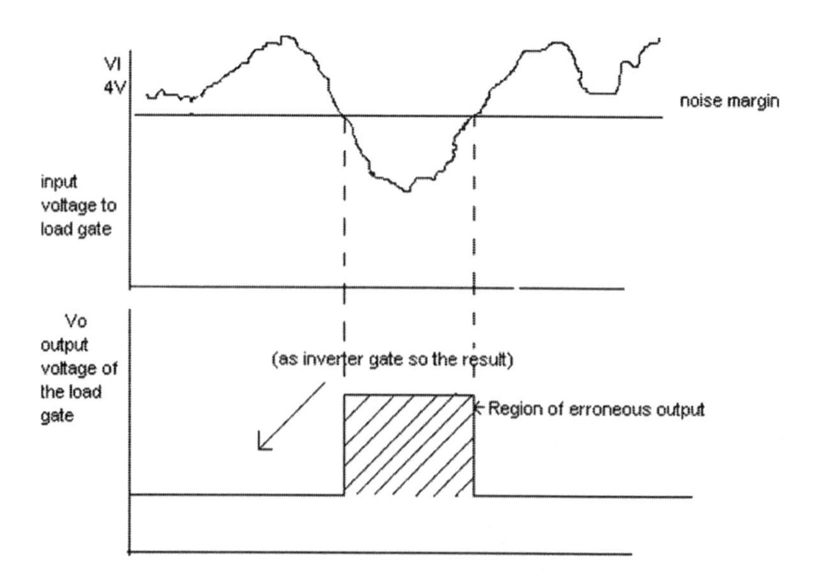

Figure 5.32 The input–output waveform of load gate.

or propagation delay of the logic gate. Thus, we can say that the AC noise margin for a gate is very large as compared to DC noise margin (so it is not important).

5.17.2.1 Fan-out

This is defined as the maximum number of inputs (Fig. 5.33) of the same logic family that the gate can drive without deviating from its specified output analogous to the situation.

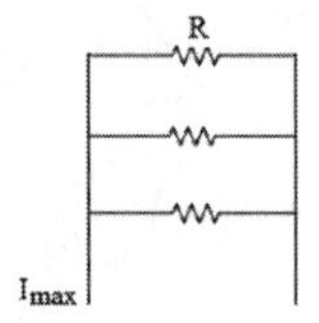

Figure 5.33 The diagram of fan-out.

$$I = \frac{V}{R_{eq}} = \frac{V}{R_{eq}} = I_{(max)}$$

with increase of R, $I_{(max)}$ will also increase.

5.17.3 Output of a TTL gate

The output of a TTL gate (Fig. 5.34) is high if V_{int} = high. In this case, the input load gate is reversing biased and the TTL driver can source a 40 μA of reverse saturation current.

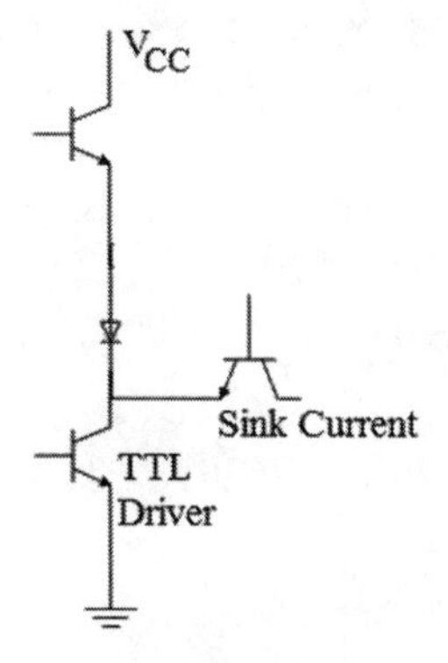

Figure 5.34 The output circuit diagram of TTL gate.

5.18 Major Comparative Features of Various Logic Families

5.18.1 Gates of TTL Logic family

Advantages

1. TTL gates require a typical supply voltage of (+/-)5%. However, currently TTL gates with lower supply voltage like 3 V,2.4 V, 1.5 V etc are also available.
2. Several types of TTL gates are now available like low power, low power Schottky, high speed, high speed Schottky and a variety of other types.
3. Amongst the saturation logic family TTL is the fastest.
4. Noise margin is better with a typical value of around 0.4 V
5. TTL gates have wide power dissipation range lying between megawatt to milliwatt.
6. TTL gates are compactable with gates of other logic fmilies.
7. TTL gates are much populr among the logic families and have easily available.
8. Fanout capabilities is upto 10 gates.
9. All logic function TTL gates are available.
10. Output impedance is low at high and low status

Disadvantages

1. TTL gates canot be used in VLSi/VLSI circuits because of their isolation problems.
2. Cost of TTL gates is more compared to NMOS/CMOS gates.
3. TTL gates produce transient volatages at the time of switching.
4. TTL gates consume more power compared to NMOS/CMOS gates.
5. TTL gates have low noise margin and hence not suitable for use in industrial atmosphere.
6. Standard TTL gates do not have wired-OR capabilities.

5.18.2 Gates of ECL logic family

Adavantages

1. ECL gates require a typical supply voltage of -5.2 V. Power supply loads do not have current switching spikes.

2. ECL gates have the highest speed among all logic families as in ECL gates transistors operate in the active region.
3. The number of function available is high but less than TTL family.
4. ECL gates have complementary outputs (NOR-OR) and Wired-OR can be achieved.

Disadvantages

1. ECL gates have resistors and hence ECL based VLSI design is difficult and costly.
2. ECL gates have limited fanout because of intensive capacitive load.
3. Inorder to interface ECL logic gates with other logic family gates, we require level shifters.
4. Among all the logic families ECL gates consume the highest power.
5. ECL gates have very noise margin like (+/-)200 mV and not suitable for use in industrial environment.

5.18.3 Gates of NMOS logic family

Adavantages

1. Fabrication cost is the lowest because NMOS gates require less number of diffusion.
2. NMOS gates are capable of using variable power supply ranging from 5 V to 15 V.
3. NMOS gates can be used as capacitor as well as resistor and can be used in charge-coupled devices.
4. NMOS gates have the highest packing density among all the logic families as which do not require isolation islands. Hence NOMS gates are suiatables for VLSI/VLSI circuits.
5. Power dessipation is very low in the order of nW, but more than CMOS gates.
6. NMOS gates have large fan-out capabilities of 20 gates.
7. NMOS gates have comparatively higher noise margin and have suitable for use in industrial environment.

Disadvantages

1. As the capacitive load of NMOS gates is very high, speed of operation is lower. Propagation delay is also very high.

2. NMOS gates are not very popular because CMOS gates have higher speed and low power dissipation.

5.18.4 Gates of IIL logic family

Advantages

1. As IIL gates are realized using BJTs, they have high speed of operation.
2. IIL gates have high packing density as they are made up of transistors only and hence they are suitable for VLSI/ULSI circuits.
3. IIL gates require very low power supply and dissipate low power.
4. Cost of production of IIL gates is less and hence several functions possible on the same chip.
5. IIL gates are compatible with gates of other logic families.

Disadvantages

1. Currently these gates are dormant because of its technology.
2. For proper functioning IIL gates need external resistors.
3. Supply voltage is in the range of 0.8 to 0.2 V (for Silicon based IIL gates) and noise margin is very low.
4. Packing density of IIL gates is lower than nMOS gates.

5.18.5 Gates of CMOS logic families

Advantages

1. CMOS gates require single power supply and logic swing is large (= Vdd). Power dissipation is in the order of nano watts.
2. CMOS gates have lower propagation delay compared to nMOS gates and higher speed(in range of GHz)
3. CMOS gates have higher fanout capability even more than 50.
4. Noise margin of CMOS gates is high (0.5Vdd) and hence suitable for use in industrial environment.
5. CMOS gates directly compatible with TTL gates.
6. CMOS gates have very good temperature stability.

Disadvantages

1. CMOS gates require additional powering steps and hence production cost is higher.
2. Packing density of CMOS gates is lower compared to nMOS logic gates. But such problem can be avoided by CMOS gates with pass

transistors logic structure. Under such situation packing density can even exceed that of nMOS gates.

3. CMOS gates may be damaged because of acquired static charge in the leads. Hence nMOS/CMOS gate based IC chips should be protected from acquiring static charges by using shorted leads.

Problems

1. What is unique about TTL devices such as the 74SXX?
2. What should be done to unused inputs on TTL gates?
3. How the data can be changed from special code to temporal code?
4. Which logic family is the fastest among the TTL, ECL, CMOS, and why?
5. Which digital logic family has the lowest propagation delay time and why?
6. In digital IC, why Schottky transistors are preferred over normal transistors?
7. What is the range of invalid TTL output voltage?
8. Why is a decoupling capacitor needed for TTL ICs and where should it be connected?
9. What is fan-in and fan-out? Give an example.
10. What is noise margin? What are the different type of noise margin ? Give the proper expression for it.
11. Which of the following summarizes the important features of emitter-coupled logic (ECL)?
12. What is the word 'interfacing' as applied to digital electronics means?
13. How the problem of interfacing IC logic families that have different supply voltages (VCCs) can be solved?
14. What are the different TTL families?
15. What are the advantages of Totem Pole Output?
16. What are the disadvantages of Totem Pole Output?
17. What are the advantages of Integrated Injection Logic?
18. What are the disadvantages of Integrated Injection Logic?
19. What are the advantages of Integrated Injection Logic?
20. What are the disadvantages of Integrated Injection Logic?

Chapter 6

Application

6.1 Introduction

In today's world digital electronics is essential in the design and working of a wide range of applications, from electronics, communication, computers, control, industrial to security, and military equipments. The digital ICs used in these applications decrease in size and more complex technology is used day by day. It is essential for engineers and students to fully understand both the fundamentals and also the implementation and application principles of digital electronics, devices, and integrated circuits, thus enabling them to use the most appropriate and effective technique to suit their technical needs. This chapter presents different types of digital components. Their internal details and applications are also presented in an elaborate manner.

6.2 Digital to Analog Converter

The signals in current use are mostly generated in analogous form where the amplitudes of these analogous signals vary continuously with respect to time. The most common real time example of an analog signal is the growth of any living being like human or plant.

Foundation of Digital Electronics and Logic Design
Subir Kumar Sarkar, Asish Kumar De, and Souvik Sarkar

ISBN 978-981-4364-58-4 (Hardcover), 978-981-4364-59-1 (eBook)
www.panstanford.com

Unfortunately, noise produces easy effect on analog signal as noise is created mainly I the form of amplitude variations. This basically indicates that we are to restrict and reduce noise and hence, amplitude is to be limited thereby creating distortion of signals because signals will be clipped due to the limit imposed on the amplitude of the signal. In order to avoid distortion and limit noise, the current technique is to use digital signals which are fixed width pulses, and take only one of the two levels of amplitude 0 or 1 at any instant. The circuit which provides an interface between the digital signals of computer system and continuous signals of the analog world is called the Digital to Analog Converter (D/A Converter or DAC) which accepts digital signals (digital codes) as input and produces constant output voltages or currents. Figure 6.1 shows a typical Analog to digital conversion system. Analog converter itself is used as a subsystem. To achieve this converter, a number of techniques can be used. But we will describe only the most common techniques here.

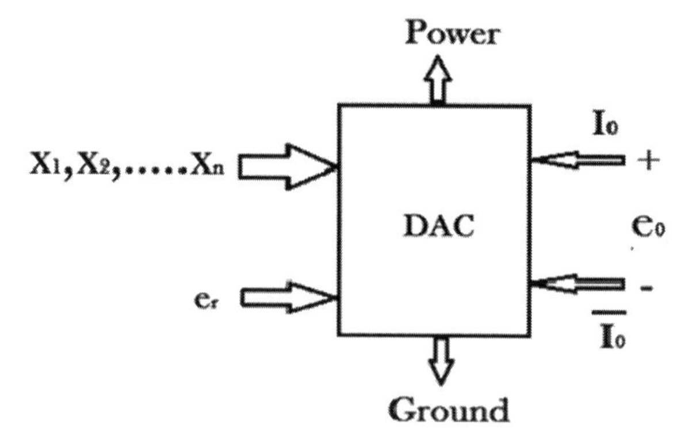

Figure 6.1 Digital to Analog Converter.

There are practically two types of DACs:

1. weighted resistor DAC
2. R-2R Ladder type DAC

An N-bit weighted resistor DAC is shown in Fig. 6.2. In both the cases we use OP-AMP as an operational device.

R_F = feedback resistor

$V_a(t)$ = analog output

V_R = reference voltage

We consider n bit weighted resistor D/A converter; each resistance can be connected to GND or V_R depending upon the type of digital input.

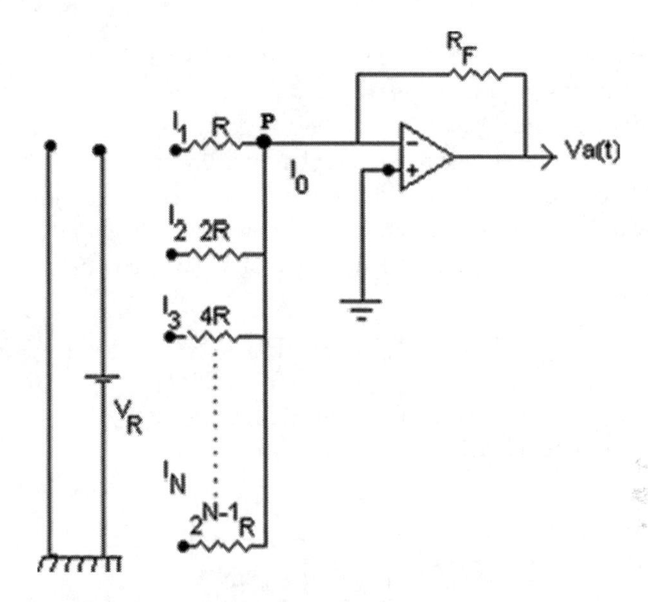

Figure 6.2 Digital to analog converter.

V_R = fixed reference voltage.

Connection to V_R means logical 1 and GND means logical 0.

The OP-AMP is assumed to be an ideal OP-AMP so the input impedance of OP-AMP is infinity.

Applying KCL at P we can write

$$I_1 + I_2 + I_3 + \ldots + I_N = I_0$$

Let $I_1 = V_{N-1}/R, I_2 = V_{N-2}/2R, I_3 = V_{N-3}/4R, \cdots, I_N = V_0/2^{N-1}R$

So, $V_{N-1}/R + V_{N-2}/2R + V_{N-3}/4R + \ldots + V_0/2^{N-1}R = I_0$

V_{N-1} is either V_R or 0 depending upon logical input 1 or 0 where $I = 0, 1, 2 \ldots N-1$

$$V_i = b_i V_R$$

where $b_i = 1$ for logical 1 input

$b_i = 0$ for logical 0 input

$$b_{N-1}V_R/R + b_{N-2}V_R/2R + b_{N-3}V_R/4R + \ldots + b_0V_R/2^{N-1}R = I_0$$

The output analog voltage is given by

$$V_a = -I_0 R_F$$

$$V_a = -R_F(V_R/2^{N-1}R)\,[b_{N-1}2^{N-1} + b_{N-2}2^{N-2} + \ldots + b_1 2^1 + b_0 2^0]$$

$$V_a = K\,[b_{N-1}2^{N-1} + b_{N-2}2^{N-2} + \ldots + b_1 2^1 + b_0 2^0] \qquad (6.1)$$

$$\text{where } K = -R_F V_R/2^{N-1}R = \text{constant} \qquad (6.2)$$

We cannot use carbon resistances R and $2R$. So, having K = constant with changes in the physical parameter such as temperature, humidity, all the resistors must be precision resistors so that its tolerance will be very small and variation is limited.

Table 6.1 depicts the analog voltage and its corresponding 3 bit digital input. This can be implemented by a 3-bit weighted resistor DAC circuit as shown on Fig. 6.3. With 3 bit weighted resistor (W–R) DACs ($N = 3$).

$$K = -(R_F/R) \times (V_R/2^{-1}) = -(R_F/R) \times (V_R/4)\ \text{V} \quad (6.3)$$

So, $V_a = K\,[2^2 b_2 + 2^1 b_1 + 2^0 b_0]$

We apply the three digital inputs b_2, b_1, b_0.

The circuit becomes:

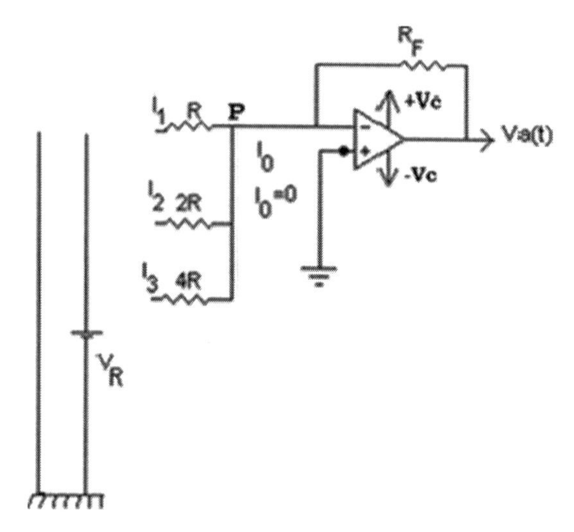

Figure 6.3 3 Bit weighted resistor (W-R) DAC.

$K = 1$ if $R_F = R$ and $V_R = (-4)$ V

with $K = 1$, $V_a = [2^2 b_2 + 2^1 b_1 + 2^0 b_0]$ V

For digital input $b_2 = b_1 = b_0 = 1$

$V_a = 2^2(1) + 2^1(1) + 2^0(1) = 7$ V

That is when all the three inputs are high connected to reference voltage V_R

$b_2 b_1 b_0 = 101$

$V_a = 2^2\,(1) + 2^1\,(0) + 2^0\,(1) = 5$ V

So, the decimal equivalent of the binary number will be the magnitude of the analog voltage. This is called unipolar W-R D/A converter.

For N bit W-R DAC, $K = 1$ if $R_F = R$ and $V_R = -2^{N-1}$

The 3 bit bipolar DAC with 1's complement analog output that is analog output voltage may be negative also.

To obtain the negative analog voltage in output, the 3 bit digital input should be applied as given in Table 6.2, and there should be some modification in the circuit. The modified circuit is depicted in Fig. 6.4. The modification in the circuit

When $b_2 = 0 \Rightarrow V_{off} = 0$ V

When $b_2 = 1 \Rightarrow V_{off} = +7$ V

$$V_a = (1)\,[2^2\, b_2 + 2^2\, b_1 + 2^0\, b_0] \tag{6.4}$$

$$V_a = \left[-\frac{V_{off}}{R_{off}} \times R_F\right]$$

By using superposition theorem, the final output for 3 bit bipolar W-R DAC will be

Table 6.1 Analog voltage and its corresponding 3 bit digital input

V_a	3 bit digital input
0 V	000
1 V	001
2 V	010
3 V	011
4 V	100
5 V	101
6 V	110
7 V	111

Table 6.2 The 3 bit digital input and corresponding analog output in 1's complement

3 bit digital input	Analog output in 1's complement
000	0 V
001	1 V
010	2 V
011	3 V
100	-3 V
101	-2 V
110	-1 V
111	0 V

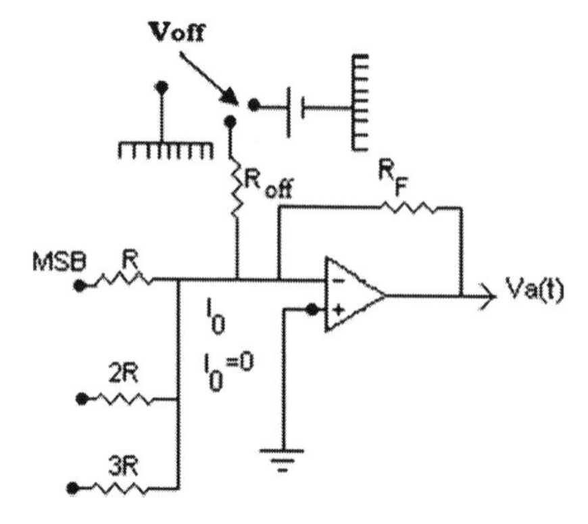

Figure 6.4 Modified circuit.

$$V_a = \left[\left(2^2 b_2 + 2^1 b_1 + 2^0 b_0\right) - \frac{V_{off}}{R_{off}} \times R_F \right] \quad (6.5)$$

If $R_F = R_{off}$, then $V_a = [2^2\, b_2 + 2^1\, b_1 + 2^0\, b_0 - V_{off}]$ (6.6)

= Final analog output voltage with the presence of offset voltage.

for $b_2b_1b_0 = 100$ from Eq. 6.6 we get:

$V_a = [2^2(1) + 2^1(0) + 2^0(0) - 7]$

$= 4 - 7$

$= -3$ V

$V_{off} = 7$ volts since $b_2 = 1$.

This gives 1's complement as if we are inputting the digital input in the 1's complement form.

Since, + 3 in 1's complement is 011 and + 3 in 1's complement is = 000 and since, it is the sign magnitude form becomes 100.

When $b_2b_1b_0 = 011$

Then $V_a = [2^1(1) + 2^0\,(1) - 0]$

= 3 volts

As $V_{off} = 0$ when $b_2 = 0$

Table 6.3 shows the 3 bit digital input and corresponding analog output in 2's complement form.

So, for 2's complement 3 bit DAC $V_{off} = 8$ V when $b_2 = 1$ and $V_{off} =$ 0 V when $b_2 = 0$

When $b_2b_1b_0 = 111$

$V_a = (2^2 \times 1 + 2^1 \times 1 + 2^0 \times 1 - 8)$

$= (-1)$ V

as $V_{off} = 8$ V when $b_2 = 1$

Table 6.3 3 bit digital input and corresponding analog output in 2's complement

3 bit digital input	Analog output in 2's complement
000	0 V
–001	1 V
–010	2 V
–011	3 V
–100	–4 V
–101	–3 V
–110	–2 V
–111	–1 V

The resistances are in GP with common ratio of 2 from top to bottom.

6.2.1 Disadvantages of W-R DAC

1. For N bit W-R DAC we require N wide spread values of resistances whose values are R, $2R$, $4R$,...$2^{N-1}R$ which are very difficult to achieve or realize. (This exact ratio of resistors is difficult to get in carbon resistors)
2. Wide spread values of precision resistors are also difficult to realize.
3. The variation of V_R is not allowed that is reference voltage source must be highly stable. Due to this W-R DAC are used not beyond 4 bit DAC.

6.2.2 The R-2R Ladder Type DAC

If we extend the ladder to ∝, the input impedance remains fixed.

Figure 6.5 shows the circuit of R-2R ladder type DAC, and its equivalent circuit is given in Fig. 6.6. It uses R-2R ladder type network whose input impedance are always constant [for all digital inputs (LSB, MSB)].

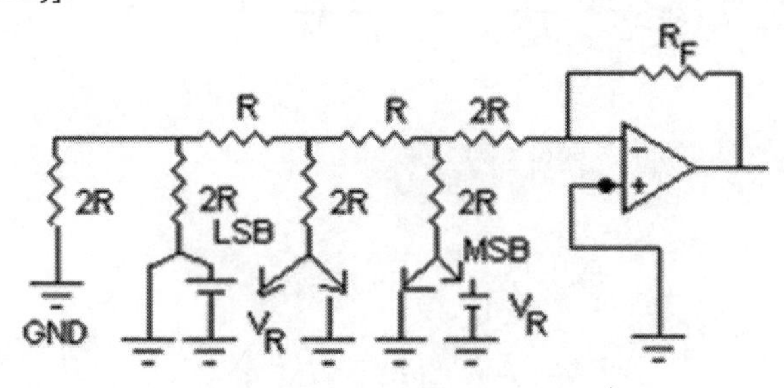

Figure 6.5 Circuit of R-2R ladder type DAC.

Let us consider, 3 bit R-2R ladder type DAC:

The OP-AMP is at the end used to convert current into voltage. The MSB will be near to OP-AMP and LSB farthest from OP-AMP.

Let the input is 001.

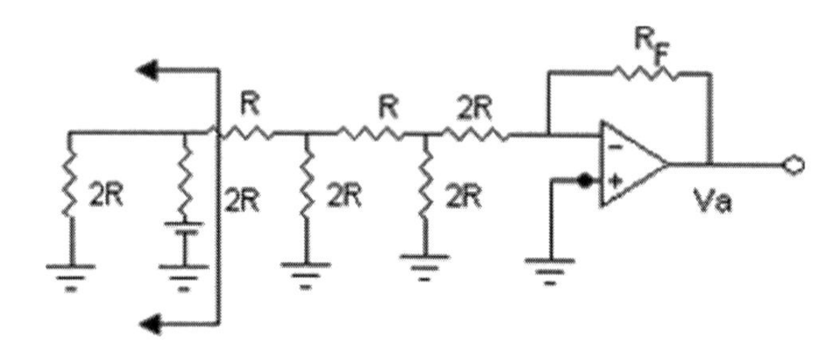

Figure 6.6 Equivalent circuit of R-2R ladder type DAC.

Using Thevenin's theorem here and the equivalent voltage source

$$V_{XY} = \left(\frac{V_R}{4R}\right) \times (2R) = \left(\frac{V_R}{2}\right) \tag{6.7}$$

and so the Thevenin equivalent circuit as shown in Fig. 6.7 is:

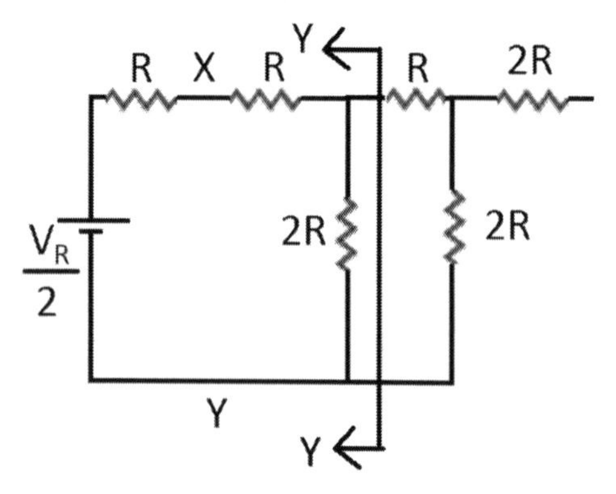

Figure 6.7 Thevenin's equivalent circuit of Fig. 6.6.

$$V_{XY} = \left(\frac{\frac{V_R}{2}}{4R}\right) \times 2R = \left(\frac{V_R}{2^2}\right) \tag{6.8}$$

So, Thevenin equivalent of circuit which is shown in Fig. 6.8 is:

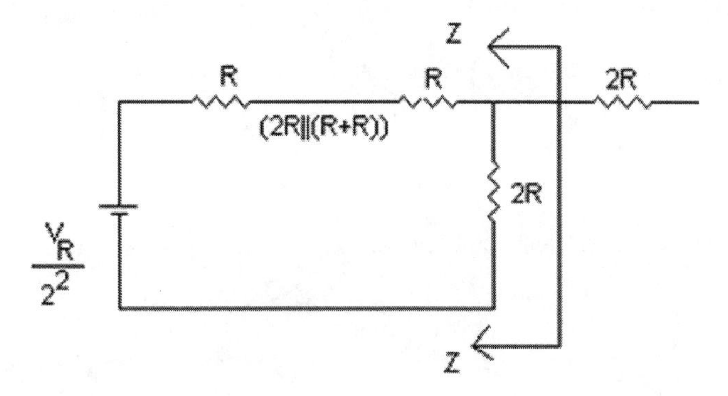

Figure 6.8 Thevenin's equivalent circuit of Fig. 6.7.

So, the final Thevenin equivalent circuit is Fig. 6.9:

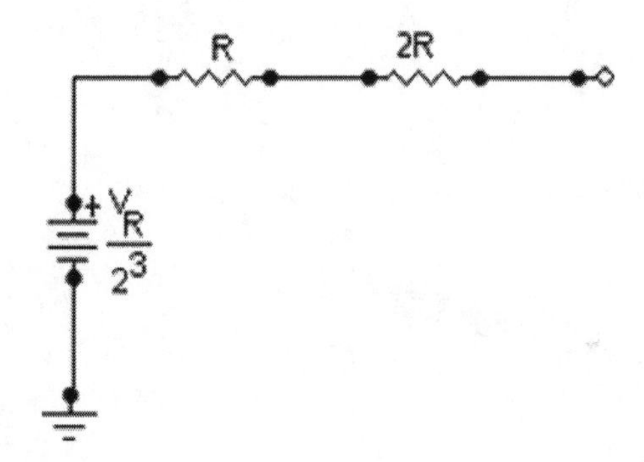

Figure 6.9 Final Thevenin equivalent circuit of Fig. 6.8.

Now, let input be 010. The circuit, equivalent circuit, Thevenin equivalent circuit and final Thevenin equivalent circuit of R-2R ladder type DAC for input 010 are shown in Figs. 6.10, 6.11, 6.12 and 6.13 respectively.

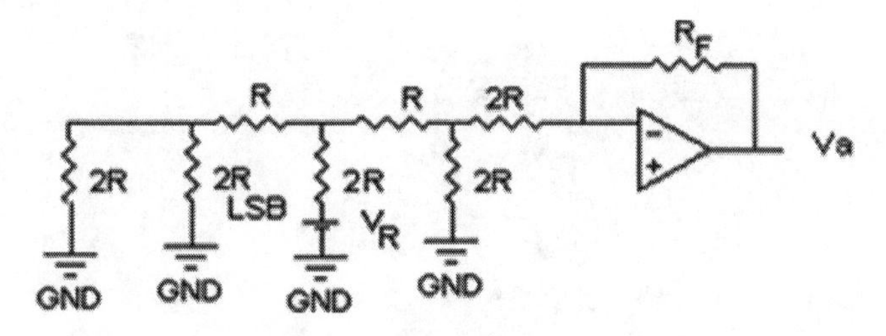

Figure 6.10 Circuit of R-2R ladder type DAC for input 010.

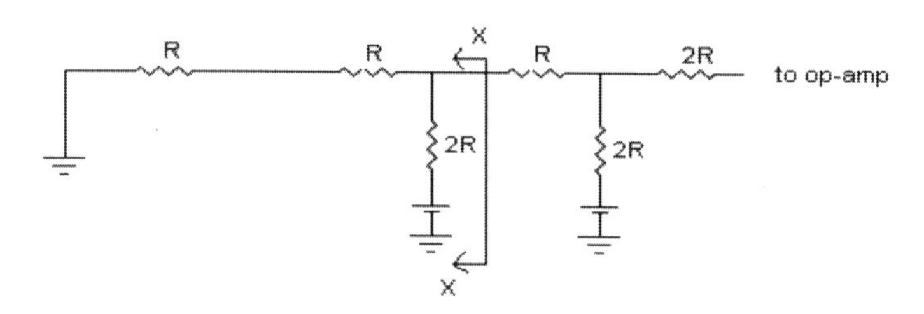

Figure 6.11 Equivalent circuit of R-2R ladder type DAC.

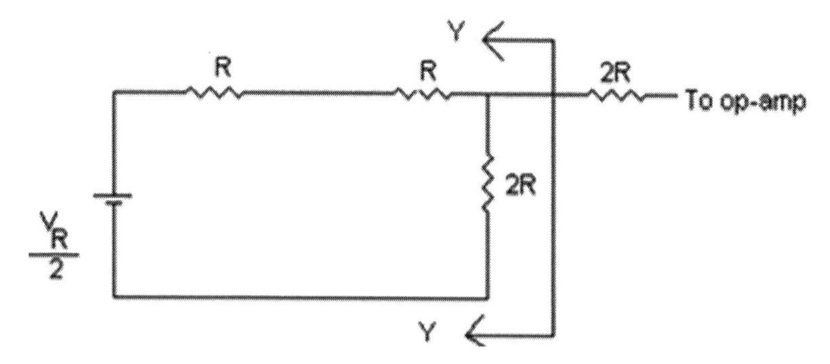

Figure 6.12 Thevenin equivalent circuit of Fig. 6.11.

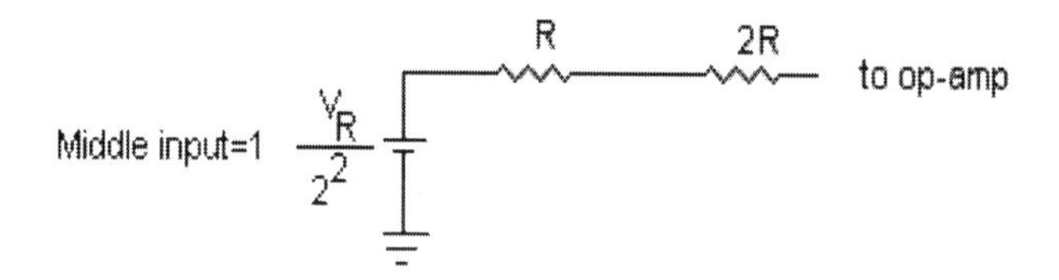

Figure 6.13 Thevenin equivalent circuit of Fig. 6.12.

Let input now be 100, the circuit, equivalent circuit and Thevenin equivalent circuit of R-2R ladder type DAC for input 100 are shown in Figs. 6.14, 6.15 and 6.16 respectively.

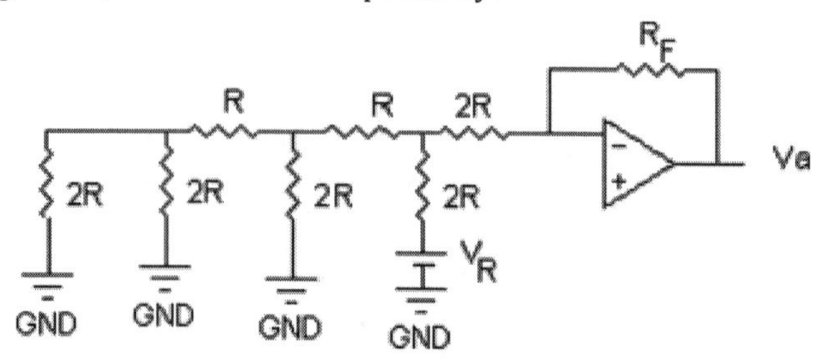

Figure 6.14 Circuit of R-2R ladder type DAC for input 100.

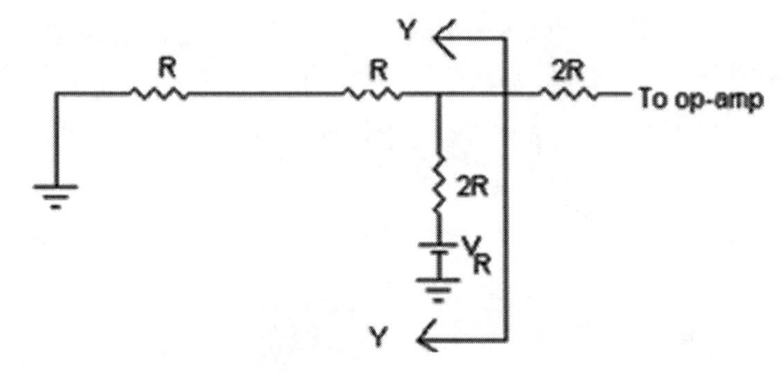

Figure 6.15 Equivalent circuit of R-2R ladder type DAC.

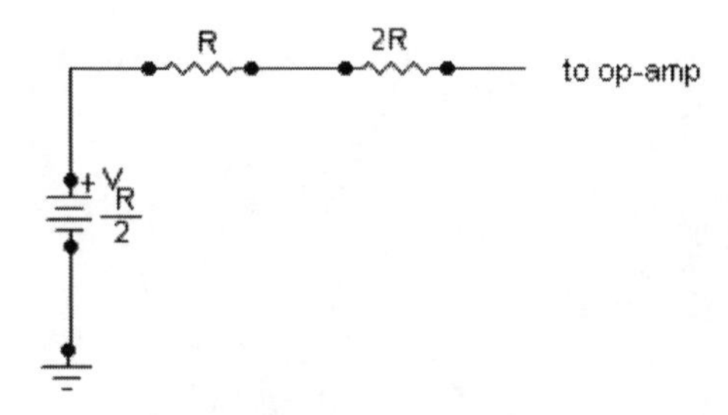

Figure 6.16 Thevenin equivalent circuit of R-2R ladder type DAC.

Combining above figures, we get the equivalent circuit for 3 bit R-2R ladder type DAC as shown in Fig. 6.17:

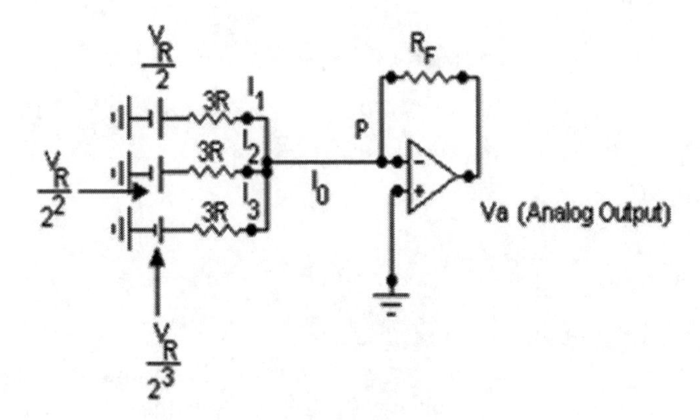

Figure 6.17 Equivalent circuit for 3 bit R-2R ladder type DAC.

The weighted resistors are same and not wide spread values of resistors are required. So, it overcomes all disadvantages of W-R DA converters.

By applying KCL at P:

$$I_0 = I_1 + I_2 + I_3$$

$$I_o = \frac{V_R}{2(3R)} + \frac{V_R}{2^2 3R} + \frac{V_R}{2^3 3R}$$

$$I_o = \frac{1}{3R}\left[\frac{V_R}{2} + \frac{V_R}{2^2} + \frac{V_R}{2^3}\right]$$

$$I_o = \frac{1}{3R} V_R \left[\frac{2^2 + 2 + 1}{2^3}\right]$$

$$I_o = \frac{V_R}{3R}\left[\frac{2^2 b_2 + 2^1 b_1 + 2^0 b_0}{2^3}\right] \quad (6.9)$$

where $b_i = 1$ or 0

$$I_o = \frac{V_R}{3R(2^3)}[2^2 b_2 + 2^1 b_1 + 2^0 b_0]$$

$$V_a = -I_0 R_F$$

$$V_a = -\frac{V_R R_F}{3R(2)^3}[2^2 b_2 + 2^1 b_1 + 2^0 b_0]$$

$$V_a = K[2^2 b_2 + 2^1 b_1 + 2^0 b_0] \quad (6.10)$$

where $K = -V_R \times \dfrac{R_F}{3R(2^3)}$

$K = (1) \Rightarrow R_F = 3R$ and $V_R = -2^3\,\text{V} = -8$ V and when $K = 1$

$$V_a = [2^2 b_2 + 2^1 b_1 + 2^0 b_0] \quad (6.11)$$

For 1's and 2's complement method, the same procedure is followed.

6.2.3 Advantage

Only three different values of precision resistors are required whose values are R, 2R, and 3R (not widespread values) and so, even a single resistance R can be used sufficiently so produce 2R and 3R by connecting them in series.

6.3 8 Bit D/A Counter IC DAC 0808/DAC 1408

8 bit D/A counter IC DAC 0808/DAC 1408 and its circuit are shown in Figs. 6.18 and 6.19 respectively. This is also called microprocessor compatible DA converter.

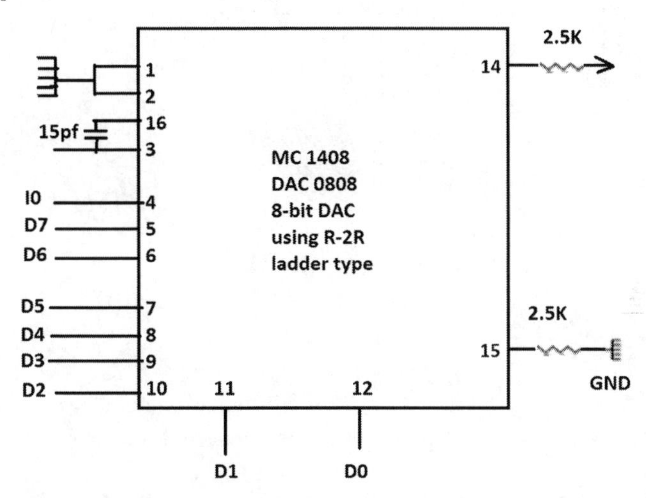

Figure 6.18 8 bit D/A counter IC DAC 0808/DAC 1408.

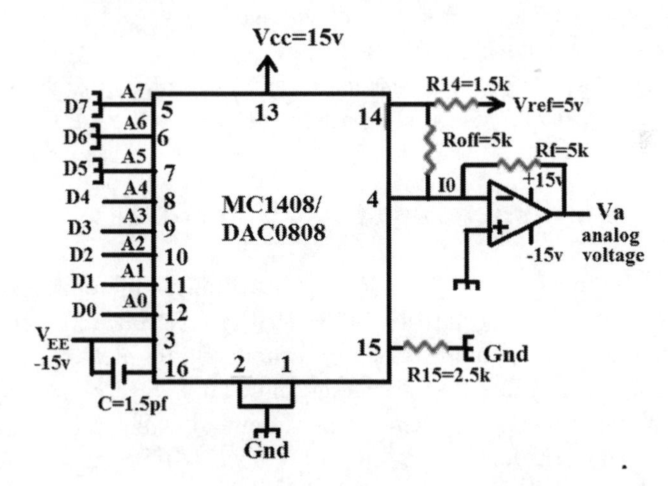

Figure 6.19 Circuit diagram of 8 bit D/A counter IC DAC 0808/DAC 1408.

$$I_0 = \frac{V_{ref}}{R_{14}}\left[\frac{A_7}{2}+\frac{A_6}{4}+\frac{A_5}{8}+\frac{A_4}{16}+\frac{A_3}{32}+\frac{A_2}{64}+\frac{A_1}{128}+\frac{A_0}{256}\right] \quad (6.12)$$

A_7 to A_0 are the digital inputs applied through D_7 to D_0.

Output analog voltage = $-I_0 R_F$

$$= \frac{V_{ref}}{R_{14}} \times R_F \left[\frac{A_7}{2} + \frac{A_6}{4} + \frac{A_5}{8} + \frac{A_4}{16} + \frac{A_3}{32} + \frac{A_2}{64} + \frac{A_1}{128} + \frac{A_0}{256} \right]$$

Maximum current I_0 will be obtained when $A_7, A_6, A_5, A_4, A_3, A_2, A_1, A_0$ all are 1.

and $(I_{0\,MAX}) = \frac{V_{ref}}{R_{14}} \times \left(\frac{255}{256} \right)$

$$= \frac{5}{2.5\text{k}} \times \left(\frac{255}{256} \right)$$

$$\approx 1.992 \text{ mA}$$

$(V_a)_{MAX}$ output analog voltage $= I_0 \times R_F$

$$= (1.992 \text{ mA}) \times 5 \times 10^3$$

$$= 1.992 \times 10^{-3} \times 5 \times 10^3$$

$$= 9.960 \text{ V}$$

= full scale voltage

Since, there are tolerances of the resistors, hence, exact 9.960 will not be obtained.

In unipolar mode, the output analog voltage varies from 0 to 10 V. In bipolar mode, the output analog voltage varies from –5 to + 5 V. The range or swing is 10 V.

We can have different logical operation by selecting different S_3 S_2 S_1 S_0, and for *A* or *L* operation, the *M* is useful.

6.4 Bit Arithmetic and Logic Unit (74181)

So, ALU is the heart of the CPU = (ALU + control unit) = microprocessor. Fig. 6.20 shows a 24 pin DIP 74181 (LSI) chip, 4 bit ALU. Logical operations corresponding to different combinations of S_3, S_2, S_1, and S_0 are given in Table 6.4 Cascading 2 bit ALU we can realize 8 bit ALU. When $M = 1$ we get logical operation and when $M = 0$ we get arithmetic operation.

Output = 1 if $A = B$

= 0 if $A \neq B$.

If carry is generated *G*, output will glow and if carry is propagated P will glow. The *G* and *P* are used with carry look ahead as cascading outputs for faster additional subtraction.

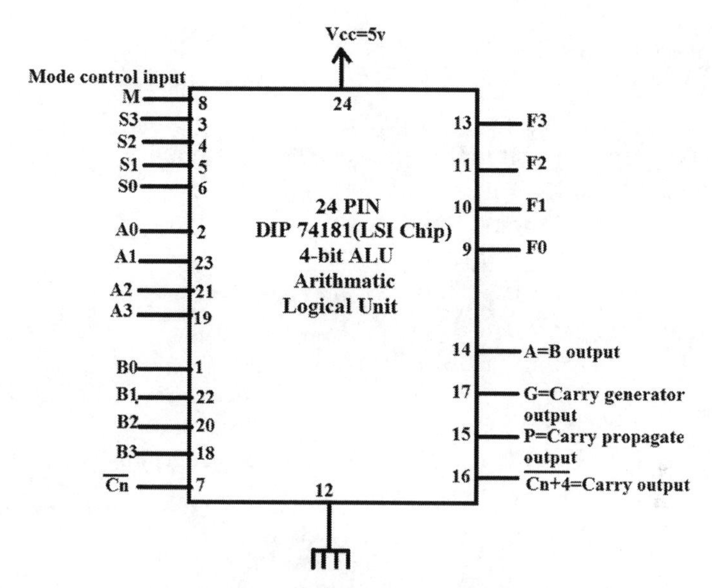

Figure 6.20 A 24 pin DIP 74181 (LSI) chip, 4 bit ALU.

Table 6.4 Logical operations corresponding to different combinations of S_3, S_2, S_1, and S_0

$S_3\ S_2\ S_1\ S_0$	Logical operation
0000	$\overline{A}$
0001	$\overline{A+B}$
0010	$\overline{A}B$
0011	0
0100	$\overline{AB}$
0101	$\overline{B}$
0110	$A \oplus B$
0111	$A\overline{B}$
1000	$\overline{A}+B$
1001	$\overline{A \oplus B}$
1010	B
1011	AB
1100	1
1101	$A+\overline{B}$
1110	A + B
1111	A*4

$M = 0$ (arithmetic)

$\bar{c}_n$ =1 (with no carry) as c_n = 0. $\bar{c}_n$ = 0 means with carry

Cascading of two 4 bit ALU to perform 8 bit addition/subtraction that is to realize 8 bit ALU as shown in Figs. 6.21 and 6.22:

For addition $M = 0$ for arithmetic operation.

Say we want to add $A = (45)_{10}$

$$B = (89)_{10}$$

$(45)_{10} = (2D)_{16}$, $(89)_{10} = (59)_{16}$

$(2D)_{16} = (00101101)_2$, $(59)_{16} = (01011001)_2$

MSB LSB MSB LSB

$\bar{C}_n = 1$	$\bar{C}_n = 0$
01 A plus B	A plus B plus 1
10 A minus B minus 1	A minus B

$\bar{C}_{n+4} = 1$ when result is negative

$\bar{C}_{n+4} = 0$ " " " positive

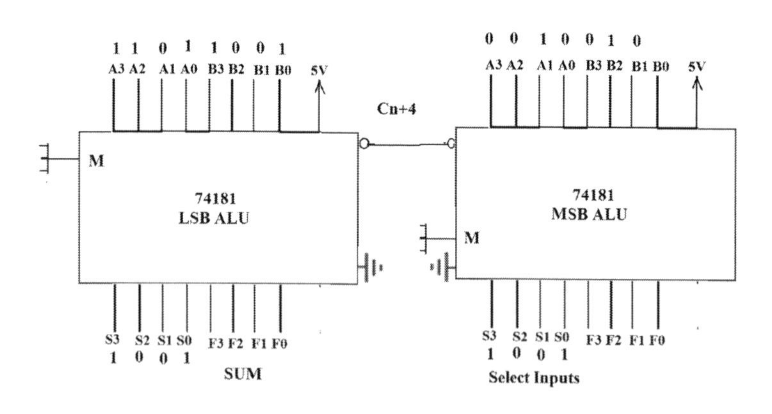

Figure 6.21 Cascading two 74181 ALU units.

M	$\overline{C_{n\text{LSB}}}$	A_{LSB}	B_{LSB}	S_{LSB}	$\overline{C_{n+4\text{ LSB}}}$	$\bar{C}_{n\text{ MSB}}$	A_{MSB}	S_{MSB}	$\overline{C_{n+4}}$
0	1	1101	1001	0110	0	0	001001	1000	1

$$\begin{array}{r} 1101 \\ 1001 \\ \hline 10110 \end{array}$$

Final sum $= S_{\text{MSB}}\, S_{\text{LSB}}$

$= 1000\ 0110 = 1 \times 2^7 \times 2^2 \times 1 + 2^1$

$= (134)_{10}$

This is for addition.

Now consider:

S_3	S_2	S_1	S_0	M	C_{nLSB}
0	1	1	0	0	0

$$\begin{array}{r} 1101 \\ -1001 \\ \hline 0100 \end{array} \qquad \begin{array}{c} C_{n+4} = 0 \\ \hline C_{n+4} = 1 \end{array} \qquad \begin{array}{r} 0010 \\ 0101 \\ \hline 1101 \end{array}$$

Here also $A = (45)_{10}$ and $B = (89)_{10}$

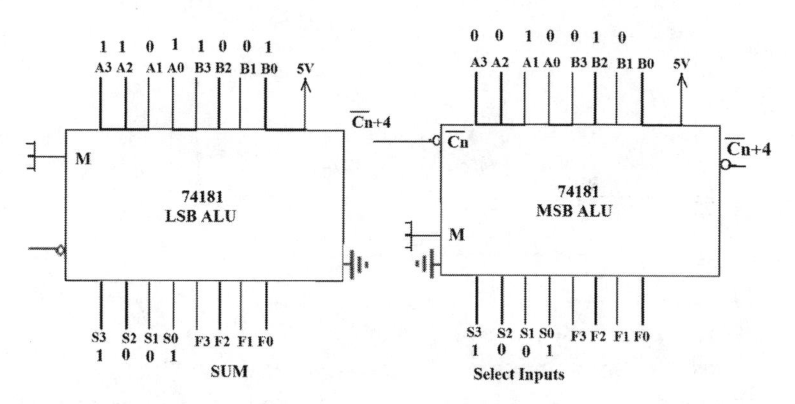

Figure 6.22 Circuit representations of two cascaded ALU units.

Final result:

$S_{\text{MSB}}\, S_{\text{LSB}}$

1101 0100

$= -128 + 64 + 16 + 4$

$= (-128 + 84)$

$= -44$

which is (45–89)

6.5 Carry Look Ahead Adder (CLA Adder)

Parallel full adder has got the disadvantage of carry propagation delay. In a CLA adder the carry propagation delay is reduced by reducing the number of gates through which the carry propagates. For an *n* bit CLA adder there is 4 bit propagation delay and the delay will not increase with increase of *n*.

We consider the truth table of a full adder as given in Table 6.5, and the corresponding circuit diagram is given in Fig. 6.23. The SOP form and the POS form of this circuit diagram are given in Figs. 6.24 and Fig. 6.25 respectively:

Table 6.5 Truth table of CLA adder

Inputs			Outputs		
A	B	C_{i-1}	S	C_i	
0	0	0	0	0	
0	0	1	1	0	
0	1	0	1	0	P_i
0	1	1	0	1	P_i
1	0	0	1	0	P_i
1	0	1	0	1	P_i
1	1	0	0	1	G_i
1	1	1	1	1	G_i

If $A = B = 1$, irrespective of c_{i-1} and $c_i = 1$.

These two conditions are called carry generate conditions.

When at least A or B is 1, the four C_is are called the carry propagate stage P_i.

$$G_i = A_i\, B_i \tag{6.13}$$

$$P_i = A_i \oplus B_i \tag{6.14}$$

Carry will generate at '*i*'th stage if that stage is a carry generator stage, or if that stage is carry propagate stage and previous carry bit is 1.

So

$$C_i = G_i + P_i C_{i-1} \tag{6.15}$$

$$S_i = A_i \oplus B_i \oplus C_{i-1}$$

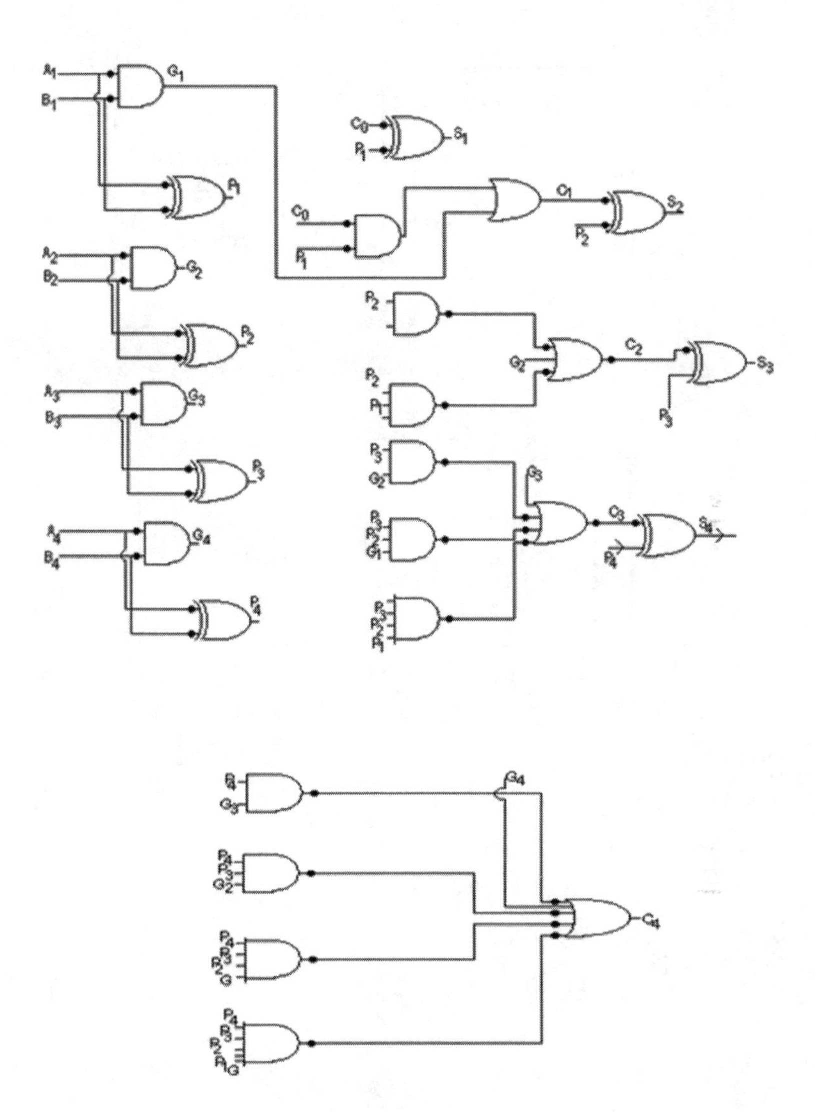

Figure 6.23 Circuit diagram of CLA adder.

$$S_i = P_i \oplus C_{i-1} \tag{6.16}$$

Let us now consider, 4 bit CLA adder.

4 bit binary $A_4 A_3 A_2 A_1 A_0 \leftarrow$ Augends bits

$+ B_4 B_3 B_2 B_1 B_0 \leftarrow$ Addend bits

Final carry $\leftarrow C_4 S_4 S_3 S_2 S_1$

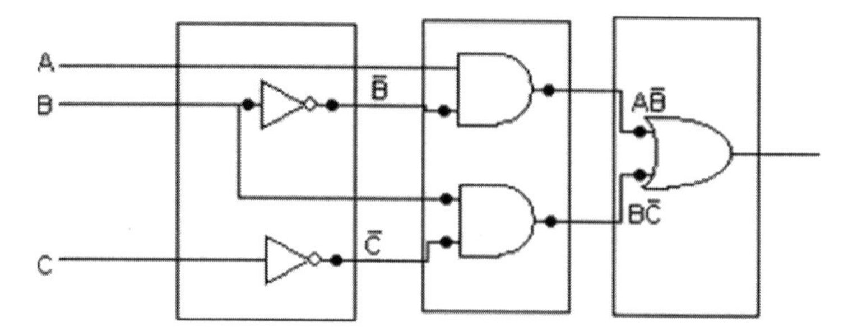

Figure 6.24 The SOP form circuit representation.

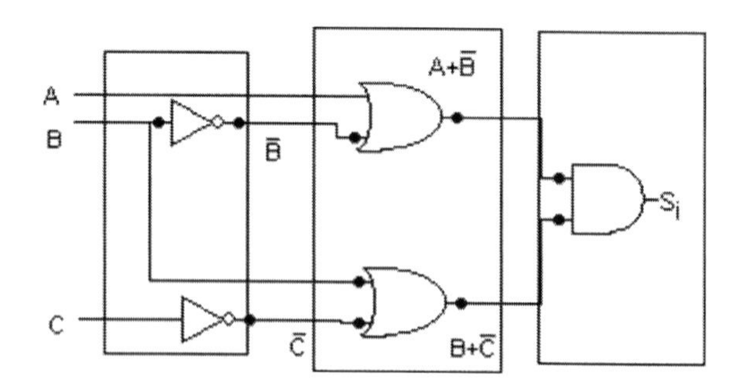

Figure 6.25 The POS form circuit representation.

So from Eq. (6.16) we get $S_1 = P_1 \oplus C_0$
In Eqs. (6.15) and (6.16) put $i = 1$

$$C_1 = G_1 + P_1C_0 \tag{6.17}$$

Putting $I = 2$

$$S_2 = P_2 \oplus C_1$$
$$C_2 = G_2 + P_2C_1$$
$$= G_2 + P_2\,(G_1{+}P_1C_0)$$
$$= G_2 + P_2G_1 + P_2P_1C_0 \tag{6.18}$$

Put $i = 3$

$$S_3 = P_3 \oplus C_2$$
$$C_3 = G_3 + P_3C_2$$
$$= G_3 + P_3G_2 + P_3P_2G_1 + P_3P_2P_1C_0 \tag{6.19}$$

Put $I = 4$

$$S_4 = P_4 \oplus C_4$$
$$C_4 = G_4 + P_4C_3$$

$$= G_4 + P_4P_3G_2 + P_4G_3 + P_4P_3P_2G_1 + P_4P_3P_2P_1C_0 \qquad (6.20)$$

Now we can realize Eqs. 6.17, 6.18, 6.19, 6.20 by logic gates,

If there is no restriction for the number of inputs for a gate then the 4 bit CLA adder can be realized by maximum of 4 gate propagation delay and if provided we assume that propagation delay for OR, AND, and XOR gate are same. For n bit CLA adder also we can realize it by 4 gate propagation delay if gates with any number of inputs are available. [Here maximum no of inputs = 5, for 4 bit, for n bit, we require $(n + 1)$ input AND gate]. But the hardware complexity is increased. There is a gain in speed over parallel addition.

The logical expressions for S_i and C_i can be minimized by using any suitable minimizing technique (K map or QM technique). The minimized expressions for S_i and C_i are in sum of product (SOP) form or product of sum (POS) form and any one of these forms can be realized by three gate propagation delay.

In sum of product form, they are:

$$C_i = AB + BC_{i-1} \qquad (6.21)$$

$$S_i = A\overline{BC}_{i-1} + \bar{A}B\bar{C}_{i-1} + \overline{AB}C_{i-1} + ABC_{i-1} \qquad (6.22)$$

6.6 Analog to Digital Converter

A simple A/D conversion scheme is shown in Fig. 6.26. This is a circuit which accepts an unknown continuous analog signal at its input and converts it into an 'n'- bit binary number which can be easily manipulated by a computer.

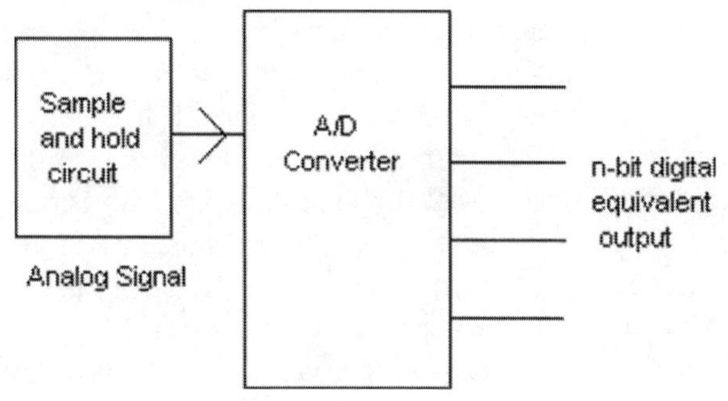

Figure 6.26 A/D Conversion scheme.

The analog signal should remain constant for an instant of time, because to get the digital equivalent at the output, the A/D converter needs some finite time. So, generally a signal is not applied directly but through sample and hold circuit.

6.6.1 Sample and Hold Circuit

A typical diagram of Sample and hold circuit. Is shown in Fig. 6.27. This configuration is called unity gain amplifier, as output = input, gain = unity.

There is isolation between input and output. Hence, it is called a buffer.

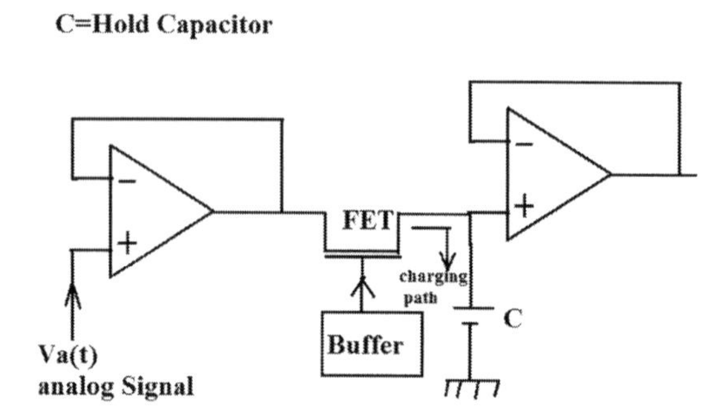

Figure 6.27 Sample and hold circuit.

6.6.2 Sampling Pulse

Figure 6.28 shows the waveform of sampling pulse for sample and hold circuit. During next T sec sampling pulse is low, Q is off. The capacitor cannot charge through Q nor does it discharge through amplifier. So, during this interval when V_S = low, capacitor holds the charge and the voltage across it will be constant. This is called hold time. So, this is the purpose of sample and hold circuit. After this again during T, capacitor will charge and then again hold for some T seconds. Hence, A/D converter works in the interval 2 or converts during this interval, as analog voltage is constant at this time interval and it requires no A/D conversion error.

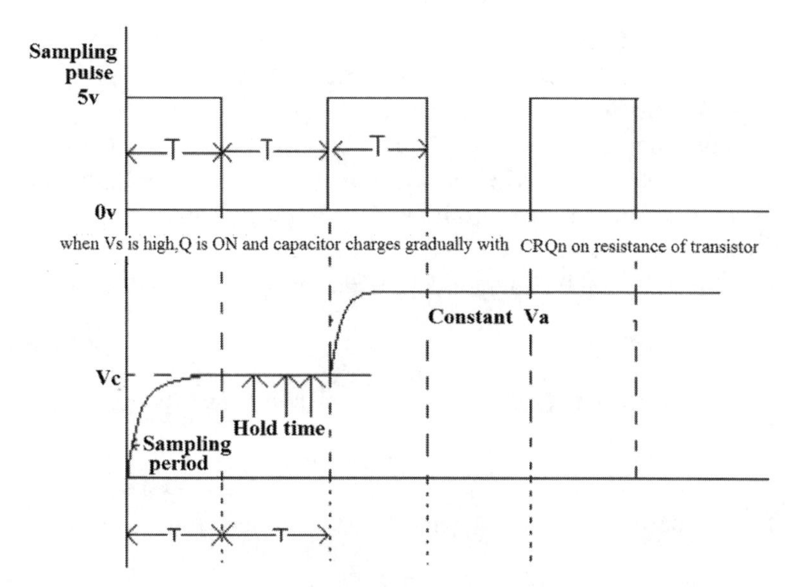

Figure 6.28 Waveform of sampling pulse.

A/D Conversion time ≤ Hold time.

Commercially available sample and hold amplifier IC is LF-398. shown in Fig. 6.29.

The value of C is in the range of 100 PF to 0.1 μF. Supply voltage is in between 5 to 18 V (dual supply voltage).

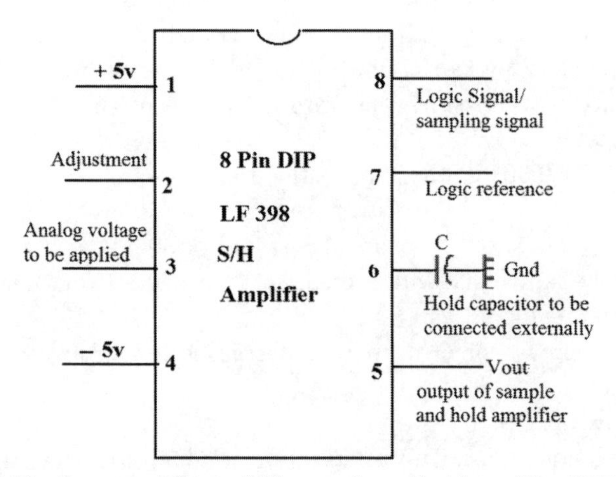

Figure 6.29 Commercially available sample and hold amplifier IC (LF-398).

The offset is used to get zero output with zero input. With zero input, if V_{out} is finite, so resistance is attached with pin number 2. As this offset error is very small so pin number 2 is left unconnected.

At pin number 7, the logic reference is normally connected to zero or ground. When logic signal at pin number 8 is high, the device takes samples of input signal at pin number 3 and thus holds capacitor charges during this time (sampling interval). When logic signal is low, the output at pin number 5 is held constant.

Parameters of this IC:

Acquisition time

It is the time needed for getting proper sample. Ideally, it should be zero that is we expect sampling to be instantaneous but practically it is finite determined by charging time constant of C and other constants.

Aperture Time

It is the time required to open the switch for holding the capacitor voltage constant (As switching is done by transistor from on to off stage of transistor, so that time is aperture time).

Drop rate

This is determined by the decrease of output voltage at pin number 5 with time during hold period (Even under the off condition of the transistor, the capacitor will slowly discharge through the off resistance of transistor. That rate of change of voltage is called drop rate).

When $C = 0.001\ \mu F$ then acquisition time = 4 sec
aperture time = 150 ms
and drop rate = 30 mV/sec

It can be shown that with decrease of C, acquisition time decreases but drop rate increases.

With increase of C, drop rate decreases but acquisition time increases.

So how to select C?

Acquisition time and drop rate has to be small so we have to make a trade-off between drop rate and acquisition time.

6.7 A/D Converter Types

Parallel type A/D converter, Counter type ADC, Successive approximation type ADC and Dual slope type ADC are some of the A/D converters which are frequently used. From the point of view we may have the following A/D converters:

1. fastest ADC having greatest hardware complexity.
2. slowest ADC, but with simplest hardware complexity.
3. most accurate and quite fast and (BLANK) reasonably complex hardware.
4. slowest, not so accurate, less complicated hardware.
5. voltage to time and voltage to frequency ADC.

6.7.1 2 Bit Parallel/Simultaneous Type ADC

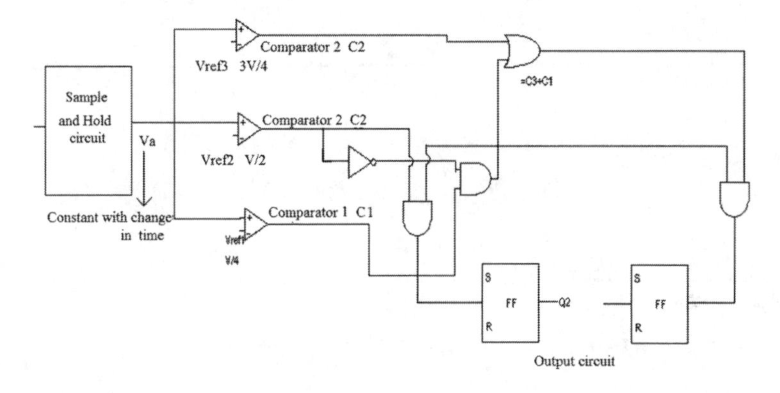

Figure 6.30 2 Bit parallel/simultaneous type ADC.

If we want to convert 3 V analog into 2 bit digital output, we use this circuit as shown in Fig. 6.30 and the comparator circuit that is used in 2 bit parallel/simultaneous type ADC is shown in Fig. 6.31.

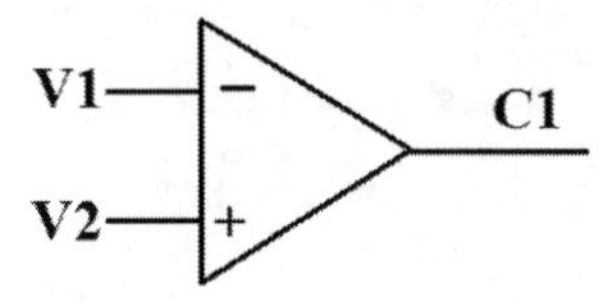

Figure 6.31 A comparator.

Comparator Output:

$$C_1 = 1 \text{ if } V_1 > V_2$$
$$= 0 \text{ if } V_1 <= V_2 \tag{6.23}$$

6.7.2 A/D Converter

Table 6.6 Analog to digital conversion

Analog voltage input	Dec	C_1	C_2	C_3	b_1	b_2
$0 < V_a < \frac{V}{4}$ (min analog voltage that can be measured)	0	0	0	0	0	0
$\frac{V}{4} < V_a < \frac{V}{2}$	4	1	0	0	0	1
$\frac{V}{2} < V_a < \frac{3V}{4}$	6	1	1	0	1	0
$\frac{3V}{4} < V_a < 0V$	7	1	1	1	1	1

$$b_1 = f(C_1, C_2, C_3) = \sum m\,(4, 7) + \sum d\,(1, 2, 3, 5) \tag{6.24}$$
$$b_2 = f(C_1, C_2, C_3) = \sum m\,(6, 7) + \sum d\,(1, 2, 3, 5) \tag{6.25}$$

A typical Analog to digital conversion chart is given in Table 6.6 and the corresponding K maps to find logical expressions of b_2 and b_1 are given in Fig. 6.32. After solving the K- maps and implementing them, we get the circuit diagram of Analog to digital converter as

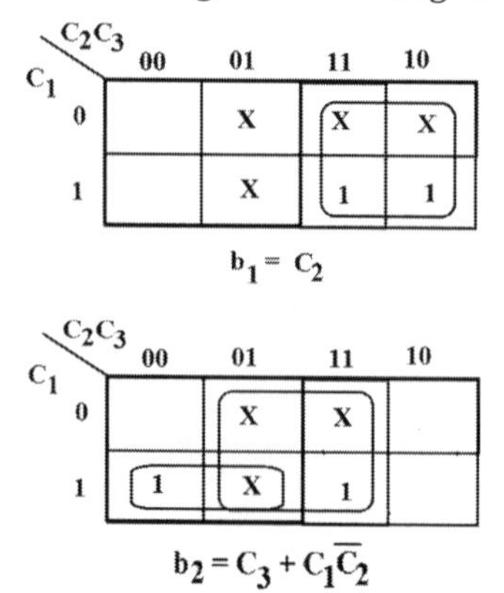

Figure 6.32 K Maps to find logical expressions of b_2 and b_1.

shown in Fig. 6.33 By observing the table we can write the logical expression for b_2 and b_1.

We apply the analog voltage simultaneously. So, it is called the parallel type. The maximum analog voltage that can be converted is 3 V.

For 2 bit parallel type ADC we require = $(3 \times \frac{V}{4})$

$2^2 - 1 = 3$ comparators

For n bit parallel type ADC we require $(2^n - 1)$ comparators.

For 3 bit parallel type ADC we require $2^3 - 1 = 7$ comparators. (With seven comparators, we can measure eight possible voltages.).

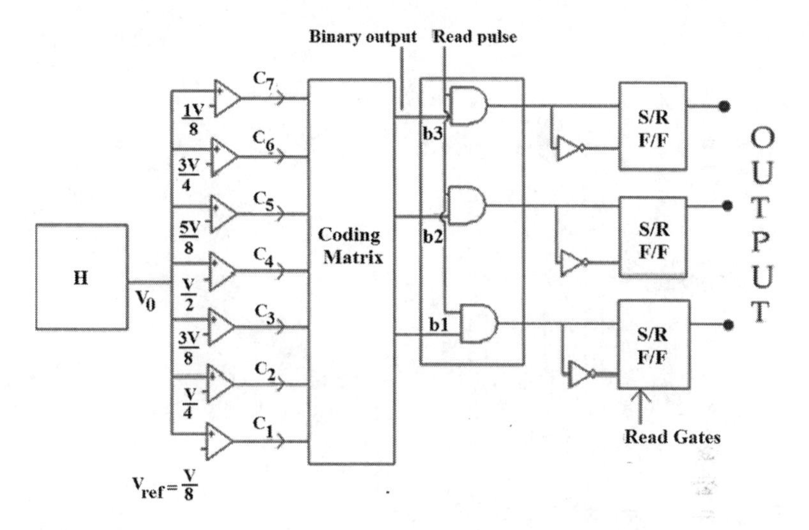

Figure 6.33 Analog to digital converter.

$V_{ref1} = \frac{V}{8} = \frac{7}{8} = 0.875$ ($V = 7$ volts in analogy with 2 bit parallel type ADC or 3 bit parallel type ADC, we can write $b_3 = C_4$ (output of middle comparator)

$$\overline{b_2} = \overline{C_4}\overline{C_5} + \overline{C_6} \tag{6.26}$$

$$b_1 = C_1C_2 + C_3C_4 + C_5C_6 + C_7 \tag{6.27}$$

6.7.2.1 Disadvantage

Since, hardware complexity increases tremendously, generally not more than 3 bit ADC are used as parallel type.

6.7.2.1.1 *Main advantage* Fastest ADC and propagation delay is small, conversion time is extremely small.

6.7.3 Counter Type ADC (3 bit CTADC)

Figure 6.34 shows a counter type A/D converter and the corresponding conversion chart is given in Table 6.7. The maximum conversion time for an n bit counter type ADC = $2n \times T$ sec where T is the time period of the clock pulse used in the counter type ADC. 3 bit counter type ADC requires:

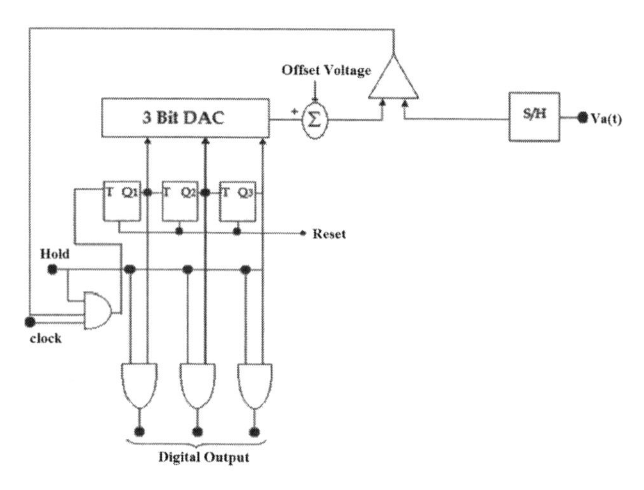

Figure 6.34 Counter type A/D converter.

1. a sample and hold circuit.
   A typical hold time diagram is shown in Fig. 6.35
2. a comparator.
3. a summer or adder.
4. one 3 bit DAC.
5. 3 bit ripple counter.
6. one clock pulse generator.
7. three output AND gates.
8. one AND gate for enabling clock pulse.

6.7.3.1 Disadvantage

It is the slowest.

With $H = 1$, the circuit takes the sample to start new conversion and counter output is read by three AND gates to give digital output

Table 6.7 Counter type A/D conversion

H	Q_3	Q_2	Q_1	V_s	V_o	V_a	C_0	Clk
0	0	0	0	0	0.5	6.3 V		
0	0	0	1	1	1.5	6.3 V		
0	0	1	0	2	2.5	6.3 V		
0	0	1	1	3	3.5	6.3 V		
0	1	0	0	4	4.5	6.3 V		
0	1	0	1	5	5.5	6.3 V		
0	1	1	0	6	6.5	6.3 V	0	no clock

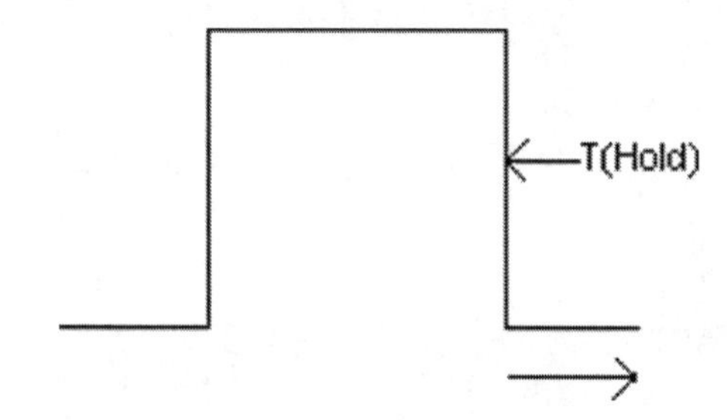

Figure 6.35 Hold time.

for previous A/D conversion. For comparator, $C_0 = 1$ if $V_a > V_o$ and $C_0 = 0$ if $V_a < V_o$.

$H = 1$ at the beginning of the conversion and at the end of conversion the output of the counter gives the digital outputs.

The counter should be reset that is $Q_3Q_2Q_1 = 000$ at the beginning.

Error in this case = 6.3 V – 6 V = 0.3 V

When clock is disabled, H becomes 1 automatically.

$H = 1$ as soon as 1 counter stops. $H = 1$ to read the counter output and also to take samples for next generation.

Without using the offset we get V_0 as V_S.

So	0	110	6 V	6 V	6.3 V	0.7
	0	111	7 V	7 V	6.3 V	0

Error here is 0.7 V.

Digital output without offset = 111 = 7 V

with offset = 110 = 6 V

Use of offset voltage reduces the quantization error. Due to use of this offset voltage, the error in the output is never greater than ± 0.5 V.

All the ADCs discussed so far are unipolar (that is we can only convert the positive analog voltage to digital) that is it gives only magnitude, but sign is not found out.

$$\text{The maximum conversion time of clock pulse} = 2^3 \times T_{\text{sec}} \qquad (6.28)$$
$$= 8\ T\text{sec}$$
$$f_{\text{clock}} = 10\ \text{MHz}$$
$$T = 10^{-7}\ \text{secs}$$
$$\text{Maximum conversion time} = 8 \times 10^{-7}\ \text{sec}$$
$$= 0.8\ \mu\text{sec}$$

If S/H LF398 is used,
Aperture time is 150 ms.
Since, 0.8 μsec > 150 ms so there is error.
So, if the hold time ≥ 0.8 μsec then there is no error.

1. one 5 bit ring counter
2. three clock S–R F/F
3. six 2 input AND gates
4. one 3 bit D/A converter
5. one comparator offset $= \frac{1}{2}$ LSB = 0.5 V (001) → LSB
6. one S/H circuit
7. one Adder.

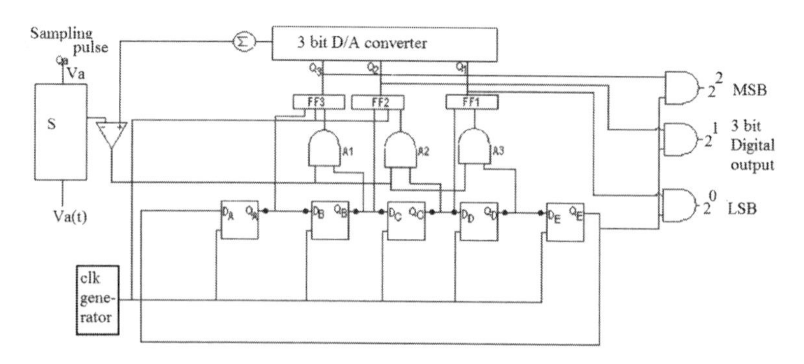

Figure 6.36 3 Bit successive approximation type ADC.

Figure 6.36 shows a 3-bit successive approximation type ADC. The 3 bit successive approximation type ADC requires five clock cycles for A/D conversion where the first 3 clock cycles are required for A/D conversion and the 4th clock cycle is required to read the digital output when $QE = 1$ and the fifth clock cycle is required to reset the SAR and to take new samples for next A/D conversion. For an n bit successive type ADC we require $(n + 2)$ clock cycles.

It is much faster than counter type. It is much more complex and accurate also. The truth table of 3-bit successive approximation type ADC is given in Table 6.8. Initially before A/D conversion, the ring counter is set so

Q_A	Q_B	Q_C	Q_D	Q_E
0	0	0	0	0

and FF_2 and FF_1 are reset and FF_3 is set to 1 so that the first trial $\omega t = 100$. The offset voltage is subtracted from V_S and C_0 accordingly varies.

Let $V_0 = (V_S - 0.5\ V)$

Table 6.8 Successive approximation type truth table

Clk	Q_A	Q_B	Q_C	Q_D	Q_E	Q_3	Q_2	Q_1	V_S	V_0	V_a	
0	1	0	0	0	0	1	0	0	4 V	3.5 V	5.3 V	0
1	0	1	0	0	0	1	1	0	6 V	5.5 V	5.3 V	1
2	0	0	1	0	0	1	0	1	5 V	4.5 V	5.3 V	0
3	0	0	0	1	0	1	0	1	5 V	4.5 V	5.3 V	0

Conversion ends after three clock pulses as after first 3 clock pulse, conversion is complete.

So, the S–R F/F output is read as 3 bit digital equivalent of analog input $V_a = 5.3$ V.

Initial State

5 10000 100 4 V 3.5 V

Next conversion starts after resetting the S–R flip-flop to 100.

So the equivalent digital output is (5.3–5.0) = 0.3 less than the analog voltage. It can be shown that without offset voltage, error is more. After first comparison, it finds unknown analog voltage is greater than $Q_3Q_2Q_1$ and after second comparison it finds that unknown analog voltage is less than $Q_3Q_2Q_1$.

6.7.4 Dual Slope A/D Converter

When $Q_n = 0$, the input to the OP-AMP R-C integrator is = V_a = the unknown analog input (of the output of S/H circuit).

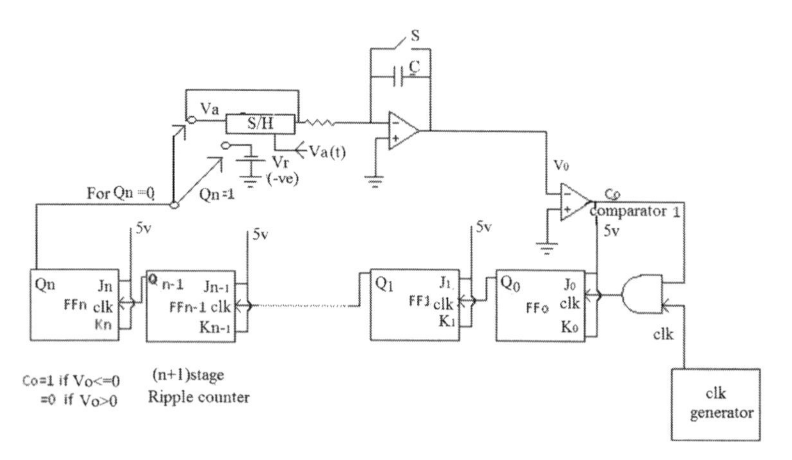

Figure 6.37 A dual slope A/D converter.

When $Q_n = 1$, the input to the integrator is negative.

$$\text{Output of the integrator} = V_0(t) = \frac{1}{RC}\int V_i(t)\mathrm{d}t$$

$$V_0(t) = -\frac{1}{\tau}V_a\int dt$$

$$V_0(t) = -\frac{V_a}{\tau}\int dt \tag{6.29}$$

So, initially reset the $(n + 1)$ stage ripple counter.
So, the counter output is all 0000 ... 0 = $(n + 1)$ bit output.

	Q_n Q_0	
Clk	$Q_n \cdots Q_0$	Count
0	0000 … 0	0
1	0000 … 1	1
2	0000 … 10	2

$$V_0(t) = -\frac{V_a}{\tau}\times t \tag{6.30}$$

So, $V_0(t)$ increases linearly in the negative direction.
After N th clock pulse $\quad Q_n \cdots Q_0$

1000000000000 …

and count $= 2^N$

and output voltage

$$V_0 = -\frac{V_a}{\tau} \times T_1 \text{ where } T_1 = 2^N \times T_C \text{ sec} \tag{6.31}$$

Summarizing:

$$V_0(t) = -\frac{V_a}{\tau} \times T_1 \text{ for } Q_n = 0 \tag{6.32}$$

$$V_0(t) = \frac{V_a}{\tau} \times T_1 + \frac{V_r(T_2 - T_1)}{\tau} \tag{6.33}$$

And say, after $t = T_2$sec, comparator output C_0 becomes 0 which disables the clock and the counter stops counting and when $C_0 = 0$, as $V_0(t) \geq 0$ then read the the first N stage counter.

So,

$$\frac{V_r(T_2 - T_1)}{\tau} = \frac{V_a T_1}{\tau} \tag{6.34}$$

During $(T_2 - T_1)$ = let the count on the n stage counter is λ.

T_1 is taken as 0.

So, $(T_2 - T_1) = \lambda T_C$

$$\text{So, } \frac{V_r(\lambda T_C)}{\lambda} = \frac{V_a}{\lambda} 2^N T_C \tag{6.35}$$

$$V_a = \frac{\lambda V_r}{2^N} \tag{6.36}$$

where V_r = reference voltage.

If V_r is set 2^N then $V_a = \lambda$

Where λ = count in the N stage ripple counter Fig. 6.38 shows the dual slope scheme.

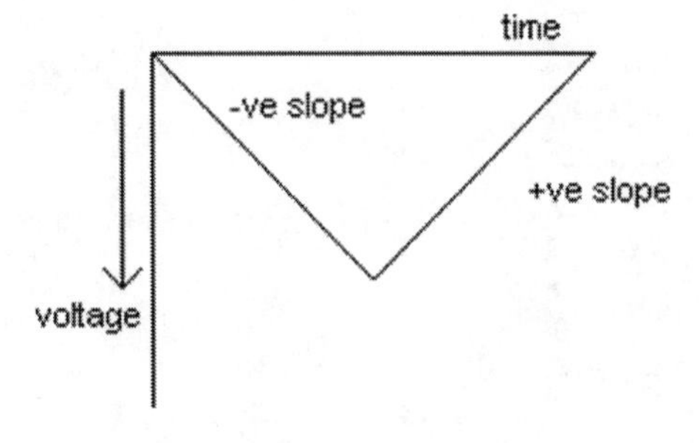

Figure 6.38 Dual slope scheme.

The dual slope is used to determine (V_a) to nullify the effect of the resistance and capacitance on analog voltage (which are temperature sensitive.) due to a variation of temperature which will affect the value of V_a increased by using single slope ADC. This is widely used in the digital voltmeter where analog voltage is converted to digital signal and this digital signal is displayed.

6.8 Microprocessor Compatible A/D Converter 0809

In this A/D converter, there are eight equivalent digital outputs for any one of the analog voltage out of eight inputs.

It has address latch enable (ALE) and output enable (OE).

It has eight input channels.

The particular analog input which will be selected to give corresponding digital output depends on channel select input.

SC = Start conversion input

EOC = End of conversion output

Clk = Clock input

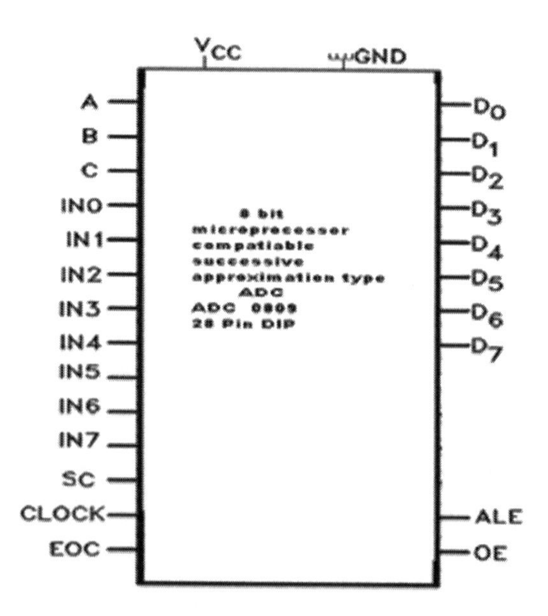

Figure 6.39 Microprocessor compatible A/D converter.

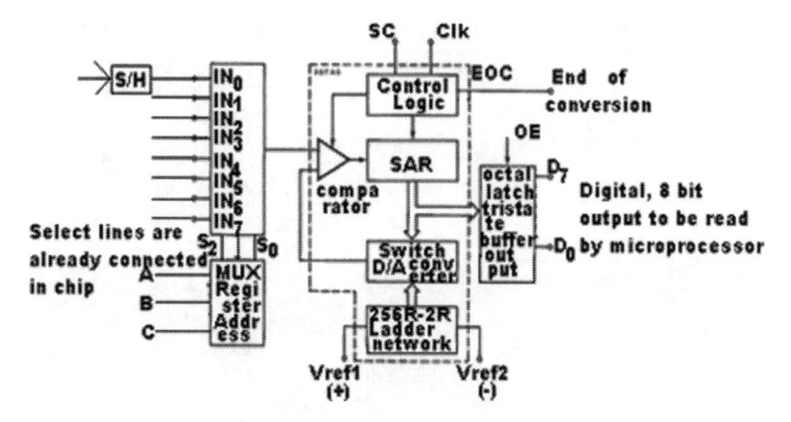

Figure 6.40 Internal block diagram.

ALE = Address latch enable input coming from ALE output of a μp.
OE = Output enable
Unless OE = 1, no output is available

Figure 6.39 shows a Microprocessor compatible A/D converter 0809. & Fig. 6.40 shows its internal diagram.

Table 6.10 shows the truth table of this A/D converter.

Besides the normal two states (0 and 1), it also a high impedance state. The truth table of microprocessor compatible A/D converter is given in Table 6.10. High impedance means it will neither offer nor accept anything.

The internal block diagram of microprocessor compatible A/D converter is shown in Fig. 6.40.

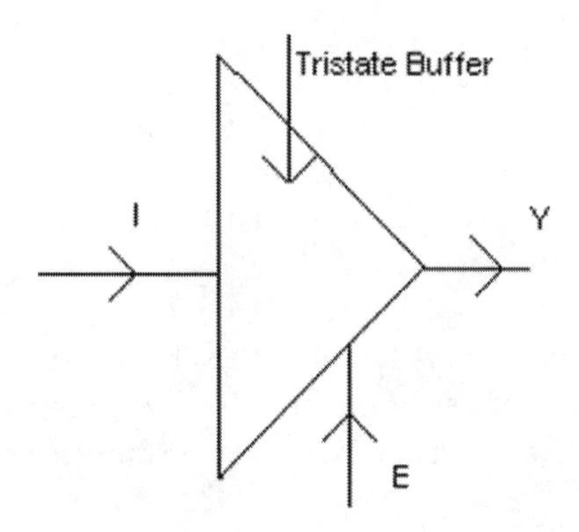

Figure 6.41 A tristate buffer.

Table 6.9 Truth table of a tristate buffer

EN	*I*	*Y*
1	0	0
1	1	1
0	X	High impedance means it will neither offer nor accept anything

Table 6.10 Truth table of microprocessor compatible A/D converter

Clock	*SC*	*OE*	*EOC*	Operation
1	1	X	X	SAR is cleared or Reset.
1	1	X	X	A/D conversion begins and it requires time to complete conversion.
1	X	X	1	When EOC = 1 A/D conversion is completed
1	X	1	1	μp can now read the equivalent digital output via some input port.

Figure 6.41 shows a tristate buffer that has similar output states with this A/D converter & Table 6.9 gives its truth table.

6.9 Specifications of D/A Converter

6.9.1 Resolution

This is the smallest possible voltage that can be measured or converted by a DAC. A 10 bit D/A converter has $\frac{1}{2^{10}}$ bit resolution. If the full scale reading or maximum output of a D/A converter is 10 V then its resolution in volts is:

$$\begin{aligned}\text{Volts} &= \frac{1}{2^{10}} \times \text{FSR} \\ &= \frac{1}{1024} \times 10 \text{ volts} \\ &\approx \frac{1}{100} = 0.01 \text{ volt} \\ &\approx 10\text{mV}\end{aligned} \tag{6.37}$$

Greater the number of bits, the resolution will be greater. So, with increase of number of bits in a D/A converter, the resolution increases that is we can measure or convert much smaller voltage.

$$\%\,\text{resolution} = \frac{1}{2^{10}} \times 100\% = \frac{1}{10} \approx 0.1 \tag{6.38}$$

6.9.2 Linearity of a DAC

If the relationship between digital input and analog output is linear then we can say that the characteristics is linear and thus, it is called DAC is linear. Figure 6.42 shows linearity of a DAC.

The magnitude of the error = actual output–expected output

V_2 = expected output

V_1 = actual output

$$\text{So, } |\in| = |V_2 - V_1| \tag{6.39}$$

If $\in$ is small, that DAC is following linearity. If $\in$ is large then DAC deviates from linearity. Closer the actual and expected analog output, the better is the linearity.

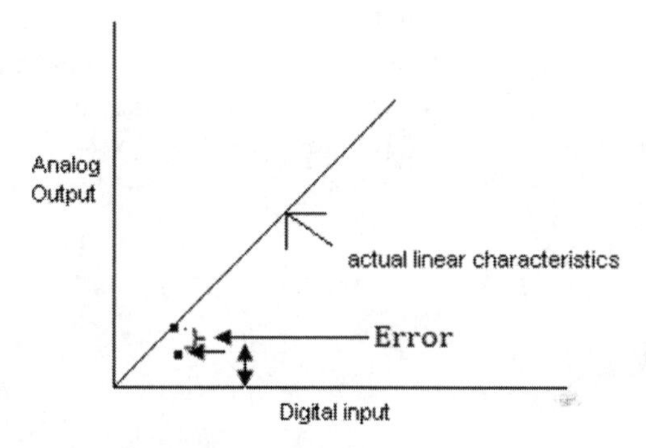

Figure 6.42 Linearity of a DAC.

6.9.3 Accuracy

Error $|\in|$ = Actual analog output–expected analog output $|\in|$ is very small then the accuracy of DAC will be larger.

Actually accuracy of a DAC is defined mathematically as

$$\frac{|\varepsilon|}{\text{FSR}} \tag{6.40}$$

$$\% \text{ accuracy} = \frac{|\varepsilon|}{\text{FSR}} \times 100\% \tag{6.41}$$

$$= \left(\frac{|\varepsilon|}{\text{FSR}} \times 100 \right)\%$$

For example:

For a DAC having FSR = 10 V if its accuracy is 0.20%. What is the maximum possible error?

$$\frac{|\varepsilon|}{\text{FSR}} \times 100 = 0.2$$
$$|\varepsilon| = 0.02\text{V}$$
$$= 20\text{ mV}$$

6.9.4 Settling Time

This is defined as the time required for a DAC to produce a settled final analog output for a digital input before the input is changed. Figure 6.43 shows settling time of a DAC. DAC uses switches, reference voltages, capacitances, and inductances and this produces transients in a DAC. That is why output requires some time to settle to a final value and that time is called settling time.

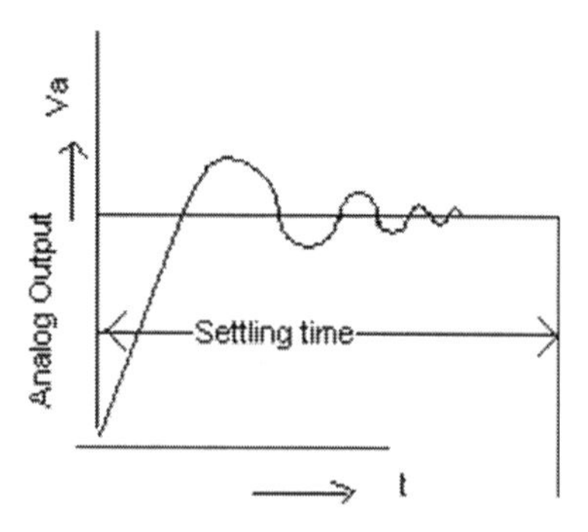

Figure 6.43 Settling time.

6.9.5 Temperature Sensitivity

The DAC uses resistances, OP-AMPs, and reference voltage source which vary with temperature. Manufacturer for a DAC specify the temperature sensitivity by a term called parts per million/°C. Temperature sensitivity of a DAC is specified by = PPM/°C (Out of 10^6 parts, how many parts vary with temperature).

6.10 Specification of A/D Converter

6.10.1 Input Voltage Range

For ADC generally input voltage range is 0 to 10 V or ±5 V or ±10 V, and so on depending on the type of ADC.

6.10.2 Input Impedance

It is generally in the range of 1 KΩ to 1 MΩ and input capacitance is few tens of Pico Farad.

6.10.3 Conversion Time

For a moderately fast ADC the conversion time is 50 μsec and for a fast ADC the conversion time is 50 nsec.

6.10.4 Format of ADC

An ADC can be unipolar, bipolar, or 1's complement or 2's complement.

6.10.5 Accuracy

Accuracy of ADC depends upon the quantization error and error due to any external noise and all other sources of error affect the accuracy of ADC.

$$\text{Quantization error} = \pm\frac{1}{2}\text{LSB bit} \tag{6.42}$$

For a 10 bit ADC, the quantization error =

$$\left(\frac{1}{2}\times\frac{1}{2^{10}}\right)=\left(\frac{1}{2}\times\frac{1}{2^{10}}\times 100\right)\%$$

$$\text{Accuracy} = \frac{|\varepsilon|}{\text{FSR}}\times 100 = \left(\frac{\text{maximum possible error}}{\text{FSR}}\right)\times 100\% \tag{6.43}$$

6.10.6 Stability or Temperature Sensitivity

Stability or temperature sensitivity is usually specified by PPM/°C of FSR.

If for an ADC with PPM = 20 PPM/°C, what is the error at 125°C? Generally PPM usually specified at a temperature of 25°C.

$$\begin{aligned} &\text{FSR} = 20\text{V} \\ &\text{PPM}\,(125-25) = \frac{(\Delta t) \rightarrow \text{error}}{\text{FSR}} \end{aligned} \quad (6.44)$$

$$20 \times 10^{-6} \times 100 = \frac{(\Delta \text{E})}{20}$$

6.11 Applications of Counters

(a) Counters can be used for direct counting of events.
(b) Counter can be used as divide by 10 circuits. A mod N counter can be used as a $\div N$ circuit.
(c) Counter can be used as a waveform generator.
(d) Counter can be used for serial to parallel conversion by using demultiplexer.
(e) Counter can be used for parallel to serial conversion by using multiplexer.
(f) Counter can be used for designing sequence generator.
(g) Counters are used as a subsystem in A/D converter.
(h) Counters can be used in a digital computer.
(i) Counter is used to find the frequency of a signal.
(j) Counter can be used to determine the time interval between two events.
(k) Counter can be used to determine the velocity of an object.
(l) Counter can be used to determine the distance between two points.
(m) Counter can be used in an automatic parking control system.
(n) Counter can be used in a digital weighing machine.
(o) Counter can be used to simulate a digital clock.
(p) Counter can be used in code generator.

6.11.1 Generation of Square Wave by ZCD

Any analog signal whose f_r is required, is passed through a zero crossing detector and the square wave is the output.

Figure 6.44 shows square wave generation by ZCD. The AND gate is enabled for 1 sec. So, the counter counts a frequency of the unknown signal.

Since, crystal oscillator gives very stable output, so it is used along with $\div 10^6$ decade counter so frequency gives square waves of 2 secs duration. Figure 6.45 gives the circuit representation of counter.

If the $Q = 0.1$ sec for which count is λ

$\therefore$ for $Q = 1$ sec for which count is $10\ \lambda$.

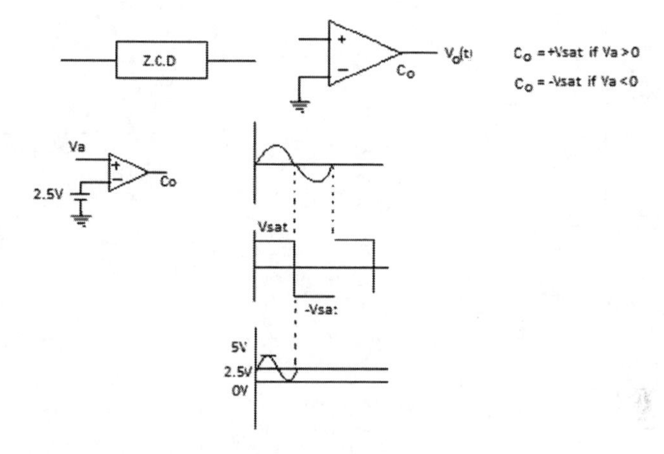

Figure 6.44 Square wave generation by ZCD.

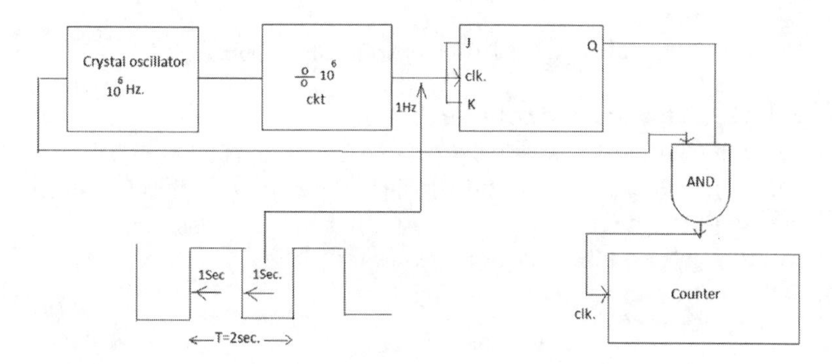

Figure 6.45 Circuit representation.

6.11.2 Counter Application

The counter can be used to determine time interval between two events. The circuit diagram of counter with pulse inputs is given in Fig. 6.46. Here we use a photocell arrangement.

The AND gate is enabled for T_1 sec and during this T_1 sec, the number of square waves appearing at input of AND gate and the count in the counter αT_1.

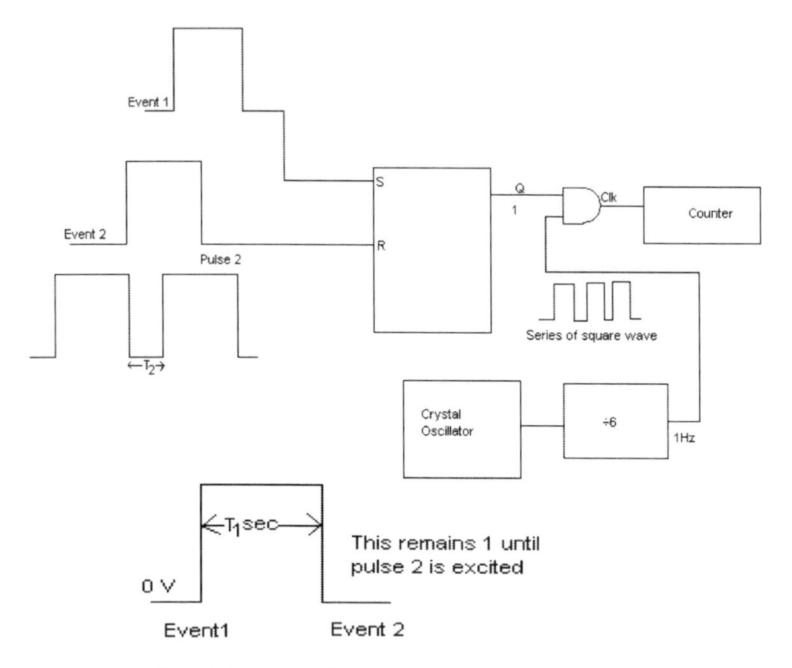

Figure 6.46 Circuit diagram of counter with pulse inputs.

6.11.2.1 Velocity of a counter

As soon the object passes here, pulse is generated at A, since photo cell arrangement is there and similarly the pulse is generated at B. We measure the time interval T by using counter technique (time interval between the two pulse generated).

Velocity of a counter is given by

$$v = \text{velocity} = (S/T) \tag{6.45}$$

as given in Fig. 6.47.

We can measure the distance between the transmitter and receiver and this is done by radar as given in Fig. 6.48.

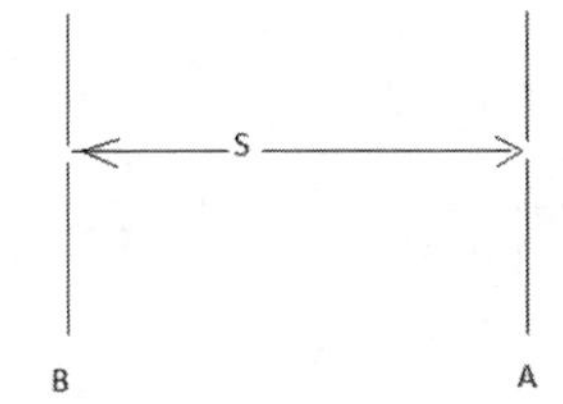

Figure 6.47 Velocity of the counter.

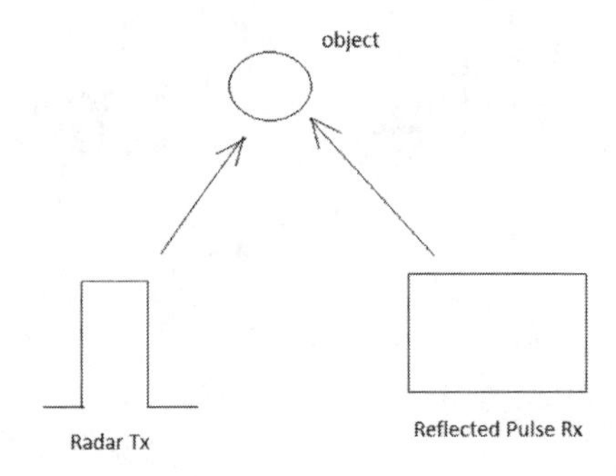

Figure 6.48 Distance measurement by radar.

Here, we measure the time interval between incident and reflected pulse from an object at a distance '*d*' from T_x and ending at R_x.

S = *vt*, *t* = since interval between two events that is incident and reflected part and $2d = c \times T$

$$\text{and } d = \left(\frac{CT}{2}\right) \tag{6.46}$$

So, the accuracy of measurement of '*d*' depends upon the measurement of time interval *T* and for this, the crystal oscillator is used which gives very stable output frequency.

6.11.3 Automatic Parking Control System

Figure 6.49 shows the block diagram of an automatic parking control system scheme and Fig. 6.50 shows the detailed circuit diagram of it. One component is a counter. Other components are sensors which are LDR (Light depended register). When light is incident on the sensor, the resistance of the register changes.

When the garage will be full, interface = 1. It activates a relay and this in turn activates gate and the gate is closed and as soon as any vehicle leaves, the gate is raised and down count starts.

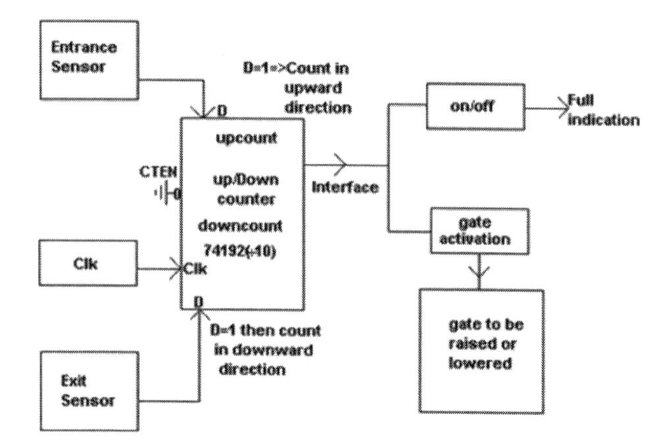

Figure 6.49 Automatic parking control system scheme.

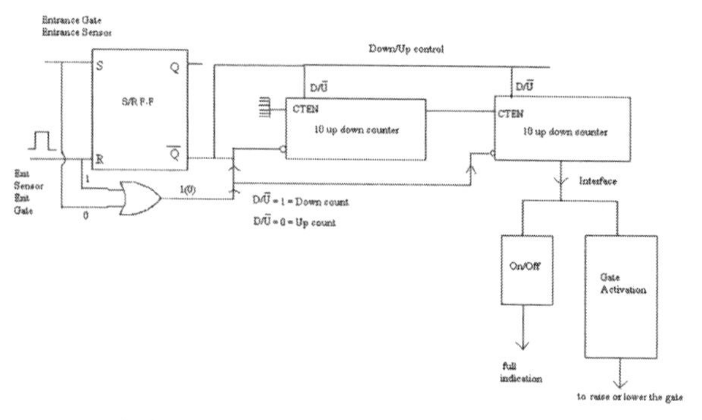

Figure 6.50 Detailed circuit diagram of the automatic parking control system.

6.12 Schmitt Trigger Circuit

6.12.1 Applications

1. It can be used as a bistable multivibrator. It has two stable states. It can switch from one stable state to other depending upon inputs.
2. It can be used as voltage comparator as it has got sharp reference voltage for comparison.
3. Any time varying signal can be converted into a square wave or a sine wave can be converted into the square wave.

6.12.2 Transistorized Schmitt Trigger Circuit

Figure 6.51 shows the circuit of a transistorized Schmitt trigger and Table 6.11 gives its truth table. For this circuit, when Q_1 is off then

Q_2 is on and then $V_0(t)$ will be small (logical 0) and when Q_1 is on Q_2 is off, and $V_0(t)$ is high (logical 1). These log 0 for one input voltage range when Q_1 and log 1 are the two stable states.

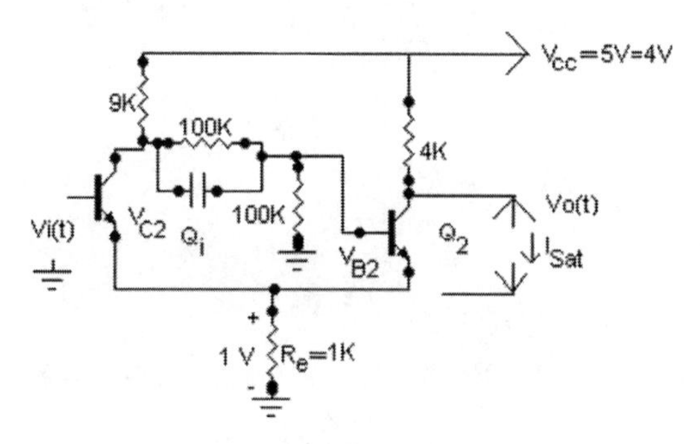

Figure 6.51 Transistorized Schmitt trigger circuit.

Table 6.11 Truth table of Schmitt trigger

$V_i(t)$	Q_1	V_{C1}	V_{B2}	Q_2	V_{C2} or $V_o(t)$
$V_i(t) < 1.7\ V$	Off	5 V	R-C circuit couples this large voltage to the base of Q_2. So V_{B2} is high.	On	0.2 V = Logical 0. Low stable state.
$V_i(t) \geq 1.7$ V	On	$V_{CE\ ON\ Sat} \approx 0.2$ V	Small	Off	5 V = high stable state

To make the output $V_0(t)$ low again, we have to reduce the input voltage $V_i(t)$ when Q_1 is on.

Here

$$I_{SatQ1} = \frac{V_{CC}}{9K + 1K} \tag{6.47}$$

$$V_{RE|Q1\ ON} = I_{SatQ1} \times 1K = \frac{V_{CC}}{10K} \times 1K\Omega = \frac{5}{10} = 0.5 \text{ volts} \tag{6.48}$$

If transistor is an ideal one then voltage across it ≈ 0 V when it is on.

Actually $V_0(t) \approx V_{CE\ Sat} = 0.2$ V

So,

$$I_{Sat} = \frac{V_{CC}}{9K + 1K} \tag{6.49}$$

$$V_0(t) = I_{cSat}\left[R_{ON} + 1K\Omega\right] \tag{6.50}$$

$$= \frac{V_{CC}}{5K\Omega}\left[0 + 1K\right]$$

$$= \frac{V_{CC}}{5K\Omega} \times 1K\Omega$$

$$= \frac{5 \text{ volts}}{5}$$

$$= 1 \text{ volts}$$

The voltage drop across B-E is 1 V so to forward bias the B-E junction of Q_1 at least voltage applied = 1.7 V.

Now we reduce $V_i(t)$ gradually from 5 to 0 V. When $V_i(t) \leq (1.2)$ V = $(0.5 + 0.7)$ V then only Q_1 come out of saturation and will be off again and the output again rises towards high voltage.

6.12.3 Characteristics of Schmitt Trigger Circuit

This type of characteristics is called S type characteristics and the transition of output occurs for two input voltages. Figure 6.52 shows characteristics of a Schmitt trigger circuit. This type of circuit is a noninverter circuit as when input is low, output is low and vice versa.

The input voltage at which the output abruptly changes from low state to high state is called positive going threshold voltage or upper trip point for Schmitt trigger noninverter circuit.

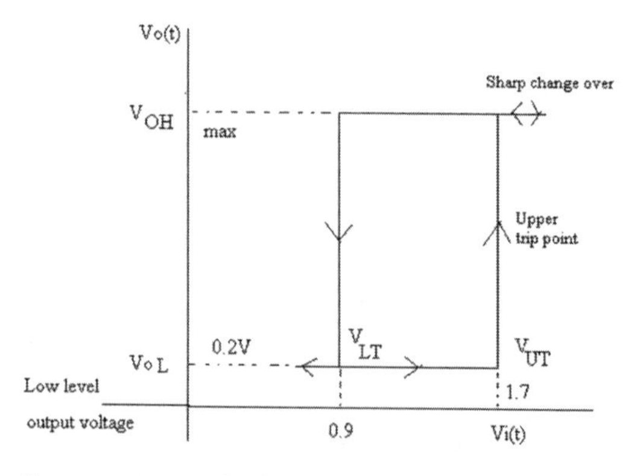

Figure 6.52 Characteristics of Schmitt trigger circuit.

$$V_{UT} = V_{T+} = 1.7\ \text{V} \tag{6.51}$$

The input voltage at which the output changes from high state to low state is called the negative going threshold voltage or lower trip point for Schmitt trigger noninverter circuit.

$$\text{So, } V_{LT} = -V_{T} = 1.2\ \text{V} \tag{6.52}$$

V_H = Hysteresis is defined as

$$V_H = V_{UT}–V_{LT} = 1.7–1.2 = 0.5 \tag{6.53}$$

The smaller is the hysteresis, the greater is the efficiency of converting sine wave to square wave. Figure 6.53 shows Input–output characteristics of ST inverter.

7414 = Hex S-T inverter circuit $V_{TT} = 1.2$ V and $V_T = 0.9$ V

$$V_H = 1.2–0.9 = 0.3 \tag{6.54}$$

$$V_{OH} = 3.4\ \text{V and} \tag{6.55}$$

$$V_{OL} = 0.2\ \text{V (Compatible)} \tag{6.56}$$

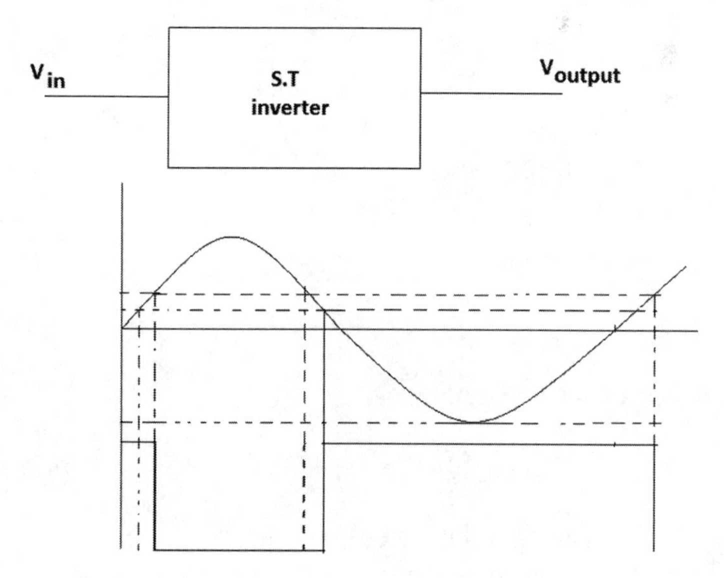

Figure 6.53 Input–output characteristics of ST inverter.

6.13 Schmitt Trigger (ST) Using OP-AMP

Figure 6.54 shows a ST inverter using an OP-Amp

$$I = \frac{V_0(t)}{(R_1 + R_2)} \tag{6.57}$$

and

$$V_f = I \times R_2 = \frac{V_0(t)R_2}{(R_1 + R_2)} \tag{6.58}$$

If $V_i(t) < V_f$
then $V_0(t) = V_{Sat}$ = Control
So long $V_i(t)$ remains $< V_f$
and $V_{Sat} \cong \pm 2$ V of the supply voltage.
$V_{Sat} = (15–2) = 13$ V

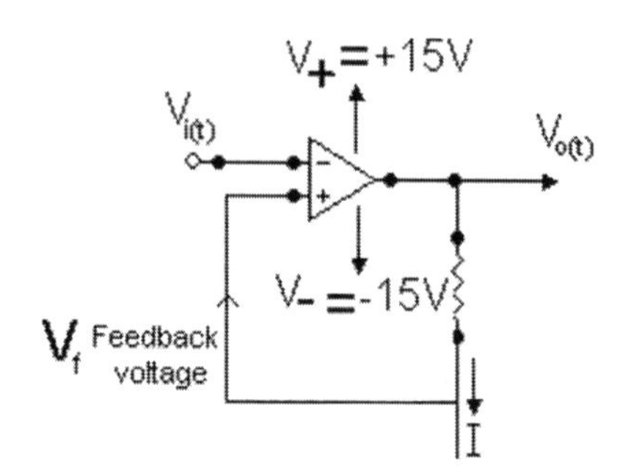

Figure 6.54 ST using OP-AMP.

6.13.1 Characteristics

Now,

$$V_f = \frac{V_{sat} \times R_2}{R_1 + R_2} \tag{6.59}$$

Upper trip point for ST inverter

$$V_{UT} = V_{T+} = \frac{R_2\left(V_{\text{Sat inverter}}\right)}{R_1 + R_2} \tag{6.60}$$

Inverter as when input voltage is small output is high and vice versa.
Now as $V_i(t)$ is increased and when $V_i(t) > V_F > V_{UT}$.
So output $V_0(t) = -V_{Sat}$

$$V_F = \frac{-V_{Sat} \times R_2}{R_1 + R_2} \tag{6.61}$$

$$V_{LT} = V_T = \frac{-V_{Sat} \times R_2}{R_1 + R_2}$$

Figure 6.55 shows the transfer characteristics of input and output of the ST inverter.

How to get the reverse condition?

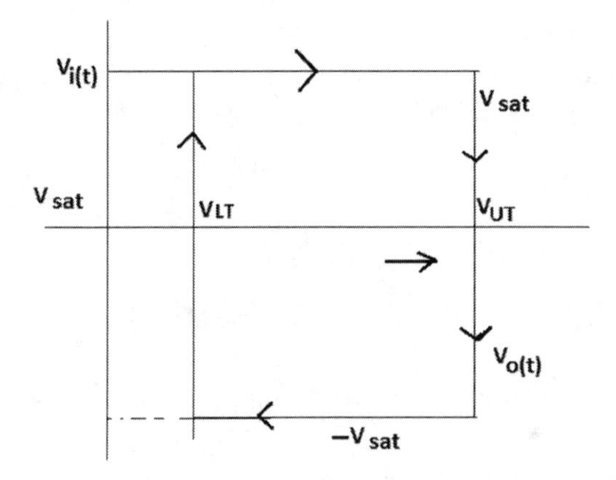

Figure 6.55 Transfer characteristics of input and output.

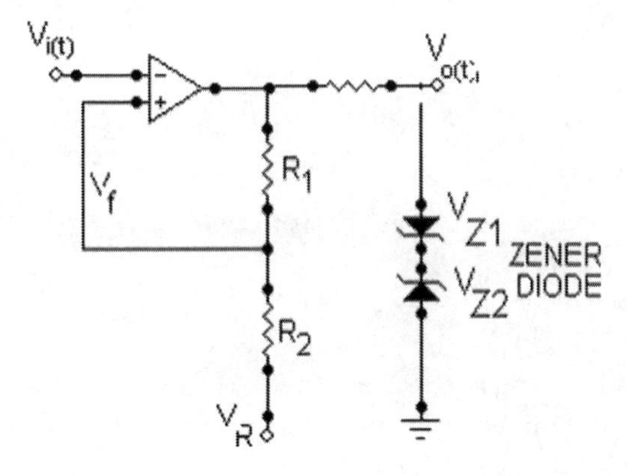

Figure 6.56 ST inverter.

Now the input voltage is reduced.

So as $V_i(t)$ is decreased when $V_i(t) < V_{LT}$ then output become high again.

$V_{Sat} \approx 13$ V of OP-AMP with supply voltage = 15 V.

$R_2 = 100\ \Omega = 0.1\ K\Omega$

$$V_{T+} = \frac{13V \times (0.1K\Omega)}{(100+1)K\Omega} = \frac{13 \times 1}{1001} \cong 13\,\text{mV}$$

$V_{LT} = (-13\ \text{mV})$

So V_{UT} and V_{LT} are very small of the order of ± 13 mV which is not TTL compatible. To make the output $V_0(t)$ TTL compatible we have to modify the circuit as follows for ST inverter.

The ST inverter with TTL compatible circuit is shown in Fig. 6.56.

V_R = reference voltage

$$\text{V}_{Sat} = V_{D1} + V_{Z2} \tag{6.62}$$

$$V_{UT} = V_F = \frac{R_2}{R_1 + R_2}\left(V_{D1} + V_{Z2}\right) \tag{6.63}$$

V_{Z1} is forward biased and V_{Z2} is reverse biased.

$$V_F = \frac{V_R}{R_1 + R_2} \times R_1 \text{ with } V_{Sat} = 0 \tag{6.64}$$

By superposition theorem,

The total feedback voltage

$$V_F = V_{UT} = V_{T+} = \frac{R_2}{R_1 + R_2}\left(V_{D1} + V_{Z2}\right) + \frac{V_R \times R_1}{R_1 + R_2} \tag{6.65}$$

$$V_f = V_{LT} = V_{T-} = \frac{-\left(V_{Z1} + V_{D2}\right)}{R_1 + R_2} \times R_2 + \frac{V_R \times R_1}{R_1 + R_2} \tag{6.66}$$

As V_{Z1} is reverse biased and V_{02} is forward biased.

$V_{D1} = V_{D2} = 0.7$ V

$V_{Z1} = V_{Z2} = 6.9$ V

$V_{UT} = 1.2$ V and $V_{LT} = 0.9$ V

Substituting in Eqs. (6.65) and (6.66) we get V_R.

So, realize symmetrical noninverter using OP-AMP V_i and V_R.

$V_{Sat} = (6.9 + 0.7)$ V = 7.6 V > 5 V

So $V_{Sat} = 7.6$ V and this is logica 1 and $-V_{Sat}$ that is logic 0.

So we mostly use TTL compatible circuits.

As for TTL compatible $V_{OH} = 3.4$ V and $V_{OL} \leq 0.2$ V.

It remains in the high state for one set of input voltage and vice versa, so it is called bistable multivibrator. It can be used as a voltage comparable circuit and the two sharp reference voltages V_{UT} and V_{LT} at which the output changes abruptly.

The 7474 is a Hex Schmitt Trigger circuit is used in frequency counter which converts any time varying signal into square wave.

6.14 Digital ICs and Characteristics

Digital ICs are available in two forms:

(i) flat package
(ii) dual in line package

The number of pins in a digital IC may vary from 14 to 64 depending upon the type, characteristics, and function of these ICs. Whatever is inside cannot be visible as every digital IC is covered or insulated with plastic or ceramic material. Digital ICs are produced by different vendors with different series such as 74XX. This series uses TTL logic family. Its operating temperature is from 0° to 50°C. These ICs are commercially used in 54XX series. This also uses TTL logic family but has operating temperature range—50°C to 150°C. These ICs are used in military application.

Both these series are pin to pin compatible (replacing one by other, the function remain unchanged).

74SXX → 74 series IC using TTL with Schottky transistor
74LSXX → 74 series IC using low power Schottky transistor
74HSXX → 74 series IC using high power Schottky transistor
74ALSXX → 74 series IC using advanced power Schottky
↑
This is called the designation number of IC.

Similarly, we have 74XXX series.

Similarly, we have 1000/8000 series IC that also uses TTL logic family by some vendors. 10,000 series ICs provided by some vendors uses emitter coupled logic (ECL) logic family. This is the fastest unsaturated logic gate.

The 4000 series uses complementary metal oxide semiconductor (CMOS) (both *N* and *P* channels are used) ICs.

6.14.1 How to Identify a Digital IC

Digital ICs are identified by the following designation:

1. the type of logic family used by the IC.
2. the type of logic circuit function, provided by the IC.
3. the type of gate complexity of the IC.
4. the number of pins in the IC and the type of package.
5. the identification number and name of the vendor.

If the IC is 74LS00 NEC

1. 74 indicates that this IC uses TTL logic family using low power Schottky.
2. this IC provides Quad 2 input NAND gate.
3. gate complexity is SSI (< 10 gates).
4. 14 pin IC and DIP package.
5. Identification number is 74LS00.

Name of vendor NEC (a Japanese company).

6.15 Programmable Logic Array (PLA)

The PLA is similar to ROM but its architecture is different from ROM and it can be used to realize random logic circuitry in SOP form that is in AND-OR logic or an AND-OR-NOT logic. The main difficulty with ROM is that if any combinational circuit contain large number of do not care term's and if we try to realize it by PROM then for do not care term's same address, we have to keep same location of ROM unaltered and this makes a wastage of some ROM locations corresponding to the do not care terms. For cases where do not care conditions are excessive it is more economical to use a second type of LSI. Figure 6.57 shows a PLA in the form of a 3:8 decoder & the truth table of the PLA is given in Table 6.12.

$$f(A, B, C, D) = \sum m\ (0, 2) + \sum d\ (1, 3, 4, 5, 6, 7) \qquad (6.67)$$

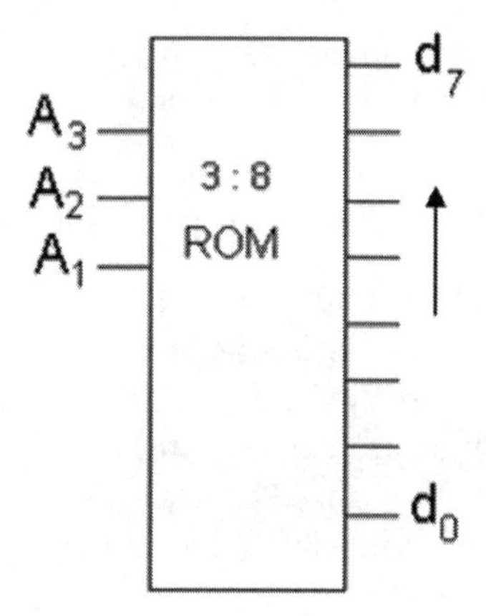

Figure 6.57 A 3: 8 decoder.

Table 6.12 Truth table of PLA

A_3	A_2	A_1	d_7	d_6	d_5	d_4	d_3	d_2	d_1	d_0
0	0	0	0	0	0	0	0	0	0	0
0	0	1								
0	1	0								
0	1	1								
1	0	0								
1	0	1								
1	1	0								
1	1	1								

The PLA is similar to a ROM in concept. Thus, a combinational circuit with large number of do not care terms can be efficiently realized by using PLA because the PLA does not decode all input variables and also does not produce minimum terms for all variables and hence efficiency of the design will be higher.
The PLAs are of two types:

1. Mask programmable logic array which is analogous to ROM. It is not programmed but is programmed by manufacturer according to its own specification using special mask according to user specification. The PLA is similar to ROM. Once it is programmed the program cannot be changed but only be read.
2. Field Programmable Logic Array (FPLA): It is analogous to PROM. The FPLA can be programmed only once like PROM. This requires

a suitable FPLA programmer. Once programmed, its contents are fixed or unaltered and hence it is a nonvolatile.

6.15.1 General Architecture of PLA

General PLA will have n inputs, K product term, and m outputs. As number of outputs = number of sum terms

$\therefore m$ output = m number of sum terms

Sum terms can be realized by OR gates.

Product terms can be realized by AND gates.

If a combinational circuit contains n inputs, K product terms, and m outputs then such a combinational circuit can be realized by using PLA with maximum number of buses which is equal to

$$2(n \times k) + (m \times k) + m \tag{6.68}$$

If we realize the same circuit by using ROM

The maximum number of fused required

$$\text{in a PROM is} = (2^n \times m) \tag{6.69}$$

Let, $n = 8$, $K = 36$ and $m = 8$

$$\text{For PROM} = (2^8 \times 8) = (256 \times 8) = (2048) \text{ number of buses.} \tag{6.70}$$

$$\begin{aligned} \text{For PLA} &= [2(8 \times 36) + (8 \times 36) + 8] \\ &= 8 \times 36 \times 3 + 8 \\ &= 864 + 8 \\ &= 872 \end{aligned} \tag{6.71}$$

So, the number of buses required is less. So, we can realize the same combinational circuit, by using a more compact circuit in PLA. Hence, PLA is used more. Figure 6.58 shows general architecture of PLA & Fig. 6.59 shows the expanded diagram. Figure 6.60 shows an input buffer used at the input of PLA in the Fig. 6.61.

$\overline{CS}$	I_0	O_0
0	0	1
0	1	0
1	X	High Z (High impedance)

The DM7575 is a PLA containing 14 inputs, 96 product terms, and 8 outputs (8 or gates).

How to realize a given logical function by using PLA?

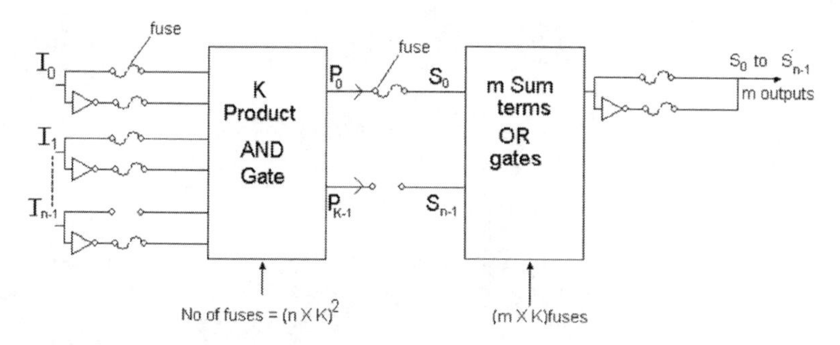

Figure 6.58 General architecture of PLA.

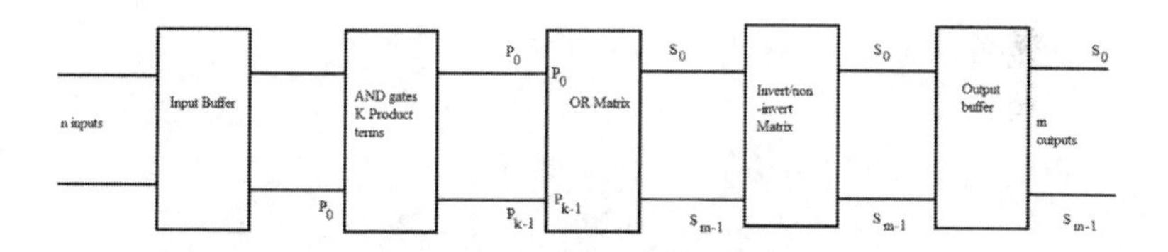

Figure 6.59 Expanded diagram of PLA.

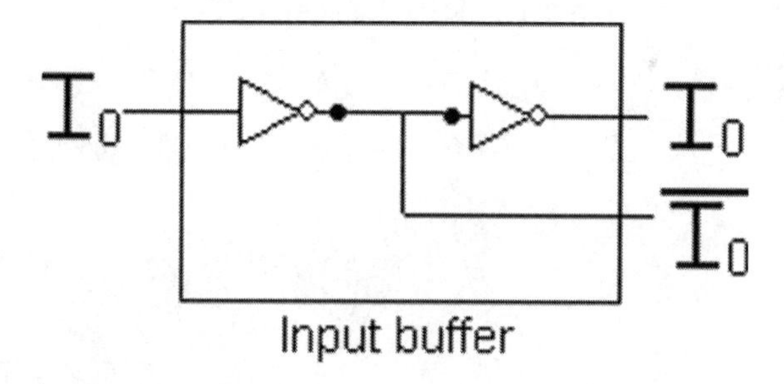

Figure 6.60 The input buffer.

1. First the TT of a logical function in min-terms should be available.
2. The TT or given logical function in min-term form is minimized by using minimizing technique.
3. Express the minimized form in SOP form.
4. Then draw the PLA program table containing the product terms, the inputs and the outputs in three successive columns.

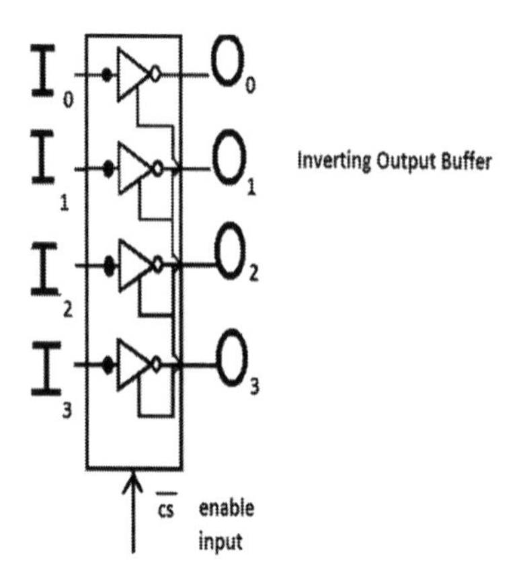

Figure 6.61 A 4 bit inverting input buffer.

5. Realize each of the product term by programming the PLA.
6. Finally realize the outputs by using OR gates.
7. The output of the OR gate should be either inverted or no-inverted.

(If the fuses are blown out, it is equivalent to logical 1 and when fuses are not connected, no electrical signal will pass.)

$$\text{For example: } f_1\,(A, B, C) = \sum m\,(0, 6) + \sum d\,(4, 7) \qquad (6.72)$$

$$f_2\,(A, B, C) = \sum m\,(3, 4) + \sum d\,(0, 2) \qquad (6.73)$$

$$= (-00) + (11-)$$

$$f_1 = (\bar{B}\bar{C} + AB) \qquad (6.74)$$

$$= (-00) + (01-)$$

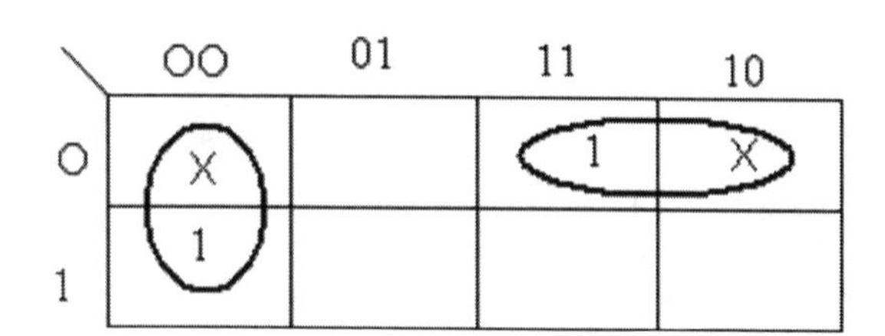

$$f_2 = (\bar{B}\bar{C} + AB) \tag{6.75}$$

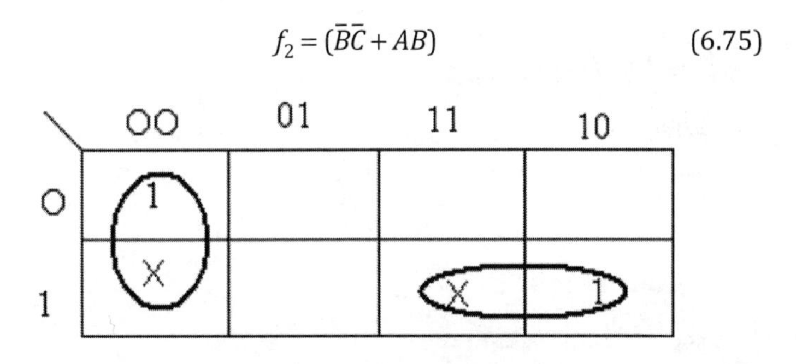

Three product terms are required to realize the above multiple output function. The programmable table is given in Table 6.13 & the implemented circuit diagram is given in Fig. 6.62.

Table 6.13 PLA programmable table

Product term	Inputs			Outputs	
	A	B	C	f_1	f_2
$P_1 = \bar{B}\bar{C}$	–	0	0	1	1
$P_2 = AB$	1	1	–	1	–
$P_1 = \bar{A}\bar{B}$	0	1	–	1	T T

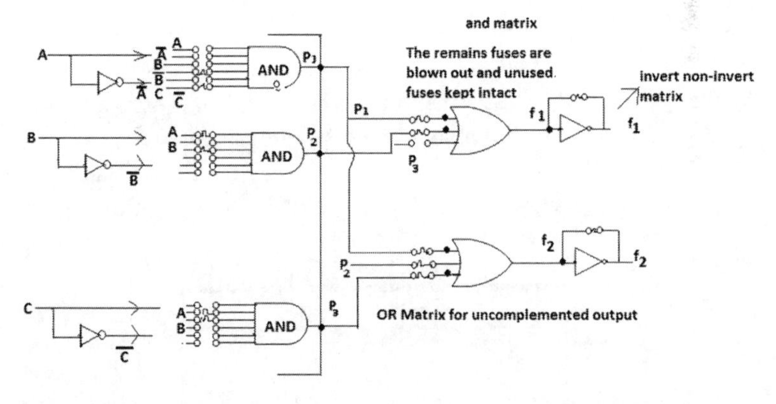

Figure 6.62 Implemented circuit diagram.

For complemented path, the fuses should be blown out.

Application of PLA is

1. It can be used to realize any random logic circuit
2. It can be used to realize code converters, character generator, look up tables, and so on.
   (We can only read the output, but cannot program it)

Main advantages of PLA:

1. It is quite compact.
2. The switching speed is quite high.
3. It is specially suitable for realizing combinational circuit when the number of do not care terms are very large.

6.16 Charge Coupled Device (CCD)

It is used as sequentially accessible memory using as FIFO memory. The data stored first in a CCD can be retrieved first. It is not random access memory.

6.16.1 Main Advantages

1. It is quite simple.
2. It is very versatile.
3. It is low cost memory.

The CCD can be considered as an array of set of MOS capacitors. Figure 6.63 shows a CCD structure, where the charges are stored not only in conductor but also in the depletion region.

At first V_1 = 5 V and parallel plate is capacitor formed and this repels the holes of the P-type substrate and forms a depletion region.

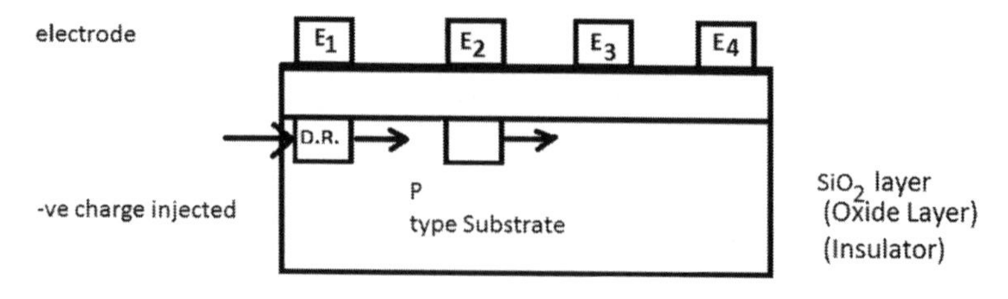

Figure 6.63 CCD structure.

The holes will be repelled downwards by the positive potential of the electrode 1. As a result, the depletion region is formed below the electrode 1 and this depletion region contains immobile negative ions. If we now inject negative charge externally and this charge is trapped by the depletion region and negative charge increases around the depletion region. So, due to the injection of negative charge externally, which will be trapped in the depletion region, under the electrode 1 the information bit 1 is stored and when the depletion region is free from any negative charge zero is stored. The depletion region can also be serially shifted if E_2 = 5 V and negative charge moves towards right and this can be continued for successive serial shifting. This is the essence of charge coupled device.

6.17 Binary Multiplication and Division Algorithm

Binary multiplication:

1. ordinary method of binary multiplication
2. boots algorithm of binary multiplication in 2's complement multiplication rule for binary number

MPD bites = Multiplicand bit = 0 0 1 1
Multiplicand bit MPR bit = × 0 1 0 1

0 0 0 1

(1) $\text{MPD} = (11)_{10}$
$\text{MPR} = (10)_{10}$

Represent the magnitude bit of MPD and MPR in binary

1011 MPD
1010 MPR

0000
1011X
0000XX
1011XXX

1101110 = Final magnitude of the product

$$(1101110) = 64 + 32 + 8 + 4 + 2$$
$$= (110)_{10}$$

(2) $11111 = (31)_{10}$ = MPD = Multiplicand
$1111 = (15)_{10}$ = MPR = Multiplier

$(31)_{10}$
$(15)_{10}$
___
$(465)_{10}$

```
 11111
  1111
 _________
  11111
  11111X
 11111XX
11111XXX
_________
111010001
11111XXXX
_________
1111000001
```

If MPD is of n bits, if MPR is of m bits then magnitude of product (result) $\leq (m + n)$ bit.

$$16 + 64 + 128 + 128 \times 2 + 1$$
$$= (465)_{10}$$

Represent the MPD and MPR with magnitude bits in binary and get the product magnitude bit. If sign bit of MPD = SMPD and sign bit of MPR = SMPR then sign bit of product = $SMPR \oplus SMPD$.

(As multiplication of two +ve numbers is +ve and two –ve numbers is –ve.) If MPD = –ve, MPR = +ve then sign bit of the product = 1.

11011
0111

1. Compare 4 bit division with 4 MSB bit of dividend.
   As dividend > divisor, so subtract divisor from dividend.
   1 = quotient bit = 1 as dividend > divisor

```
 11011
-0111↓
 01101
```

This is the new dividend. Now right shift divisor by one bit.

$$\begin{array}{lr} & 01101 \\ \text{after right shift} & \underline{00111} \\ & 00110 \end{array}$$

Again dividend > divisor

So subtract divisor from dividend and set quotient bit = 1

This is the magnitude bit of the final remainder.

And quotient = 11

The comparison bit can be realized by using a hardware ckt.

$$\text{Dividend} = (13)_{10} \rightarrow, \text{Divisor} = (3)$$
$$\text{Dividend} = \text{X} \qquad \text{Divisor} = \text{Y}$$
$$X = 1101 \qquad Y = 00101$$

Hardware circuit for dividing a 4 bit dividend

Initial counter is loaded with 0, which is initial quotient. Final remainder will be obtained at the output of 4 flip-flops.

Initially the dividend is loaded at the output of 4 F/F and initially we set $C_0 = 1$ because in the circuit we use 1's complement addition.

When C_4 = 1, C_0 = 1 is added with the LSB. When C_4 = 1 then only *Clk* = 1 When C_4 = 0 then *Clk* = 0 and counter stops under this condition. Whatever counter reads, it gives the magnitude of the

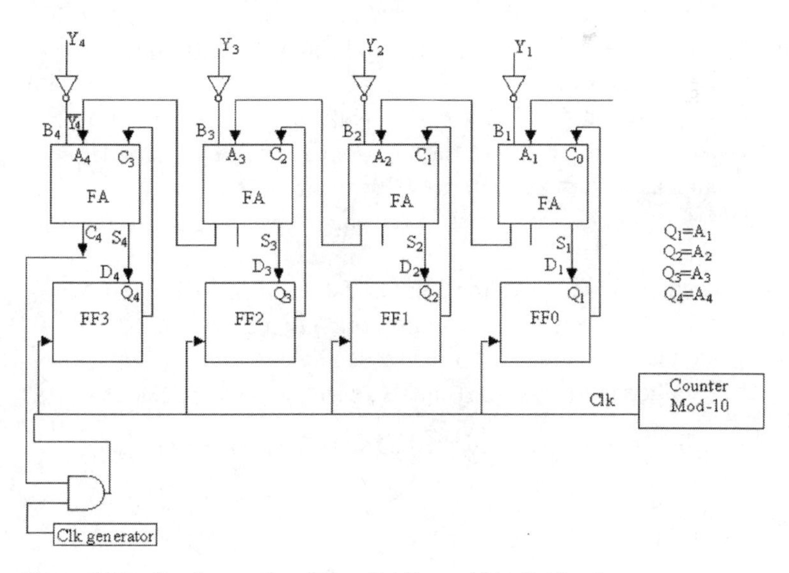

Figure 6.64 Hardware circuit for dividing a 4 bit dividend.

quotient and the final contents of the 4 F/F gives magnitude bit for the final remainder. The hardware circuit for dividing a 4 bit dividend is shown in Fig. 6.64.

6.18 Booth's Multiplication Algorithm

The one of the most interesting 2's complement multiplication algorithm is Booth's algorithm. This algorithm is proposed by Andrew D. Booth in the 1950s. The advantage of this algorithm is that it treats positive and negative numbers in the same manner and is different from Robertson's method.

6.18.1 Algorithm

1. Initially represent both multiplicand and multiplier in signed 2's complement form.
2. There is no partial product at first so accumulator (AC) is cleared to zero.
3. Least significant bit (Q_n) and its presiding bit (Q_{n+1}) of multiplier is cleared to zero.
4. Sequence counter (SC) holds the number of multiplier bits 'n'.
5. If $Q_n = 1$ and $Q_{n+1} = 0$ then it means '1' in the string of 1's has been encountered and the multiplicand is to be subtracted from the partial product value in the AC register.
6. If $Q_n = 0$ and $Q_{n+1} = 1$ then it means '0' in the string of 0's has been encountered and the multiplicand is to be added to the partial product value in the AC register.
7. If Q_n and Q_{n+1} are equal then partial product value remains unchanged.
8. Then both accumulator (AC) and multiplier (QR) are shifted right. However, signed bit in AC remained unchanged.
9. Sequence counter (SC) is decreased by one.
10. If sequence counter (SC) equal to zero then stop else go to step 5.

Figure 6.65 shows the pictorial diagram of this algorithm.

Example: Multiply (–9) and (–13)

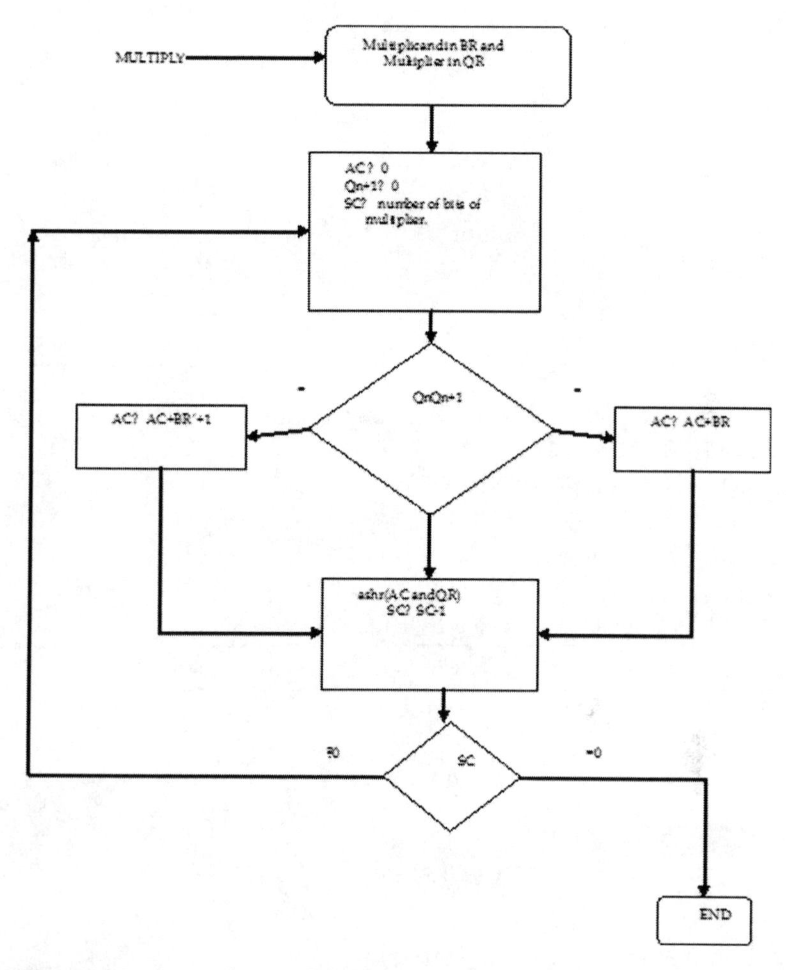

Figure 6.65 Pictorial view of algorithm.

$BR = 10111$ and $QR = 10011$ (Both in 2's complement form).

Table 6.14 Tabular representation of Booth's algorithm

Q_n	Q_{n+1}	AC	QR	Q_{n+1}	SC	Comment
0	0	00000	10011	0	101	Initial
1	0	+01001				Subtracted
		01001				BR
		00100	11001	1	100	Arithmetic right shift
1	1	00010	01100	1	011	Arithmetic right shift
0	1	+10111				Add
		11001				BR
		11100	10110	0	010	Arithmetic right shift
0	0	11110	01011	0	001	Arithmetic right shift
1	0	01001				Subtracted
		00111				BR
		00011	10101	1	000	Arithmetic right shift

Therefore, 10 bit final product appears in AC and QR and is = 0001110101 (+117).

Example: Multiply (9) and (13)

$BR = 01001$, $BR' + 1 = 10111$

and $QR = 01101$

Table 6.15 Tabular representation of Booth's algorithm

Q_n	Q_{n+1}	AC	QR	Q_{n+1}	SC	Comment
0	0	00000	01101	0	5	Initial
1	0	+10111				Subtracted
		10111				BR
		11011	10110	1	4	Arithmetic right shift
0	1	+01001				Add
		00100				BR
		00010	01011	0	3	Arithmetic right shift
1	0	+10111				Add
		11001				BR
		11100	10101	1	2	Arithmetic right shift
1	1	11110	01010	1	1	Arithmetic right shift
0	1	+01001				Add
		00111				BR
		00011	10101	0	0	Arithmetic right shift

Therefore, 10 bit final product appears in AC and QR and is = 0001110101 (+ 117).

Example: Multiply (–9) and (13)

$BR = -9 = 10111$, $BR' + 1 = 01001$

and $QR = 13 = 01101$

Table 6.16 Tabular representation of Booth's algorithm

Q_n	Q_{n+1}	AC	QR	Q_{n+1}	SC	Comment
0	0	00000	01101	0	5	Initial
1	0	+ 01001				Subtracted
		01001				BR
		00100	10110	1	4	Arithmetic right shift
0	1	+ 10111				Add
		11011				BR
		11101	11011	0	3	Arithmetic right shift
1	0	+ 01001				Subtracted
		00110				BR
		00011	01101	1	2	Arithmetic right shift
1	1	00001	10110	1	1	Arithmetic right shift
0	1	+ 10111				Add
		11000				BR
		11100	01011	0	0	Arithmetic right shift

So, the final answer is 1110001011 but the MSB bit is "1" so it indicates that the answer is in negative.

So, the final answer will be 2's complement of the final answer, that is (0001110100 + 1) = 1110101.

Problems

1. What do you mean by DAC? Why is it required?
2. Name two most frequently used DAC.
3. What are the disadvantages of W-R type DAC?
4. What is the major advantage of R-2R type DAC?
5. Which IC is used as 8-bit D/A Counter?
6. How two ALU units are cascaded? Describe with example.
7. What is CLA Adder? What are its advantages?
8. What do you mean by ADC? What is the role played by sample and hold circuit in ADC?
9. Give an example of waveform of sampling pulse.
10. What are the parameters of IC LF 398?
11. What is resolution of a DAC?
12. What is accuracy of a DAC?
13. Explain what is linearity of a DAC.
14. What is settling time of a DAC?
15. Name some of the counter applications.

16. Name some of the Schmitt trigger applications.
17. What is hysteresis of a Schmitt Trigger?
18. What are the available packages of Digital ICs?
19. How to identify a digital IC?
20. What is Charge Coupled Device (CCD)? Name some of its advantages.

Index